Math in Society

Edition 2.4

Contents

David Lippman
Pierce College Ft Steilacoom

About the Author/Editor

David Lippman received his master's degree in mathematics from Western Washington University and has been teaching at Pierce College since Fall 2000.

David has been a long time advocate of open learning, open materials, and basically any idea that will reduce the cost of education for students. It started by supporting the college's calculator rental program, and running a book loan scholarship program. Eventually the frustration with the escalating costs of commercial text books and the online homework systems that charged for access led to action.

First, David developed IMathAS, open source online math homework software that runs WAMAP.org and MyOpenMath.com. Through this platform, he became an integral part of a vibrant sharing and learning community of teachers from around Washington State that support and contribute to WAMAP. These pioneering efforts, supported by dozens of other dedicated faculty and financial support from the Transition Math Project, have led to a system used by thousands of students every quarter, saving hundreds of thousands of dollars over comparable commercial offerings.

David continued further and wrote the first edition of this textbook, *Math in Society*, after being frustrated by students having to pay $100+ for a textbook for a terminal course. Together with Melonie Rasmussen, he co-authored *PreCalculus: An Investigation of Functions* in 2010.

Acknowledgements

David would like to thank the following for their generous support and feedback.

- Jeff Eldridge, Lawrence Morales, and Mike Kenyon, who were kind enough to license me use of their works.

- The community of WAMAP users and developers for creating some of the homework content used in the online homework sets.

- Pierce College students in David's online Math 107 classes for helping correct typos and identifying portions of the text that needed improving, along with other users of the text.

- The Open Course Library Project for providing the support needed to produce a full course package for this book.

Preface

The traditional high school and college mathematics sequence leading from algebra up through calculus could leave one with the impression that mathematics is all about algebraic manipulations. This book is an exploration of the wide world of mathematics, of which algebra is only one small piece. The topics were chosen because they provide glimpses into other ways of thinking mathematically, and because they have interesting applications to everyday life. Together, they highlight algorithmic, graphical, algebraic, statistical, and analytic approaches to solving problems.

This book is available online for free, in both Word and PDF format. You are free to change the wording, add materials and sections or take them away. I welcome feedback, comments and suggestions for future development. If you add a section, chapter or problems, I would love to hear from you and possibly add your materials so everyone can benefit.

New in This Edition

Edition 2 has been heavily revised to introduce a new layout that emphasizes core concepts and definitions, and examples. Based on experience using the first edition for three years as the primarily learning materials in a fully online course, concepts that were causing students confusion were clarified, and additional examples were added. New "Try it Now" problems were introduced, which give students the opportunity to test out their understanding in a zero-stakes format. Edition 2.0 also added four new chapters.

Edition 2.1 was a typo and clarification update on the first 14 chapters, and added 2 additional new chapters. No page or exercise numbers changed on the first 14 chapters. Edition 2.2 was a typo revision. A couple new exploration exercises were added. Edition 2.3 and 2.4 were typo revisions.

Supplements

The Washington Open Course Library (OCL) project helped fund the creation of a full course package for this book, which contains the following features:

- Suggested syllabus for a fully online course
- Possible syllabi for an on-campus course
- Online homework for most chapters (algorithmically generated, free response)
- Online quizzes for most chapters (algorithmically generated, free response)
- Written assignments and discussion forum assignments for most chapters

The course shell was built for the IMathAS online homework platform, and is available for Washington State faculty at www.wamap.org and mirrored for others at www.myopenmath.com.

The course shell was designed to follow Quality Matters (QM) guidelines, but has not yet been formally reviewed.

Problem Solving

In previous math courses, you've no doubt run into the infamous "word problems." Unfortunately, these problems rarely resemble the type of problems we actually encounter in everyday life. In math books, you usually are told exactly which formula or procedure to use, and are given exactly the information you need to answer the question. In real life, problem solving requires identifying an appropriate formula or procedure, and determining what information you will need (and won't need) to answer the question.

In this chapter, we will review several basic but powerful algebraic ideas: percents, rates, and proportions. We will then focus on the problem solving process, and explore how to use these ideas to solve problems where we don't have perfect information.

Percents

In the 2004 vice-presidential debates, Edwards's claimed that US forces have suffered "90% of the coalition casualties" in Iraq. Cheney disputed this, saying that in fact Iraqi security forces and coalition allies "have taken almost 50 percent" of the casualties[1]. Who is correct? How can we make sense of these numbers?

Percent literally means "per 100," or "parts per hundred." When we write 40%, this is equivalent to the fraction $\frac{40}{100}$ or the decimal 0.40. Notice that 80 out of 200 and 10 out of 25 are also 40%, since $\frac{80}{200} = \frac{10}{25} = \frac{40}{100}$.

Example 1

243 people out of 400 state that they like dogs. What percent is this?

$\frac{243}{400} = 0.6075 = \frac{60.75}{100}$. This is 60.75%.

Notice that the percent can be found from the equivalent decimal by moving the decimal point two places to the right.

Example 2

Write each as a percent: a) $\frac{1}{4}$ b) 0.02 c) 2.35

a) $\frac{1}{4} = 0.25 = 25\%$ b) $0.02 = 2\%$ c) $2.35 = 235\%$

---

[1] http://www.factcheck.org/cheney_edwards_mangle_facts.html

---

Percents

If we have a *part* that is some *percent* of a *whole*, then

$$\text{percent} = \frac{\text{part}}{\text{whole}}, \text{ or equivalently, } \text{part} = \text{percent} \cdot \text{whole}$$

To do the calculations, we write the percent as a decimal.

Example 3

The sales tax in a town is 9.4%. How much tax will you pay on a $140 purchase?

Here, $140 is the whole, and we want to find 9.4% *of* $140. We start by writing the percent as a decimal by moving the decimal point two places to the left (which is equivalent to dividing by 100). We can then compute:

$$\text{tax} = 0.094(140) = \$13.16 \text{ in tax.}$$

Example 4

In the news, you hear "tuition is expected to increase by 7% next year." If tuition this year was $1200 per quarter, what will it be next year?

The tuition next year will be the current tuition plus an additional 7%, so it will be 107% of this year's tuition:
$1200(1.07) = $1284.

Alternatively, we could have first calculated 7% of $1200: $1200(0.07) = $84.

Notice this is *not* the expected tuition for next year (we could only wish). Instead, this is the expected *increase*, so to calculate the expected tuition, we'll need to add this change to the previous year's tuition:
$1200 + $84 = $1284.

Try it Now 1

A TV originally priced at $799 is on sale for 30% off. There is then a 9.2% sales tax. Find the price after including the discount and sales tax.

Example 5

The value of a car dropped from $7400 to $6800 over the last year. What percent decrease is this?

To compute the percent change, we first need to find the dollar value change: $6800-$7400 = -$600. Often we will take the absolute value of this amount, which is called the **absolute change**: $|-600| = 600$.

Since we are computing the decrease relative to the starting value, we compute this percent out of $7400:

$$\frac{600}{7400} = 0.081 = 8.1\% \text{ decrease. This is called a } \textbf{relative change}.$$

Absolute and Relative Change

Given two quantities,

Absolute change = $|\text{ending quantity} - \text{starting quantity}|$

Relative change: $\dfrac{\text{absolute change}}{\text{starting quantity}}$

Absolute change has the same units as the original quantity.
Relative change gives a percent change.
The starting quantity is called the **base** of the percent change.

The base of a percent is very important. For example, while Nixon was president, it was argued that marijuana was a "gateway" drug, claiming that 80% of marijuana smokers went on to use harder drugs like cocaine. The problem is, this isn't true. The true claim is that 80% of harder drug users first smoked marijuana. The difference is one of base: 80% of marijuana smokers using hard drugs, vs. 80% of hard drug users having smoked marijuana. These numbers are not equivalent. As it turns out, only one in 2,400 marijuana users actually go on to use harder drugs[2].

Example 6

There are about 75 QFC supermarkets in the U.S. Albertsons has about 215 stores. Compare the size of the two companies.

When we make comparisons, we must ask first whether an absolute or relative comparison. The absolute difference is $215 - 75 = 140$. From this, we could say "Albertsons has 140 more stores than QFC." However, if you wrote this in an article or paper, that number does not mean much. The relative difference may be more meaningful. There are two different relative changes we could calculate, depending on which store we use as the base:

Using QFC as the base, $\dfrac{140}{75} = 1.867$.

This tells us Albertsons is 186.7% larger than QFC.

Using Albertsons as the base, $\dfrac{140}{215} = 0.651$.

This tells us QFC is 65.1% smaller than Albertsons.

---

[2] http://tvtropes.org/pmwiki/pmwiki.php/Main/LiesDamnedLiesAndStatistics

Notice both of these are showing percent *differences*. We could also calculate the size of Albertsons relative to QFC: $\frac{215}{75} = 2.867$, which tells us Albertsons is 2.867 times the size of QFC. Likewise, we could calculate the size of QFC relative to Albertsons: $\frac{75}{215} = 0.349$, which tells us that QFC is 34.9% of the size of Albertsons.

Example 7

Suppose a stock drops in value by 60% one week, then increases in value the next week by 75%. Is the value higher or lower than where it started?

To answer this question, suppose the value started at $100. After one week, the value dropped by 60%:
$100 - $100(0.60) = $100 - $60 = $40.

In the next week, notice that base of the percent has changed to the new value, $40. Computing the 75% increase:
$40 + $40(0.75) = $40 + $30 = $70.

In the end, the stock is still $30 lower, or $\frac{\$30}{\$100} = 30\%$ lower, valued than it started.

Try it Now 2

The U.S. federal debt at the end of 2001 was $5.77 trillion, and grew to $6.20 trillion by the end of 2002. At the end of 2005 it was $7.91 trillion, and grew to $8.45 trillion by the end of 2006[3]. Calculate the absolute and relative increase for 2001-2002 and 2005-2006. Which year saw a larger increase in federal debt?

Example 8

A Seattle Times article on high school graduation rates reported "The number of schools graduating 60 percent or fewer students in four years – sometimes referred to as "dropout factories" – decreased by 17 during that time period. The number of kids attending schools with such low graduation rates was cut in half."

a) Is the "decrease by 17" number a useful comparison?

b) Considering the last sentence, can we conclude that the number of "dropout factories" was originally 34?

[3] http://www.whitehouse.gov/sites/default/files/omb/budget/fy2013/assets/hist07z1.xls

a) This number is hard to evaluate, since we have no basis for judging whether this is a larger or small change. If the number of "dropout factories" dropped from 20 to 3, that'd be a very significant change, but if the number dropped from 217 to 200, that'd be less of an improvement.

b) The last sentence provides relative change which helps put the first sentence in perspective. We can estimate that the number of "dropout factories" was probably previously around 34. However, it's possible that students simply moved schools rather than the school improving, so that estimate might not be fully accurate.

Example 9

In the 2004 vice-presidential debates, Edwards's claimed that US forces have suffered "90% of the coalition casualties" in Iraq. Cheney disputed this, saying that in fact Iraqi security forces and coalition allies "have taken almost 50 percent" of the casualties. Who is correct?

Without more information, it is hard for us to judge who is correct, but we can easily conclude that these two percents are talking about different things, so one does not necessarily contradict the other. Edward's claim was a percent with coalition forces as the base of the percent, while Cheney's claim was a percent with both coalition and Iraqi security forces as the base of the percent. It turns out both statistics are in fact fairly accurate.

Try it Now 3

In the 2012 presidential elections, one candidate argued that "the president's plan will cut $716 billion from Medicare, leading to fewer services for seniors," while the other candidate rebuts that "our plan does not cut current spending and actually expands benefits for seniors, while implementing cost saving measures." Are these claims in conflict, in agreement, or not comparable because they're talking about different things?

We'll wrap up our review of percents with a couple cautions. First, when talking about a change of quantities that are already measured in percents, we have to be careful in how we describe the change.

Example 10

A politician's support increases from 40% of voters to 50% of voters. Describe the change.

We could describe this using an absolute change: $|50\% - 40\%| = 10\%$. Notice that since the original quantities were percents, this change also has the units of percent. In this case, it is best to describe this as an increase of 10 **percentage points**.

In contrast, we could compute the percent change: $\dfrac{10\%}{40\%} = 0.25 = 25\%$ increase. This is the relative change, and we'd say the politician's support has increased by 25%.

Lastly, a caution against averaging percents.

Example 11

A basketball player scores on 40% of 2-point field goal attempts, and on 30% of 3-point of field goal attempts. Find the player's overall field goal percentage.

It is very tempting to average these values, and claim the overall average is 35%, but this is likely not correct, since most players make many more 2-point attempts than 3-point attempts. We don't actually have enough information to answer the question. Suppose the player attempted 200 2-point field goals and 100 3-point field goals. Then they made $200(0.40) = 80$ 2-point shots and $100(0.30) = 30$ 3-point shots. Overall, they made 110 shots out of 300, for a $\frac{110}{300} = 0.367 = 36.7\%$ overall field goal percentage.

Proportions and Rates

If you wanted to power the city of Seattle using wind power, how many windmills would you need to install? Questions like these can be answered using rates and proportions.

> **Rates**
> A rate is the ratio (fraction) of two quantities.
> A **unit rate** is a rate with a denominator of one.

Example 12

Your car can drive 300 miles on a tank of 15 gallons. Express this as a rate.

Expressed as a rate, $\frac{300\,\text{miles}}{15\,\text{gallons}}$. We can divide to find a unit rate: $\frac{20\,\text{miles}}{1\,\text{gallon}}$, which we could also write as $20\frac{\text{miles}}{\text{gallon}}$, or just 20 miles per gallon.

> **Proportion Equation**
> A proportion equation is an equation showing the equivalence of two rates or ratios.

Example 13

Solve the proportion $\frac{5}{3} = \frac{x}{6}$ for the unknown value x.

This proportion is asking us to find a fraction with denominator 6 that is equivalent to the fraction $\frac{5}{3}$. We can solve this by multiplying both sides of the equation by 6, giving

$x = \frac{5}{3} \cdot 6 = 10$.

Example 14

A map scale indicates that ½ inch on the map corresponds with 3 real miles. How many miles apart are two cities that are $2\frac{1}{4}$ inches apart on the map?

We can set up a proportion by setting equal two $\dfrac{\text{map inches}}{\text{real miles}}$ rates, and introducing a variable, x, to represent the unknown quantity – the mile distance between the cities.

$\dfrac{\frac{1}{2}\,\text{map inch}}{3\,\text{miles}} = \dfrac{2\frac{1}{4}\,\text{map inches}}{x\,\text{miles}}$ Multiply both sides by x

and rewriting the mixed number

$\dfrac{\frac{1}{2}}{3}\cdot x = \dfrac{9}{4}$ Multiply both sides by 3

$\dfrac{1}{2}x = \dfrac{27}{4}$ Multiply both sides by 2 (or divide by ½)

$x = \dfrac{27}{2} = 13\frac{1}{2}$ miles

Many proportion problems can also be solved using **dimensional analysis**, the process of multiplying a quantity by rates to change the units.

Example 15

Your car can drive 300 miles on a tank of 15 gallons. How far can it drive on 40 gallons?

We could certainly answer this question using a proportion: $\dfrac{300\,\text{miles}}{15\,\text{gallons}} = \dfrac{x\,\text{miles}}{40\,\text{gallons}}$.

However, we earlier found that 300 miles on 15 gallons gives a rate of 20 miles per gallon. If we multiply the given 40 gallon quantity by this rate, the *gallons* unit "cancels" and we're left with a number of miles:

$40\,\text{gallons}\cdot\dfrac{20\,\text{miles}}{\text{gallon}} = \dfrac{40\,\text{gallons}}{1}\cdot\dfrac{20\,\text{miles}}{\text{gallon}} = 800\,\text{miles}$

Notice if instead we were asked "how many gallons are needed to drive 50 miles?" we could answer this question by inverting the 20 mile per gallon rate so that the *miles* unit cancels and we're left with gallons:

$50\,\text{miles}\cdot\dfrac{1\,\text{gallon}}{20\,\text{miles}} = \dfrac{50\,\text{miles}}{1}\cdot\dfrac{1\,\text{gallon}}{20\,\text{miles}} = \dfrac{50\,\text{gallons}}{20} = 2.5\,\text{gallons}$

Dimensional analysis can also be used to do unit conversions. Here are some unit conversions for reference.

> **Unit Conversions**
> **Length**
> 1 foot (ft) = 12 inches (in)   1 yard (yd) = 3 feet (ft)
> 1 mile = 5,280 feet
> 1000 millimeters (mm) = 1 meter (m)   100 centimeters (cm) = 1 meter
> 1000 meters (m) = 1 kilometer (km)   2.54 centimeters (cm) = 1 inch
>
> **Weight and Mass**
> 1 pound (lb) = 16 ounces (oz)   1 ton = 2000 pounds
> 1000 milligrams (mg) = 1 gram (g)   1000 grams = 1kilogram (kg)
> 1 kilogram = 2.2 pounds (on earth)
>
> **Capacity**
> 1 cup = 8 fluid ounces (fl oz)[*]   1 pint = 2 cups
> 1 quart = 2 pints = 4 cups   1 gallon = 4 quarts = 16 cups
> 1000 milliliters (ml) = 1 liter (L)
>
> [*]Fluid ounces are a capacity measurement for liquids. 1 fluid ounce ≈ 1 ounce (weight) for water only.

Example 16

A bicycle is traveling at 15 miles per hour. How many feet will it cover in 20 seconds?

To answer this question, we need to convert 20 seconds into feet. If we know the speed of the bicycle in feet per second, this question would be simpler. Since we don't, we will need to do additional unit conversions. We will need to know that 5280 ft = 1 mile. We might start by converting the 20 seconds into hours:

$$20\,\text{seconds} \cdot \frac{1\,\text{minute}}{60\,\text{seconds}} \cdot \frac{1\,\text{hour}}{60\,\text{minutes}} = \frac{1}{180}\,\text{hour}$$ Now we can multiply by the 15 miles/hr

$$\frac{1}{180}\,\text{hour} \cdot \frac{15\,\text{miles}}{1\,\text{hour}} = \frac{1}{12}\,\text{mile}$$ Now we can convert to feet

$$\frac{1}{12}\,\text{mile} \cdot \frac{5280\,\text{feet}}{1\,\text{mile}} = 440\,\text{feet}$$

We could have also done this entire calculation in one long set of products:

$$20\,\text{seconds} \cdot \frac{1\,\text{minute}}{60\,\text{seconds}} \cdot \frac{1\,\text{hour}}{60\,\text{minutes}} \cdot \frac{15\,\text{miles}}{1\,\text{hour}} \cdot \frac{5280\,\text{feet}}{1\,\text{mile}} = 440\,\text{feet}$$

Try it Now 4

A 1000 foot spool of bare 12-gauge copper wire weighs 19.8 pounds. How much will 18 inches of the wire weigh, in ounces?

Notice that with the miles per gallon example, if we double the miles driven, we double the gas used. Likewise, with the map distance example, if the map distance doubles, the real-life distance doubles. This is a key feature of proportional relationships, and one we must confirm before assuming two things are related proportionally.

Example 17

Suppose you're tiling the floor of a 10 ft by 10 ft room, and find that 100 tiles will be needed. How many tiles will be needed to tile the floor of a 20 ft by 20 ft room?

In this case, while the width the room has doubled, the area has quadrupled. Since the number of tiles needed corresponds with the area of the floor, not the width, 400 tiles will be needed. We could find this using a proportion based on the areas of the rooms:

$$\frac{100\,\text{tiles}}{100\,\text{ft}^2} = \frac{n\,\text{tiles}}{400\,\text{ft}^2}$$

Other quantities just don't scale proportionally at all.

Example 18

Suppose a small company spends $1000 on an advertising campaign, and gains 100 new customers from it. How many new customers should they expect if they spend $10,000?

While it is tempting to say that they will gain 1000 new customers, it is likely that additional advertising will be less effective than the initial advertising. For example, if the company is a hot tub store, there are likely only a fixed number of people interested in buying a hot tub, so there might not even be 1000 people in the town who would be potential customers.

Sometimes when working with rates, proportions, and percents, the process can be made more challenging by the magnitude of the numbers involved. Sometimes, large numbers are just difficult to comprehend.

Example 19

Compare the 2010 U.S. military budget of $683.7 billion to other quantities.

Here we have a very large number, about $683,700,000,000 written out. Of course, imagining a billion dollars is very difficult, so it can help to compare it to other quantities.

If that amount of money was used to pay the salaries of the 1.4 million Walmart employees in the U.S., each would earn over $488,000.

There are about 300 million people in the U.S. The military budget is about $2,200 per person.

If you were to put $683.7 billion in $100 bills, and count out 1 per second, it would take 216 years to finish counting it.

Example 20

Compare the electricity consumption per capita in China to the rate in Japan.

To address this question, we will first need data. From the CIA[4] website we can find the electricity consumption in 2011 for China was 4,693,000,000,000 KWH (kilowatt-hours), or 4.693 trillion KWH, while the consumption for Japan was 859,700,000,000, or 859.7 billion KWH. To find the rate per capita (per person), we will also need the population of the two countries. From the World Bank[5], we can find the population of China is 1,344,130,000, or 1.344 billion, and the population of Japan is 127,817,277, or 127.8 million.

Computing the consumption per capita for each country:

China: $\dfrac{4{,}693{,}000{,}000{,}000\,\text{KWH}}{1{,}344{,}130{,}000\,\text{people}} \approx 3491.5$ KWH per person

Japan: $\dfrac{859{,}700{,}000{,}000\,\text{KWH}}{127{,}817{,}277\,\text{people}} \approx 6726$ KWH per person

While China uses more than 5 times the electricity of Japan overall, because the population of Japan is so much smaller, it turns out Japan uses almost twice the electricity per person compared to China.

Geometry

Geometric shapes, as well as area and volumes, can often be important in problem solving.

Example 21

You are curious how tall a tree is, but don't have any way to climb it. Describe a method for determining the height.

There are several approaches we could take. We'll use one based on triangles, which requires that it's a sunny day. Suppose the tree is casting a shadow, say 15 ft long. I can then have a friend help me measure my own shadow. Suppose I am 6 ft tall, and cast a 1.5 ft shadow. Since the triangle formed by the tree and its shadow has the same angles as the triangle formed by me and my shadow, these triangles are called **similar triangles** and their sides will scale proportionally. In other words, the ratio of height to width will be the same in both triangles. Using this, we can find the height of the tree, which we'll denote by h:

$$\frac{6\,\text{ft tall}}{1.5\,\text{ft shadow}} = \frac{h\,\text{ft tall}}{15\,\text{ft shadow}}$$

Multiplying both sides by 15, we get $h = 60$. The tree is about 60 ft tall.

---

[4] https://www.cia.gov/library/publications/the-world-factbook/rankorder/2042rank.html
[5] http://data.worldbank.org/indicator/SP.POP.TOTL

It may be helpful to recall some formulas for areas and volumes of a few basic shapes.

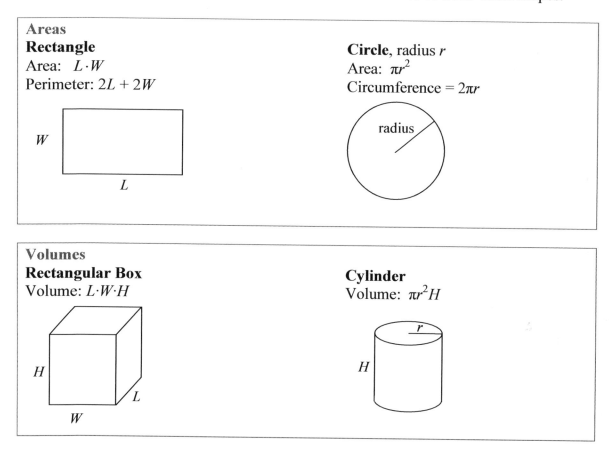

Areas
Rectangle
Area: $L \cdot W$
Perimeter: $2L + 2W$

Circle, radius r
Area: πr^2
Circumference $= 2\pi r$

Volumes
Rectangular Box
Volume: $L \cdot W \cdot H$

Cylinder
Volume: $\pi r^2 H$

Example 22

If a 12 inch diameter pizza requires 10 ounces of dough, how much dough is needed for a 16 inch pizza?

To answer this question, we need to consider how the weight of the dough will scale. The weight will be based on the volume of the dough. However, since both pizzas will be about the same thickness, the weight will scale with the area of the top of the pizza. We can find the area of each pizza using the formula for area of a circle, $A = \pi r^2$:
A 12" pizza has radius 6 inches, so the area will be $\pi 6^2 =$ about 113 square inches.
A 16" pizza has radius 8 inches, so the area will be $\pi 8^2 =$ about 201 square inches.

Notice that if both pizzas were 1 inch thick, the volumes would be 113 in^3 and 201 in^3 respectively, which are at the same ratio as the areas. As mentioned earlier, since the thickness is the same for both pizzas, we can safely ignore it.

We can now set up a proportion to find the weight of the dough for a 16" pizza:
$$\frac{10 \text{ ounces}}{113 \text{ in}^2} = \frac{x \text{ ounces}}{201 \text{ in}^2} \qquad \text{Multiply both sides by 201}$$

$$x = 201 \cdot \frac{10}{113} = \text{about 17.8 ounces of dough for a 16" pizza.}$$

It is interesting to note that while the diameter is $\frac{16}{12} = 1.33$ times larger, the dough required, which scales with area, is $1.33^2 = 1.78$ times larger.

Example 23

A company makes regular and jumbo marshmallows. The regular marshmallow has 25 calories. How many calories will the jumbo marshmallow have?

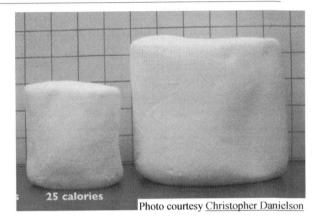
25 calories
Photo courtesy Christopher Danielson

We would expect the calories to scale with volume. Since the marshmallows have cylindrical shapes, we can use that formula to find the volume. From the grid in the image, we can estimate the radius and height of each marshmallow.

The regular marshmallow appears to have a diameter of about 3.5 units, giving a radius of 1.75 units, and a height of about 3.5 units. The volume is about $\pi(1.75)^2(3.5) = 33.7$ units^3.

The jumbo marshmallow appears to have a diameter of about 5.5 units, giving a radius of 2.75 units, and a height of about 5 units. The volume is about $\pi(2.75)^2(5) = 118.8$ units^3.

We could now set up a proportion, or use rates. The regular marshmallow has 25 calories for 33.7 cubic units of volume. The jumbo marshmallow will have:

$$118.8 \text{ units}^3 \cdot \frac{25 \text{ calories}}{33.7 \text{ units}^3} = 88.1 \text{ calories}$$

It is interesting to note that while the diameter and height are about 1.5 times larger for the jumbo marshmallow, the volume and calories are about $1.5^3 = 3.375$ times larger.

Try it Now 5

A website says that you'll need 48 fifty-pound bags of sand to fill a sandbox that measure 8ft by 8ft by 1ft. How many bags would you need for a sandbox 6ft by 4ft by 1ft?

Problem Solving and Estimating

Finally, we will bring together the mathematical tools we've reviewed, and use them to approach more complex problems. In many problems, it is tempting to take the given information, plug it into whatever formulas you have handy, and hope that the result is what you were supposed to find. Chances are, this approach has served you well in other math classes.

This approach does not work well with real life problems. Instead, problem solving is best approached by first starting at the end: identifying exactly what you are looking for. From there, you then work backwards, asking "what information and procedures will I need to find this?" Very few interesting questions can be answered in one mathematical step; often times you will need to chain together a solution pathway, a series of steps that will allow you to answer the question.

Problem Solving Process
1. Identify the question you're trying to answer.
2. Work backwards, identifying the information you will need and the relationships you will use to answer that question.
3. Continue working backwards, creating a solution pathway.
4. If you are missing necessary information, look it up or estimate it. If you have unnecessary information, ignore it.
5. Solve the problem, following your solution pathway.

In most problems we work, we will be approximating a solution, because we will not have perfect information. We will begin with a few examples where we will be able to approximate the solution using basic knowledge from our lives.

Example 24

How many times does your heart beat in a year?

This question is asking for the rate of heart beats per year. Since a year is a long time to measure heart beats for, if we knew the rate of heart beats per minute, we could scale that quantity up to a year. So the information we need to answer this question is heart beats per minute. This is something you can easily measure by counting your pulse while watching a clock for a minute.

Suppose you count 80 beats in a minute. To convert this beats per year:

$$\frac{80 \text{ beats}}{1 \text{ minute}} \cdot \frac{60 \text{ minutes}}{1 \text{ hour}} \cdot \frac{24 \text{ hours}}{1 \text{ day}} \cdot \frac{365 \text{ days}}{1 \text{ year}} = 42{,}048{,}000 \text{ beats per year}$$

Example 25

How thick is a single sheet of paper? How much does it weigh?

While you might have a sheet of paper handy, trying to measure it would be tricky. Instead we might imagine a stack of paper, and then scale the thickness and weight to a single sheet. If you've ever bought paper for a printer or copier, you probably bought a ream, which contains 500 sheets. We could estimate that a ream of paper is about 2 inches thick and weighs about 5 pounds. Scaling these down,

14

$$\frac{2 \text{ inches}}{\text{ream}} \cdot \frac{1 \text{ ream}}{500 \text{ pages}} = 0.004 \text{ inches per sheet}$$

$$\frac{5 \text{ pounds}}{\text{ream}} \cdot \frac{1 \text{ ream}}{500 \text{ pages}} = 0.01 \text{ pounds per sheet, or } 0.16 \text{ ounces per sheet.}$$

Example 26

A recipe for zucchini muffins states that it yields 12 muffins, with 250 calories per muffin. You instead decide to make mini-muffins, and the recipe yields 20 muffins. If you eat 4, how many calories will you consume?

There are several possible solution pathways to answer this question. We will explore one.

To answer the question of how many calories 4 mini-muffins will contain, we would want to know the number of calories in each mini-muffin. To find the calories in each mini-muffin, we could first find the total calories for the entire recipe, then divide it by the number of mini-muffins produced. To find the total calories for the recipe, we could multiply the calories per standard muffin by the number per muffin. Notice that this produces a multi-step solution pathway. It is often easier to solve a problem in small steps, rather than trying to find a way to jump directly from the given information to the solution.

We can now execute our plan:

$$12 \text{ muffins} \cdot \frac{250 \text{ calories}}{\text{muffin}} = 3000 \text{ calories for the whole recipe}$$

$$\frac{3000 \text{ calories}}{20 \text{ mini} - \text{muffins}} \text{ gives 150 calories per mini-muffin}$$

$$4 \text{ mini muffins} \cdot \frac{150 \text{ calories}}{\text{mini} - \text{muffin}} \text{ totals 600 calories consumed.}$$

Example 27

You need to replace the boards on your deck. About how much will the materials cost?

There are two approaches we could take to this problem: 1) estimate the number of boards we will need and find the cost per board, or 2) estimate the area of the deck and find the approximate cost per square foot for deck boards. We will take the latter approach.

For this solution pathway, we will be able to answer the question if we know the cost per square foot for decking boards and the square footage of the deck. To find the cost per square foot for decking boards, we could compute the area of a single board, and divide it into the cost for that board. We can compute the square footage of the deck using geometric formulas. So first we need information: the dimensions of the deck, and the cost and dimensions of a single deck board.

Suppose that measuring the deck, it is rectangular, measuring 16 ft by 24 ft, for a total area of 384 ft^2.

From a visit to the local home store, you find that an 8 foot by 4 inch cedar deck board costs about $7.50. The area of this board, doing the necessary conversion from inches to feet, is:

$$8 \text{ feet} \cdot 4 \text{ inches} \cdot \frac{1 \text{ foot}}{12 \text{ inches}} = 2.667 \text{ ft}^2.$$ The cost per square foot is then

$$\frac{\$7.50}{2.667 \text{ ft}^2} = \$2.8125 \text{ per ft}^2.$$

This will allow us to estimate the material cost for the whole 384 ft^2 deck

$$\$384 \text{ ft}^2 \cdot \frac{\$2.8125}{\text{ft}^2} = \$1080 \text{ total cost}.$$

Of course, this cost estimate assumes that there is no waste, which is rarely the case. It is common to add at least 10% to the cost estimate to account for waste.

Example 28

Is it worth buying a Hyundai Sonata hybrid instead the regular Hyundai Sonata?

To make this decision, we must first decide what our basis for comparison will be. For the purposes of this example, we'll focus on fuel and purchase costs, but environmental impacts and maintenance costs are other factors a buyer might consider.

It might be interesting to compare the cost of gas to run both cars for a year. To determine this, we will need to know the miles per gallon both cars get, as well as the number of miles we expect to drive in a year. From that information, we can find the number of gallons required from a year. Using the price of gas per gallon, we can find the running cost.

From Hyundai's website, the 2013 Sonata will get 24 miles per gallon (mpg) in the city, and 35 mpg on the highway. The hybrid will get 35 mpg in the city, and 40 mpg on the highway.

An average driver drives about 12,000 miles a year. Suppose that you expect to drive about 75% of that in the city, so 9,000 city miles a year, and 3,000 highway miles a year.

We can then find the number of gallons each car would require for the year.

Sonata:
$$9000 \text{ city miles} \cdot \frac{1 \text{ gallon}}{24 \text{ city miles}} + 3000 \text{ hightway miles} \cdot \frac{1 \text{ gallon}}{35 \text{ highway miles}} = 460.7 \text{ gallons}$$

Hybrid:
$$9000 \text{ city miles} \cdot \frac{1 \text{ gallon}}{35 \text{ city miles}} + 3000 \text{ hightway miles} \cdot \frac{1 \text{ gallon}}{40 \text{ highway miles}} = 332.1 \text{ gallons}$$

If gas in your area averages about $3.50 per gallon, we can use that to find the running cost:

Sonata: $460.7 \text{ gallons} \cdot \dfrac{\$3.50}{\text{gallon}} = \$1612.45$

Hybrid: $332.1 \text{ gallons} \cdot \dfrac{\$3.50}{\text{gallon}} = \$1162.35$

The hybrid will save \$450.10 a year. The gas costs for the hybrid are about $\dfrac{\$450.10}{\$1612.45} = 0.279 = 27.9\%$ lower than the costs for the standard Sonata.

While both the absolute and relative comparisons are useful here, they still make it hard to answer the original question, since "is it worth it" implies there is some tradeoff for the gas savings. Indeed, the hybrid Sonata costs about \$25,850, compared to the base model for the regular Sonata, at \$20,895.

To better answer the "is it worth it" question, we might explore how long it will take the gas savings to make up for the additional initial cost. The hybrid costs \$4965 more. With gas savings of \$451.10 a year, it will take about 11 years for the gas savings to make up for the higher initial costs.

We can conclude that if you expect to own the car 11 years, the hybrid is indeed worth it. If you plan to own the car for less than 11 years, it may still be worth it, since the resale value of the hybrid may be higher, or for other non-monetary reasons. This is a case where math can help guide your decision, but it can't make it for you.

Try it Now 6
If traveling from Seattle, WA to Spokane WA for a three-day conference, does it make more sense to drive or fly?

Try it Now Answers
1. The sale price is $799(0.70) = \$559.30$. After tax, the price is $\$559.30(1.092) = \$610.76$
2. 2001-2002: Absolute change: \$0.43 trillion. Relative change: 7.45%
   2005-2006: Absolute change: \$0.54 trillion. Relative change: 6.83%
   2005-2006 saw a larger absolute increase, but a smaller relative increase.
3. Without more information, it is hard to judge these arguments. This is compounded by the complexity of Medicare. As it turns out, the \$716 billion is not a cut in current spending, but a cut in future increases in spending, largely reducing future growth in health care payments. In this case, at least the numerical claims in both statements could be considered at least partially true. Here is one source of more information if you're interested: http://factcheck.org/2012/08/a-campaign-full-of-mediscare/
4. $18 \text{ inches} \cdot \dfrac{1 \text{ foot}}{12 \text{ inches}} \cdot \dfrac{19.8 \text{ pounds}}{1000 \text{ feet}} \cdot \dfrac{16 \text{ ounces}}{1 \text{ pound}} \approx 0.475 \text{ ounces}$

Try it Now Answers Continued

5. The original sandbox has volume 64 ft^3. The smaller sandbox has volume 24ft^3.

$$\frac{48\,\text{bags}}{64\,\text{ft}^3} = \frac{x\,\text{bags}}{24\,\text{ft}^3} \text{ results in } x = 18 \text{ bags.}$$

6. There is not enough information provided to answer the question, so we will have to make some assumptions, and look up some values.

Assumptions:

a) We own a car. Suppose it gets 24 miles to the gallon. We will only consider gas cost.

b) We will not need to rent a car in Spokane, but will need to get a taxi from the airport to the conference hotel downtown and back.

c) We can get someone to drop us off at the airport, so we don't need to consider airport parking.

d) We will not consider whether we will lose money by having to take time off work to drive.

Values looked up (your values may be different)

a) Flight cost: $184

b) Taxi cost: $25 each way (estimate, according to hotel website)

c) Driving distance: 280 miles each way

d) Gas cost: $3.79 a gallon

Cost for flying: $184 flight cost + $50 in taxi fares = $234.

Cost for driving: 560 miles round trip will require 23.3 gallons of gas, costing $88.31.

Based on these assumptions, driving is cheaper. However, our assumption that we only include gas cost may not be a good one. Tax law allows you deduct $0.55 (in 2012) for each mile driven, a value that accounts for gas as well as a portion of the car cost, insurance, maintenance, etc. Based on this number, the cost of driving would be $319.

Exercises

1. Out of 230 racers who started the marathon, 212 completed the race, 14 gave up, and 4 were disqualified. What percentage did not complete the marathon?

2. Patrick left an $8 tip on a $50 restaurant bill. What percent tip is that?

3. Ireland has a 23% VAT (value-added tax, similar to a sales tax). How much will the VAT be on a purchase of a €250 item?

4. Employees in 2012 paid 4.2% of their gross wages towards social security (FICA tax), while employers paid another 6.2%. How much will someone earning $45,000 a year pay towards social security out of their gross wages?

5. A project on Kickstarter.com was aiming to raise $15,000 for a precision coffee press. They ended up with 714 supporters, raising 557% of their goal. How much did they raise?

6. Another project on Kickstarter for an iPad stylus raised 1,253% of their goal, raising a total of $313,490 from 7,511 supporters. What was their original goal?

7. The population of a town increased from 3,250 in 2008 to 4,300 in 2010. Find the absolute and relative (percent) increase.

8. The number of CDs sold in 2010 was 114 million, down from 147 million the previous year[6]. Find the absolute and relative (percent) decrease.

9. A company wants to decrease their energy use by 15%.
   a. If their electric bill is currently $2,200 a month, what will their bill be if they're successful?
   b. If their next bill is $1,700 a month, were they successful? Why or why not?

10. A store is hoping an advertising campaign will increase their number of customers by 30%. They currently have about 80 customers a day.
    a. How many customers will they have if their campaign is successful?
    b. If they increase to 120 customers a day, were they successful? Why or why not?

11. An article reports "attendance dropped 6% this year, to 300." What was the attendance before the drop?

12. An article reports "sales have grown by 30% this year, to $200 million." What were sales before the growth?

---

[6] http://www.cnn.com/2010/SHOWBIZ/Music/07/19/cd.digital.sales/index.html

13. The Walden University had 47,456 students in 2010, while Kaplan University had 77,966 students. Complete the following statements:
    a. Kaplan's enrollment was ___% larger than Walden's.
    b. Walden's enrollment was ___% smaller than Kaplan's.
    c. Walden's enrollment was ___% of Kaplan's.

14. In the 2012 Olympics, Usain Bolt ran the 100m dash in 9.63 seconds. Jim Hines won the 1968 Olympic gold with a time of 9.95 seconds.
    a. Bolt's time was ___% faster than Hines'.
    b. Hine' time was ___% slower than Bolt's.
    c. Hine' time was ___% of Bolt's.

15. A store has clearance items that have been marked down by 60%. They are having a sale, advertising an additional 30% off clearance items. What percent of the original price do you end up paying?

16. Which is better: having a stock that goes up 30% on Monday than drops 30% on Tuesday, or a stock that drops 30% on Monday and goes up 30% on Tuesday? In each case, what is the net percent gain or loss?

17. Are these two claims equivalent, in conflict, or not comparable because they're talking about different things?
    a. "16.3% of Americans are without health insurance"[7]
    b. "only 55.9% of adults receive employer provided health insurance"[8]

18. Are these two claims equivalent, in conflict, or not comparable because they're talking about different things?
    a. "We mark up the wholesale price by 33% to come up with the retail price"
    b. "The store has a 25% profit margin"

19. Are these two claims equivalent, in conflict, or not comparable because they're talking about different things?
    a. "Every year since 1950, the number of American children gunned down has doubled."
    b. "The number of child gunshot deaths has doubled from 1950 to 1994."

20. Are these two claims equivalent, in conflict, or not comparable because they're talking about different things?[9]
    a. "75 percent of the federal health care law's taxes would be paid by those earning less than $120,000 a year"
    b. "76 percent of those who would pay the penalty [health care law's taxes] for not having insurance in 2016 would earn under $120,000"

[7] http://www.cnn.com/2012/06/27/politics/btn-health-care/index.html
[8] http://www.politico.com/news/stories/0712/78134.html
[9] http://factcheck.org/2012/07/twisting-health-care-taxes/

21. Are these two claims equivalent, in conflict, or not comparable because they're talking about different things?
    a. "The school levy is only a 0.1% increase of the property tax rate."
    b. "This new levy is a 12% tax hike, raising our total rate to $9.33 per $1000 of value."

22. Are the values compared in this statement comparable or not comparable? "Guns have murdered more Americans here at home in recent years than have died on the battlefields of Iraq and Afghanistan. In support of the two wars, more than 6,500 American soldiers have lost their lives. During the same period, however, guns have been used to murder about 100,000 people on American soil"[10]

23. A high school currently has a 30% dropout rate. They've been tasked to decrease that rate by 20%. Find the equivalent percentage point drop.

24. A politician's support grew from 42% by 3 percentage points to 45%. What percent (relative) change is this?

25. Marcy has a 70% average in her class going into the final exam. She says "I need to get a 100% on this final so I can raise my score to 85%." Is she correct?

26. Suppose you have one quart of water/juice mix that is 50% juice, and you add 2 quarts of juice. What percent juice is the final mix?

27. Find a unit rate: You bought 10 pounds of potatoes for $4.

28. Find a unit rate: Joel ran 1500 meters in 4 minutes, 45 seconds.

29. Solve: $\dfrac{2}{5} = \dfrac{6}{x}$.

30. Solve: $\dfrac{n}{5} = \dfrac{16}{20}$.

31. A crepe recipe calls for 2 eggs, 1 cup of flour, and 1 cup of milk. How much flour would you need if you use 5 eggs?

32. An 8ft length of 4 inch wide crown molding costs $14. How much will it cost to buy 40ft of crown molding?

33. Four 3-megawatt wind turbines can supply enough electricity to power 3000 homes. How many turbines would be required to power 55,000 homes?

---

[10] http://www.northjersey.com/news/opinions/lautenberg_073112.html?c=y&page=2

34. A highway had a landslide, where 3,000 cubic yards of material fell on the road, requiring 200 dump truck loads to clear. On another highway, a slide left 40,000 cubic yards on the road. How many dump truck loads would be needed to clear this slide?

35. Convert 8 feet to inches.

36. Convert 6 kilograms to grams.

37. A wire costs $2 per meter. How much will 3 kilometers of wire cost?

38. Sugar contains 15 calories per teaspoon. How many calories are in 1 cup of sugar?

39. A car is driving at 100 kilometers per hour. How far does it travel in 2 seconds?

40. A chain weighs 10 pounds per foot. How many ounces will 4 inches weigh?

41. The table below gives data on three movies. Gross earnings is the amount of money the movie brings in. Compare the net earnings (money made after expenses) for the three movies.[11]

Movie	Release Date	Budget	Gross earnings
Saw	10/29/2004	$1,200,000	$103,096,345
Titanic	12/19/1997	$200,000,000	$1,842,879,955
Jurassic Park	6/11/1993	$63,000,000	$923,863,984

42. For the movies in the previous problem, which provided the best return on investment?

43. The population of the U.S. is about 309,975,000, covering a land area of 3,717,000 square miles. The population of India is about 1,184,639,000, covering a land area of 1,269,000 square miles. Compare the population densities of the two countries.

44. The GDP (Gross Domestic Product) of China was $5,739 billion in 2010, and the GDP of Sweden was $435 billion. The population of China is about 1,347 million, while the population of Sweden is about 9.5 million. Compare the GDP per capita of the two countries.

45. In June 2012, Twitter was reporting 400 million tweets per day. Each tweet can consist of up to 140 characters (letter, numbers, etc.). Create a comparison to help understand the amount of tweets in a year by imagining each character was a drop of water and comparing to filling something up.

46. The photo sharing site Flickr had 2.7 billion photos in June 2012. Create a comparison to understand this number by assuming each picture is about 2 megabytes in size, and comparing to the data stored on other media like DVDs, iPods, or flash drives.

---

[11] http://www.the-numbers.com/movies/records/budgets.php

47. Your chocolate milk mix says to use 4 scoops of mix for 2 cups of milk. After pouring in the milk, you start adding the mix, but get distracted and accidentally put in 5 scoops of mix. How can you adjust the mix if:
    a. There is still room in the cup?
    b. The cup is already full?

48. A recipe for sabayon calls for 2 egg yolks, 3 tablespoons of sugar, and ¼ cup of white wine. After cracking the eggs, you start measuring the sugar, but accidentally put in 4 tablespoons of sugar. How can you compensate?

49. The Deepwater Horizon oil spill resulted in 4.9 million barrels of oil spilling into the Gulf of Mexico. Each barrel of oil can be processed into about 19 gallons of gasoline. How many cars could this have fueled for a year? Assume an average car gets 20 miles to the gallon, and drives about 12,000 miles in a year.

50. The store is selling lemons at 2 for $1. Each yields about 2 tablespoons of juice. How much will it cost to buy enough lemons to make a 9-inch lemon pie requiring ½ cup of lemon juice?

51. A piece of paper can be made into a cylinder in two ways: by joining the short sides together, or by joining the long sides together[12]. Which cylinder would hold more? How much more?

52. Which of these glasses contains more liquid? How much more?

In the next 4 questions, estimate the values by making reasonable approximations for unknown values, or by doing some research to find reasonable values.

53. Estimate how many gallons of water you drink in a year.

54. Estimate how many times you blink in a day.

55. How much does the water in a 6-person hot tub weigh?

56. How many gallons of paint would be needed to paint a two-story house 40 ft long and 30 ft wide?

57. During the landing of the Mars Science Laboratory *Curiosity*, it was reported that the signal from the rover would take 14 minutes to reach earth. Radio signals travel at the speed of light, about 186,000 miles per second. How far was Mars from Earth when *Curiosity* landed?

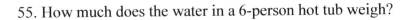

[12] http://vimeo.com/42501010

58. It is estimated that a driver takes, on average, 1.5 seconds from seeing an obstacle to reacting by applying the brake or swerving. How far will a car traveling at 60 miles per hour travel (in feet) before the driver reacts to an obstacle?

59. The flash of lightning travels at the speed of light, which is about 186,000 miles per second. The sound of lightning (thunder) travels at the speed of sound, which is about 750 miles per hour.
    a. If you see a flash of lightning, then hear the thunder 4 seconds later, how far away is the lightning?
    b. Now let's generalize that result. Suppose it takes n seconds to hear the thunder after a flash of lightning. How far away is the lightning, in terms of n?

60. Sound travels about 750 miles per hour. If you stand in a parking lot near a building and sound a horn, you will hear an echo.
    a. Suppose it takes about ½ a second to hear the echo. How far away is the building[13]?
    b. Now let's generalize that result. Suppose it takes n seconds to hear the echo. How far away is the building, in terms of n?

61. It takes an air pump 5 minutes to fill a twin sized air mattress (39 by 8.75 by 75 inches). How long will it take to fill a queen sized mattress (60 by 8.75 by 80 inches)?

62. It takes your garden hose 20 seconds to fill your 2-gallon watering can. How long will it take to fill
    a. An inflatable pool measuring 3 feet wide, 8 feet long, and 1 foot deep.[14]
    b. A circular inflatable pool 13 feet in diameter and 3 feet deep.[15]

63. You want to put a 2" thick layer of topsoil for a new 20'x30' garden. The dirt store sells by the cubic yards. How many cubic yards will you need to order?

64. A box of Jell-O costs $0.50, and makes 2 cups. How much would it cost to fill a swimming pool 4 feet deep, 8 feet wide, and 12 feet long with Jell-O? (1 cubic foot is about 7.5 gallons)

65. You read online that a 15 ft by 20 ft brick patio would cost about $2,275 to have professionally installed. Estimate the cost of having a 18 by 22 ft brick patio installed.

66. I was at the store, and saw two sizes of avocados being sold. The regular size sold for $0.88 each, while the jumbo ones sold for $1.68 each. Which is the better deal?

[13] http://vimeo.com/40377128
[14] http://www.youtube.com/watch?v=DIkwefReHZc
[15] http://www.youtube.com/watch?v=p9SABH7Yg9M

24

67. The grocery store has bulk pecans on sale, which is great since you're planning on making 10 pecan pies for a wedding. Your recipe calls for 1¾ cups pecans per pie. However, in the bulk section there's only a scale available, not a measuring cup. You run over to the baking aisle and find a bag of pecans, and look at the nutrition label to gather some info. How many pounds of pecans should you buy?

Nutrition Facts

Serving Size: 1 cup, halves (99 g)
Servings per Container: about 2

Amount Per Serving

Calories 684 Calories from Fat 596

	% Daily Value*
Total Fat 71g	110%
Saturated Fat 6g	31%
Trans Fat	
Cholesterol 0mg	0%

68. Soda is often sold in 20 ounce bottles. The nutrition label for one of these bottles is shown to the right. A packet of sugar (the kind they have at restaurants for your coffee or tea) typically contain 4 grams of sugar in the U.S. Drinking a 20 oz soda is equivalent to eating how many packets of sugar?[16]

Nutrition Facts

Serving Size: 8 fl oz (240 mL)
Servings Per Container: about 2.5

Amount Per Serving

Calories 110

	% Daily Value*
Total Fat 0g	0%
Sodium 70mg	3%
Total Carbohydrate 31g	10%
Sugars 30g	
Protein 0g	

For the next set of questions, *first* identify the information you need to answer the question, and *then* turn to the end of the section to find that information. The details may be imprecise; answer the question the best you can with the provided information. Be sure to justify your decision.

69. You're planning on making 6 meatloafs for a party. You go to the store to buy breadcrumbs, and see they are sold by the canister. How many canisters do you need to buy?

70. Your friend wants to cover their car in bottle caps, like in this picture.[17] How many bottle caps are you going to need?

71. You need to buy some chicken for dinner tonight. You found an ad showing that the store across town has it on sale for $2.99 a pound, which is cheaper than your usual neighborhood store, which sells it for $3.79 a pound. Is it worth the extra drive?

---

[16] http://www.youtube.com/watch?v=62JMfv0tf3Q

[17] Photo credit: http://www.flickr.com/photos/swayze/, CC-BY

72. I have an old gas furnace, and am considering replacing it with a new, high efficiency model. Is upgrading worth it?

73. Janine is considering buying a water filter and a reusable water bottle rather than buying bottled water. Will doing so save her money?

74. Marcus is considering going car-free to save money and be more environmentally friendly. Is this financially a good decision?

For the next set of problems, research or make educated estimates for any unknown quantities needed to answer the question.

75. You want to travel from Tacoma, WA to Chico, CA for a wedding. Compare the costs and time involved with driving, flying, and taking a train. Assume that if you fly or take the train you'll need to rent a car while you're there. Which option is best?

76. You want to paint the walls of a 6ft by 9ft storage room that has one door and one window. You want to put on two coats of paint. How many gallons and/or quarts of paint should you buy to paint the room as cheaply as possible?

77. A restaurant in New York tiled their floor with pennies[18]. Just for the materials, is this more expensive than using a more traditional material like ceramic tiles? If each penny has to be laid by hand, estimate how long it would take to lay the pennies for a 12ft by 10ft room. Considering material and labor costs, are pennies a cost-effective replacement for ceramic tiles?

78. You are considering taking up part of your back yard and turning it into a vegetable garden, to grow broccoli, tomatoes, and zucchini. Will doing so save you money, or cost you more than buying vegetables from the store?

79. Barry is trying to decide whether to keep his 1993 Honda Civic with 140,000 miles, or trade it in for a used 2008 Honda Civic. Consider gas, maintenance, and insurance costs in helping him make a decision.

80. Some people claim it costs more to eat vegetarian, while some claim it costs less. Examine your own grocery habits, and compare your current costs to the costs of switching your diet (from omnivore to vegetarian or vice versa as appropriate). Which diet is more cost effective based on your eating habits?

---

[18] http://www.notcot.com/archives/2009/06/floor-of-pennie.php

Info for the breadcrumbs question

How much breadcrumbs does the recipe call for?

It calls for 1½ cups of breadcrumbs.

How many meatloafs does the recipe make?

It makes 1 meatloaf.

How many servings does that recipe make?

It says it serves 8.

How big is the canister?

It is cylindrical, 3.5 inches across and
7 inches tall.

What is the net weight of the contents of 1 canister?

15 ounces.

How much does a cup of breadcrumbs weigh?

I'm not sure, but maybe something from the nutritional label will help.

How much does a canister cost? $2.39

Nutrition Facts	
Serving Size: 1/3 cup (30g)	
Servings per Container: about 14	
Amount Per Serving	
Calories 110 Calories from Fat 15	
	% Daily Value*
Total Fat 1.5g	2%

Info for bottle cap car

What kind of car is that?

A 1993 Honda Accord.

How big is that car / what are the dimensions? Here is some details from MSN autos:

Weight: 2800lb   Length: 185.2 in   Width: 67.1 in   Height: 55.2 in

How much of the car was covered with caps?

Everything but the windows and the underside.

How big is a bottle cap?

Caps are 1 inch in diameter.

Info for chicken problem

How much chicken will you be buying?

Four pounds

How far are the two stores?

My neighborhood store is 2.2 miles away, and takes about 7 minutes. The store
across town is 8.9 miles away, and takes about 25 minutes.

What kind of mileage does your car get?

It averages about 24 miles per gallon in the city.

How many gallons does your car hold?

About 14 gallons

How much is gas?

About $3.69/gallon right now.

Info for furnace problem

How efficient is the current furnace?

It is a 60% efficient furnace.

How efficient is the new furnace?

It is 94% efficient.

What is your gas bill?

Here is the history for 2 years:

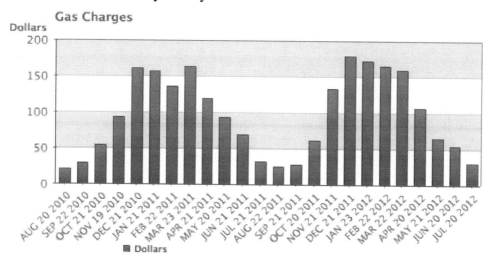

How much do you pay for gas?

There is $10.34 base charge, plus $0.39097 per Therm for a delivery charge, and $0.65195 per Therm for cost of gas.

How much gas do you use?

Here is the history for 2 years:

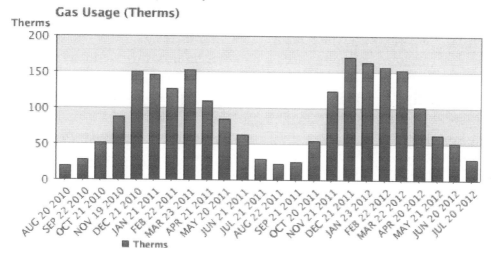

How much does the new furnace cost?

It will cost $7,450.

How long do you plan to live in the house?

Probably at least 15 years.

Info for water filter problem

How much water does Janine drink in a day?

　　　She normally drinks 3 bottles a day, each 16.9 ounces.

How much does a bottle of water cost?

　　　She buys 24-packs of 16.9 ounce bottles for $3.99.

How much does a reusable water bottle cost?

　　　About $10.

How long does a reusable water bottle last?

　　　Basically forever (or until you lose it).

How much does a water filter cost? How much water will they filter?

- A faucet-mounted filter costs about $28. Refill filters cost about $33 for a 3-pack. The box says each filter will filter up to 100 gallons (378 liters)
- A water filter pitcher costs about $22. Refill filters cost about $20 for a 4-pack. The box says each filter lasts for 40 gallons or 2 months
- An under-sink filter costs $130. Refill filters cost about $60 each. The filter lasts for 500 gallons.

Info for car-free problem

Where does Marcus currently drive? He:

- Drives to work 5 days a week, located 4 miles from his house.
- Drives to the store twice a week, located 7 miles from his house.
- Drives to other locations on average 5 days a week, with locations ranging from 1 mile to 20 miles.
- Drives to his parent's house 80 miles away once a month.

How will he get to these locations without a car?

- For work, he can walk when it's sunny and he gets up early enough. Otherwise he can take a bus, which takes about 20 minutes
- For the store, he can take a bus, which takes about 35 minutes.
- Some of the other locations he can bus to. Sometimes he'll be able to get a friend to pick him up. A few locations he is able to walk to. A couple locations are hard to get to by bus, but there is a ZipCar (short term car rental) location within a few blocks.
- He'll need to get a ZipCar to visit his parents.

How much does gas cost?

　　　About $3.69/gallon.

How much does he pay for insurance and maintenance?

- He pays $95/month for insurance.
- He pays $30 every 3 months for an oil change, and has averaged about $300/year for other maintenance costs.

How much is he paying for the car?

- He's paying $220/month on his car loan right now, and has 3 years left on the loan.
- If he sold the car, he'd be able to make enough to pay off the loan.
- If he keeps the car, he's planning on trading the car in for a newer model in a couple years.

What mileage does his car get?

　　　About 26 miles per gallon on average.

How much does a bus ride cost?

　　　$2.50 per trip, or $90 for an unlimited monthly pass.

How much does a ZipCar rental cost?

- The "occasional driving plan": $25 application fee and $60 annual fee, with no monthly commitment. Monday-Thursday the cost is $8/hour, or $72 per day. Friday-Sunday the cost is $8/hour or $78/day. Gas, insurance, and 180 miles are included in the cost. Additional miles are $0.45/mile.

- The "extra value plan": Same as above, but with a $50 monthly commitment, getting you a 10% discount on the usage costs.

Extension: Taxes

Governments collect taxes to pay for the services they provide. In the United States, federal income taxes help fund the military, the environmental protection agency, and thousands of other programs. Property taxes help fund schools. Gasoline taxes help pay for road improvements. While very few people enjoy paying taxes, they are necessary to pay for the services we all depend upon.

Taxes can be computed in a variety of ways, but are typically computed as a percentage of a sale, of one's income, or of one's assets.

Example 1

The sales tax rate in a city is 9.3%. How much sales tax will you pay on a $140 purchase?

The sales tax will be 9.3% of $140. To compute this, we multiply $140 by the percent written as a decimal: $140(0.093) = $13.02.

When taxes are not given as a fixed percentage rate, sometimes it is necessary to calculate the **effective rate.**

Effective rate
The effective tax rate is the equivalent percent rate of the tax paid out of the dollar amount the tax is based on.

Example 2

Joan paid $3,200 in property taxes on her house valued at $215,000 last year. What is the effective tax rate?

We can compute the equivalent percentage: $3200/215000 = 0.01488$, or about 1.49% effective rate.

Taxes are often referred to as progressive, regressive, or flat.

Tax categories
A **flat tax**, or proportional tax, charges a constant percentage rate.
A **progressive tax** increases the percent rate as the base amount increases.
A **regressive tax** decreases the percent rate as the base amount increases.

Example 3

The United States federal income tax on earned wages is an example of a progressive tax. People with a higher wage income pay a higher percent tax on their income.

For a single person in 2011, adjusted gross income (income after deductions) under $8,500 was taxed at 10%. Income over $8,500 but under $34,500 was taxed at 15%.

A person earning $10,000 would pay 10% on the portion of their income under $8,500, and 15% on the income over $8,500, so they'd pay:

        8500(0.10) = 850   10% of $8500
        1500(0.15) = 225   15% of the remaining $1500 of income
        Total tax: = $1075

The effective tax rate paid is 1075/10000 = 10.75%

A person earning $30,000 would also pay 10% on the portion of their income under $8,500, and 15% on the income over $8,500, so they'd pay:

        8500(0.10) = 850   10% of $8500
        21500(0.15) = 3225   15% of the remaining $21500 of income
        Total tax: = $4075

The effective tax rate paid is 4075/30000 = 13.58%.

Notice that the effective rate has increased with income, showing this is a progressive tax.

Example 4

A gasoline tax is a flat tax when considered in terms of consumption, a tax of, say, $0.30 per gallon is proportional to the amount of gasoline purchased. Someone buying 10 gallons of gas at $4 a gallon would pay $3 in tax, which is $3/$40 = 7.5%. Someone buying 30 gallons of gas at $4 a gallon would pay $9 in tax, which is $9/$120 = 7.5%, the same effective rate.

However, in terms of income, a gasoline tax is often considered a regressive tax. It is likely that someone earning $30,000 a year and someone earning $60,000 a year will drive about the same amount. If both pay $60 in gasoline taxes over a year, the person earning $30,000 has paid 0.2% of their income, while the person earning $60,000 has paid 0.1% of their income in gas taxes.

Try it Now 1

A sales tax is a fixed percentage tax on a person's purchases. Is this a flat, progressive, or regressive tax?

Try it Now Answers

1. While sales tax is a flat percentage rate, it is often considered a regressive tax for the same reasons as the gasoline tax.

Income Taxation

Many people have proposed various revisions to the income tax collection in the United States. Some, for example, have claimed that a flat tax would be fairer. Others call for revisions to how different types of income are taxed, since currently investment income is taxed at a different rate than wage income.

The following two projects will allow you to explore some of these ideas and draw your own conclusions.

Project 1: Flat tax, Modified Flat Tax, and Progressive Tax.

Imagine the country is made up of 100 households. The federal government needs to collect $800,000 in income taxes to be able to function. The population consists of 6 groups:

Group A: 20 households that earn $12,000 each
Group B: 20 households that earn $29,000 each
Group C: 20 households that earn $50,000 each
Group D: 20 households that earn $79,000 each
Group E: 15 households that earn $129,000 each
Group F: 5 households that earn $295,000 each

This scenario is roughly proportional to the actual United States population and tax needs. We are going to determine new income tax rates.

The first proposal we'll consider is a flat tax – one where every income group is taxed at the same percentage tax rate.

1) Determine the total income for the population (all 100 people together)

2) Determine what flat tax rate would be necessary to collect enough money.

The second proposal we'll consider is a modified flat-tax plan, where everyone only pays taxes on any income over $20,000. So, everyone in group A will pay no taxes. Everyone in group B will pay taxes only on $9,000.

3) Determine the total *taxable* income for the whole population

4) Determine what flat tax rate would be necessary to collect enough money in this modified system

5) Complete this table for both the plans

Group	Income per household	Flat Tax Plan		Modified Flat Tax Plan	
		Income tax per household	Income after taxes	Income tax per household	Income after taxes
A	$12,000				
B	$29,000				
C	$50,000				
D	$79,000				
E	$129,000				
F	$295,000				

The third proposal we'll consider is a progressive tax, where lower income groups are taxed at a lower percent rate, and higher income groups are taxed at a higher percent rate. For simplicity, we're going to assume that a household is taxed at the same rate on *all* their income.

6) Set progressive tax rates for each income group to bring in enough money. There is no one right answer here – just make sure you bring in enough money!

Group	Income per household	Tax rate (%)	Income tax per household	Total tax collected for all households	Income after taxes per household
A	$12,000				
B	$29,000				
C	$50,000				
D	$79,000				
E	$129,000				
F	$295,000				
				This better total to $800,000	

7) Discretionary income is the income people have left over after paying for necessities like rent, food, transportation, etc. The cost of basic expenses does increase with income, since housing and car costs are higher, however usually not proportionally. For each income group, estimate their essential expenses, and calculate their discretionary income. Then compute the effective tax rate for each plan relative to discretionary income rather than income.

Group	Income per household	Discretionary Income (estimated)	Effective rate, flat	Effective rate, modified	Effective rate, progressive
A	$12,000				
B	$29,000				
C	$50,000				
D	$79,000				
E	$129,000				
F	$295,000				

8) Which plan seems the most fair to you? Which plan seems the least fair to you? Why?

Project 2: Calculating Taxes.

Visit www.irs.gov, and download the most recent version of forms 1040, and schedules A, B, C, and D.

Scenario 1: Calculate the taxes for someone who earned $60,000 in standard wage income (W-2 income), has no dependents, and takes the standard deduction.

Scenario 2: Calculate the taxes for someone who earned $20,000 in standard wage income, $40,000 in qualified dividends, has no dependents, and takes the standard deduction. (Qualified dividends are earnings on certain investments such as stocks.)

Scenario 3: Calculate the taxes for someone who earned $60,000 in small business income, has no dependents, and takes the standard deduction.

Based on these three scenarios, what are your impressions of how the income tax system treats these different forms of income (wage, dividends, and business income)?

Scenario 4: To get a more realistic sense for calculating taxes, you'll need to consider itemized deductions. Calculate the income taxes for someone with the income and expenses listed below.

Married with 2 children, filing jointly
Wage income: $50,000 combined
Paid sales tax in Washington State
Property taxes paid: $3200
Home mortgage interest paid: $4800
Charitable gifts: $1200

Voting Theory

In many decision making situations, it is necessary to gather the group consensus. This happens when a group of friends decides which movie to watch, when a company decides which product design to manufacture, and when a democratic country elects its leaders.

While the basic idea of voting is fairly universal, the method by which those votes are used to determine a winner can vary. Amongst a group of friends, you may decide upon a movie by voting for all the movies you're willing to watch, with the winner being the one with the greatest approval. A company might eliminate unpopular designs then revote on the remaining. A country might look for the candidate with the most votes.

In deciding upon a winner, there is always one main goal: to reflect the preferences of the people in the most fair way possible.

Preference Schedules

To begin, we're going to want more information than a traditional ballot normally provides. A traditional ballot usually asks you to pick your favorite from a list of choices. This ballot fails to provide any information on how a voter would rank the alternatives if their first choice was unsuccessful.

> **Preference ballot**
> A **preference ballot** is a ballot in which the voter ranks the choices in order of preference.

Example 1

A vacation club is trying to decide which destination to visit this year: Hawaii (H), Orlando (O), or Anaheim (A). Their votes are shown below:

	Bob	Ann	Marv	Alice	Eve	Omar	Lupe	Dave	Tish	Jim
1st choice	A	A	O	H	A	O	H	O	H	A
2nd choice	O	H	H	A	H	H	A	H	A	H
3rd choice	H	O	A	O	O	A	O	A	O	O

These individual ballots are typically combined into one **preference schedule**, which shows the number of voters in the top row that voted for each option:

	1	3	3	3
1st choice	A	A	O	H
2nd choice	O	H	H	A
3rd choice	H	O	A	O

Notice that by totaling the vote counts across the top of the preference schedule we can recover the total number of votes cast: 1+3+3+3 = 10 total votes.

Plurality

The voting method we're most familiar with in the United States is the **plurality method**.

> **Plurality Method**
> In this method, the choice with the most first-preference votes is declared the winner. Ties are possible, and would have to be settled through some sort of run-off vote.

This method is sometimes mistakenly called the majority method, or "majority rules", but it is not necessary for a choice to have gained a majority of votes to win. A majority is over 50%; it is possible for a winner to have a **plurality** without having a majority.

Example 2

In our election from above, we had the preference table:

	1	3	3	3
1st choice	A	A	O	H
2nd choice	O	H	H	A
3rd choice	H	O	A	O

For the plurality method, we only care about the first choice options. Totaling them up:
Anaheim: 1+3 = 4 first-choice votes
Orlando: 3 first-choice votes
Hawaii: 3 first-choice votes

Anaheim is the winner using the plurality voting method.

Notice that Anaheim won with 4 out of 10 votes, 40% of the votes, which is a plurality of the votes, but not a majority.

Try it Now 1

Three candidates are running in an election for County Executive: Goings (G), McCarthy (M), and Bunney (B)[1]. The voting schedule is shown below. Which candidate wins under the plurality method?

	44	14	20	70	22	80	39
1st choice	G	G	G	M	M	B	B
2nd choice	M	B		G	B	M	
3rd choice	B	M		B	G	G	

Note: In the third column and last column, those voters only recorded a first-place vote, so we don't know who their second and third choices would have been.

---

[1] This data is loosely based on the 2008 County Executive election in Pierce County, Washington. See http://www.co.pierce.wa.us/xml/abtus/ourorg/aud/Elections/RCV/ranked/exec/summary.pdf

What's Wrong with Plurality?

The election from Example 2 may seem totally clean, but there is a problem lurking that arises whenever there are three or more choices. Looking back at our preference table, how would our members vote if they only had two choices?

Anaheim vs Orlando: 7 out of the 10 would prefer Anaheim over Orlando

	1	3	3	3
1st choice	A	A	O	H
2nd choice	O	H	H	A
3rd choice	H	O	A	O

Anaheim vs Hawaii: 6 out of 10 would prefer Hawaii over Anaheim

	1	3	3	3
1st choice	A	A	O	H
2nd choice	O	H	H	A
3rd choice	H	O	A	O

This doesn't seem right, does it? Anaheim just won the election, yet 6 out of 10 voters, 60% of them, would have preferred Hawaii! That hardly seems fair. Marquis de Condorcet, a French philosopher, mathematician, and political scientist wrote about how this could happen in 1785, and for him we name our first **fairness criterion**.

Fairness Criteria
The fairness criteria are statements that seem like they should be true in a fair election.

Condorcet Criterion
If there is a choice that is preferred in every one-to-one comparison with the other choices, that choice should be the winner. We call this winner the **Condorcet Winner**, or Condorcet Candidate.

Example 3

In the election from Example 2, what choice is the Condorcet Winner?

We see above that Hawaii is preferred over Anaheim. Comparing Hawaii to Orlando, we can see 6 out of 10 would prefer Hawaii to Orlando.

	1	3	3	3
1st choice	A	A	O	H
2nd choice	O	H	H	A
3rd choice	H	O	A	O

Since Hawaii is preferred in a one-to-one comparison to both other choices, Hawaii is the Condorcet Winner.

Example 4

Consider a city council election in a district that is historically 60% Democratic voters and 40% Republican voters. Even though city council is technically a nonpartisan office, people generally know the affiliations of the candidates. In this election there are three candidates: Don and Key, both Democrats, and Elle, a Republican. A preference schedule for the votes looks as follows:

	342	214	298
1st choice	Elle	Don	Key
2nd choice	Don	Key	Don
3rd choice	Key	Elle	Elle

We can see a total of 342 + 214 + 298 = 854 voters participated in this election. Computing percentage of first place votes:

Don: 214/854 = 25.1%
Key: 298/854 = 34.9%
Elle: 342/854 = 40.0%

So in this election, the Democratic voters split their vote over the two Democratic candidates, allowing the Republican candidate Elle to win under the plurality method with 40% of the vote.

Analyzing this election closer, we see that it violates the Condorcet Criterion. Analyzing the one-to-one comparisons:

Elle vs Don: 342 prefer Elle; 512 prefer Don: Don is preferred
Elle vs Key: 342 prefer Elle; 512 prefer Key: Key is preferred
Don vs Key: 556 prefer Don; 298 prefer Key: Don is preferred

So even though Don had the smallest number of first-place votes in the election, he is the Condorcet winner, being preferred in every one-to-one comparison with the other candidates.

Try it Now 2

Consider the election from Try it Now 1. Is there a Condorcet winner in this election?

	44	14	20	70	22	80	39
1st choice	G	G	G	M	M	B	B
2nd choice	M	B		G	B	M	
3rd choice	B	M		B	G	G	

Insincere Voting

Situations like the one in Example 4 above, when there are more than one candidate that share somewhat similar points of view, can lead to **insincere voting**. Insincere voting is when a person casts a ballot counter to their actual preference for strategic purposes. In the case above, the democratic leadership might realize that Don and Key will split the vote, and encourage voters to vote for Key by officially endorsing him. Not wanting to see their party lose the election, as happened in the scenario above, Don's supporters might insincerely vote for Key, effectively voting against Elle.

Instant Runoff Voting

Instant Runoff Voting (IRV), also called Plurality with Elimination, is a modification of the plurality method that attempts to address the issue of insincere voting.

> **Instant Runoff Voting (IRV)**
> In IRV, voting is done with preference ballots, and a preference schedule is generated. The choice with the *least* first-place votes is then eliminated from the election, and any votes for that candidate are redistributed to the voters' next choice. This continues until a choice has a majority (over 50%).

This is similar to the idea of holding runoff elections, but since every voter's order of preference is recorded on the ballot, the runoff can be computed without requiring a second costly election.

This voting method is used in several political elections around the world, including election of members of the Australian House of Representatives, and was used for county positions in Pierce County, Washington until it was eliminated by voters in 2009. A version of IRV is used by the International Olympic Committee to select host nations.

Example 5

Consider the preference schedule below, in which a company's advertising team is voting on five different advertising slogans, called A, B, C, D, and E here for simplicity.

Initial votes

	3	4	4	6	2	1
1st choice	B	C	B	D	B	E
2nd choice	C	A	D	C	E	A
3rd choice	A	D	C	A	A	D
4th choice	D	B	A	E	C	B
5th choice	E	E	E	B	D	C

If this was a plurality election, note that B would be the winner with 9 first-choice votes, compared to 6 for D, 4 for C, and 1 for E.

There are total of 3+4+4+6+2+1 = 20 votes. A majority would be 11 votes. No one yet has a majority, so we proceed to elimination rounds.

Round 1: We make our first elimination. Choice A has the fewest first-place votes, so we remove that choice

	3	4	4	6	2	1
1st choice	B	C	B	D	B	E
2nd choice	C		D	C	E	
3rd choice		D	C			D
4th choice	D	B		E	C	B
5th choice	E	E	E	B	D	C

We then shift everyone's choices up to fill the gaps. There is still no choice with a majority, so we eliminate again.

	3	4	4	6	2	1
1st choice	B	C	B	D	B	E
2nd choice	C	D	D	C	E	D
3rd choice	D	B	C	E	C	B
4th choice	E	E	E	B	D	C

Round 2: We make our second elimination. Choice E has the fewest first-place votes, so we remove that choice, shifting everyone's options to fill the gaps.

	3	4	4	6	2	1
1st choice	B	C	B	D	B	D
2nd choice	C	D	D	C	C	B
3rd choice	D	B	C	B	D	C

Notice that the first and fifth columns have the same preferences now, we can condense those down to one column.

	5	4	4	6	1
1st choice	B	C	B	D	D
2nd choice	C	D	D	C	B
3rd choice	D	B	C	B	C

Now B has 9 first-choice votes, C has 4 votes, and D has 7 votes. Still no majority, so we eliminate again.

Round 3: We make our third elimination. C has the fewest votes.

	5	4	4	6	1
1st choice	B	D	B	D	D
2nd choice	D	B	D	B	B

Condensing this down:

	9	11
1st choice	B	D
2nd choice	D	B

D has now gained a majority, and is declared the winner under IRV.

Try it Now 3

Consider again the election from Try it Now 1. Find the winner using IRV.

	44	14	20	70	22	80	39
1st choice	G	G	G	M	M	B	B
2nd choice	M	B		G	B	M	
3rd choice	B	M		B	G	G	

What's Wrong with IRV?

Example 6

Let's return to our City Council Election

	342	214	298
1st choice	Elle	Don	Key
2nd choice	Don	Key	Don
3rd choice	Key	Elle	Elle

In this election, Don has the smallest number of first place votes, so Don is eliminated in the first round. The 214 people who voted for Don have their votes transferred to their second choice, Key.

	342	512
1st choice	Elle	Key
2nd choice	Key	Elle

So Key is the winner under the IRV method.

We can immediately notice that in this election, IRV violates the Condorcet Criterion, since we determined earlier that Don was the Condorcet winner. On the other hand, the temptation has been removed for Don's supporters to vote for Key; they now know their vote will be transferred to Key, not simply discarded.

Example 7

Consider the voting system below.

	37	22	12	29
1st choice	Adams	Brown	Brown	Carter
2nd choice	Brown	Carter	Adams	Adams
3rd choice	Carter	Adams	Carter	Brown

In this election, Carter would be eliminated in the first round, and Adams would be the winner with 66 votes to 34 for Brown.

Now suppose that the results were announced, but election officials accidentally destroyed the ballots before they could be certified, and the votes had to be recast. Wanting to "jump on the bandwagon", 10 of the voters who had originally voted in the order Brown, Adams, Carter change their vote to favor the presumed winner, changing those votes to Adams, Brown, Carter.

	47	22	2	29
1st choice	Adams	Brown	Brown	Carter
2nd choice	Brown	Carter	Adams	Adams
3rd choice	Carter	Adams	Carter	Brown

In this re-vote, Brown will be eliminated in the first round, having the fewest first-place votes. After transferring votes, we find that Carter will win this election with 51 votes to Adams' 49 votes! Even though the only vote changes made *favored* Adams, the change ended up costing Adams the election. This doesn't seem right, and introduces our second fairness criterion:

Monotonicity Criterion
If voters change their votes to increase the preference for a candidate, it should not harm that candidate's chances of winning.

This criterion is violated by this election. Note that even though the criterion is violated in this particular election, it does not mean that IRV always violates the criterion; just that IRV has the potential to violate the criterion in certain elections.

Borda Count

Borda Count is another voting method, named for Jean-Charles de Borda, who developed the system in 1770.

Borda Count
In this method, points are assigned to candidates based on their ranking; 1 point for last choice, 2 points for second-to-last choice, and so on. The point values for all ballots are totaled, and the candidate with the largest point total is the winner.

Example 8

A group of mathematicians are getting together for a conference. The members are coming from four cities: Seattle, Tacoma, Puyallup, and Olympia. Their approximate locations on a map are shown to the right.

The votes for where to hold the conference were:

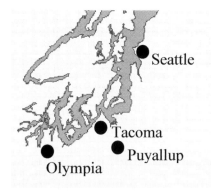

	51	25	10	14
1st choice	Seattle	Tacoma	Puyallup	Olympia
2nd choice	Tacoma	Puyallup	Tacoma	Tacoma
3rd choice	Olympia	Olympia	Olympia	Puyallup
4th choice	Puyallup	Seattle	Seattle	Seattle

In each of the 51 ballots ranking Seattle first, Puyallup will be given 1 point, Olympia 2 points, Tacoma 3 points, and Seattle 4 points. Multiplying the points per vote times the number of votes allows us to calculate points awarded:

	51	25	10	14
1st choice 4 points	Seattle $4 \cdot 51 = 204$	Tacoma $4 \cdot 25 = 100$	Puyallup $4 \cdot 10 = 40$	Olympia $4 \cdot 14 = 56$
2nd choice 3 points	Tacoma $3 \cdot 51 = 153$	Puyallup $3 \cdot 25 = 75$	Tacoma $3 \cdot 10 = 30$	Tacoma $3 \cdot 14 = 42$
3rd choice 2 points	Olympia $2 \cdot 51 = 102$	Olympia $2 \cdot 25 = 50$	Olympia $2 \cdot 10 = 20$	Puyallup $2 \cdot 14 = 28$
4th choice 1 point	Puyallup $1 \cdot 51 = 51$	Seattle $1 \cdot 25 = 25$	Seattle $1 \cdot 10 = 10$	Seattle $1 \cdot 14 = 14$

Adding up the points:
Seattle: $204 + 25 + 10 + 14 = 253$ points
Tacoma: $153 + 100 + 30 + 42 = 325$ points
Puyallup: $51 + 75 + 40 + 28 = 194$ points
Olympia: $102 + 50 + 20 + 56 = 228$ points

Under the Borda Count method, Tacoma is the winner of this vote.

Try it Now 4
Consider again the election from Try it Now 1. Find the winner using Borda Count. Since we have some incomplete preference ballots, for simplicity, give every unranked candidate 1 point, the points they would normally get for last place.

	44	14	20	70	22	80	39
1st choice	G	G	G	M	M	B	B
2nd choice	M	B		G	B	M	
3rd choice	B	M		B	G	G	

What's Wrong with Borda Count?

You might have already noticed one potential flaw of the Borda Count from the previous example. In that example, Seattle had a majority of first-choice votes, yet lost the election! This seems odd, and prompts our next fairness criterion:

> **Majority Criterion**
> If a choice has a majority of first-place votes, that choice should be the winner.

The election from Example 8 using the Borda Count violates the Majority Criterion. Notice also that this automatically means that the Condorcet Criterion will also be violated, as Seattle would have been preferred by 51% of voters in any head-to-head comparison.

Borda count is sometimes described as a consensus-based voting system, since it can sometimes choose a more broadly acceptable option over the one with majority support. In the example above, Tacoma is probably the best compromise location. This is a different approach than plurality and instant runoff voting that focus on first-choice votes; Borda Count considers every voter's entire ranking to determine the outcome.

Because of this consensus behavior, Borda Count, or some variation of it, is commonly used in awarding sports awards. Variations are used to determine the Most Valuable Player in baseball, to rank teams in NCAA sports, and to award the Heisman trophy.

Copeland's Method (Pairwise Comparisons)

So far none of our voting methods have satisfied the Condorcet Criterion. The Copeland Method specifically attempts to satisfy the Condorcet Criterion by looking at pairwise (one-to-one) comparisons.

> **Copeland's Method**
> In this method, each pair of candidates is compared, using all preferences to determine which of the two is more preferred. The more preferred candidate is awarded 1 point. If there is a tie, each candidate is awarded ½ point. After all pairwise comparisons are made, the candidate with the most points, and hence the most pairwise wins, is declared the winner.

Variations of Copeland's Method are used in many professional organizations, including election of the Board of Trustees for the Wikimedia Foundation that runs Wikipedia.

Example 9

Consider our vacation group example from the beginning of the chapter. Determine the winner using Copeland's Method.

	1	3	3	3
1st choice	A	A	O	H
2nd choice	O	H	H	A
3rd choice	H	O	A	O

We need to look at each pair of choices, and see which choice would win in a one-to-one comparison. You may recall we did this earlier when determining the Condorcet Winner. For example, comparing Hawaii vs Orlando, we see that 6 voters, those shaded below in the first table below, would prefer Hawaii to Orlando. Note that Hawaii doesn't have to be the voter's first choice – we're imagining that Anaheim wasn't an option. If it helps, you can imagine removing Anaheim, as in the second table below.

	1	3	3	3
1st choice	A	A	O	H
2nd choice	O	H	H	A
3rd choice	H	O	A	O

	1	3	3	3
1st choice			O	H
2nd choice	O	H	H	
3rd choice	H	O		O

Based on this, in the comparison of Hawaii vs Orlando, Hawaii wins, and receives 1 point.

Comparing Anaheim to Orlando, the 1 voter in the first column clearly prefers Anaheim, as do the 3 voters in the second column. The 3 voters in the third column clearly prefer Orlando. The 3 voters in the last column prefer Hawaii as their first choice, but if they had to choose between Anaheim and Orlando, they'd choose Anaheim, their second choice overall. So, altogether 1+3+3=7 voters prefer Anaheim over Orlando, and 3 prefer Orlando over Anaheim. So, comparing Anaheim vs Orlando: 7 votes to 3 votes: Anaheim gets 1 point.

All together,
Hawaii vs Orlando: 6 votes to 4 votes: Hawaii gets 1 point
Anaheim vs Orlando: 7 votes to 3 votes: Anaheim gets 1 point
Hawaii vs Anaheim: 6 votes to 4 votes: Hawaii gets 1 point

Hawaii is the winner under Copeland's Method, having earned the most points.

Notice this process is consistent with our determination of a Condorcet Winner.

Example 10

Consider the advertising group's vote we explored earlier. Determine the winner using Copeland's method.

	3	4	4	6	2	1
1st choice	B	C	B	D	B	E
2nd choice	C	A	D	C	E	A
3rd choice	A	D	C	A	A	D
4th choice	D	B	A	E	C	B
5th choice	E	E	E	B	D	C

With 5 candidates, there are 10 comparisons to make:

A vs B: 11 votes to 9 votes   A gets 1 point
A vs C: 3 votes to 17 votes   C gets 1 point
A vs D: 10 votes to 10 votes   A gets ½ point, D gets ½ point
A vs E: 17 votes to 3 votes   A gets 1 point
B vs C: 10 votes to 10 votes   B gets ½ point, C gets ½ point
B vs D: 9 votes to 11 votes   D gets 1 point
B vs E: 13 votes to 7 votes   B gets 1 point
C vs D: 9 votes to 11 votes   D gets 1 point
C vs E: 17 votes to 3 votes   C gets 1 point
D vs E: 17 votes to 3 votes   D gets 1 point

Totaling these up:
A gets 2½ points
B gets 1½ points
C gets 2½ points
D gets 3½ points
E gets 0 points

Using Copeland's Method, we declare D as the winner.

Notice that in this case, D is not a Condorcet Winner. While Copeland's method will also select a Condorcet Candidate as the winner, the method still works in cases where there is no Condorcet Winner.

Try it Now 5

Consider again the election from Try it Now 1. Find the winner using Copeland's method. Since we have some incomplete preference ballots, we'll have to adjust. For example, when comparing M to B, we'll ignore the 20 votes in the third column which do not rank either candidate.

	44	14	20	70	22	80	39
1st choice	G	G	G	M	M	B	B
2nd choice	M	B		G	B	M	
3rd choice	B	M		B	G	G	

What's Wrong with Copeland's Method?

As already noted, Copeland's Method does satisfy the Condorcet Criterion. It also satisfies the Majority Criterion and the Monotonicity Criterion. So is this the perfect method? Well, in a word, no.

Example 11

A committee is trying to award a scholarship to one of four students, Anna (A), Brian (B), Carlos (C), and Dimitry (D). The votes are shown below:

	5	5	6	4
1st choice	D	A	C	B
2nd choice	A	C	B	D
3rd choice	C	B	D	A
4th choice	B	D	A	C

Making the comparisons:
A vs B: 10 votes to 10 votes   A gets ½ point, B gets ½ point
A vs C: 14 votes to 6 votes: A gets 1 point
A vs D: 5 votes to 15 votes: D gets 1 point
B vs C: 4 votes to 16 votes: C gets 1 point
B vs D: 15 votes to 5 votes: B gets 1 point
C vs D: 11 votes to 9 votes: C gets 1 point

Totaling:
A has 1 ½ points   B has 1 ½ points
C has 2 points   D has 1 point

So Carlos is awarded the scholarship. However, the committee then discovers that Dimitry was not eligible for the scholarship (he failed his last math class). Even though this seems like it shouldn't affect the outcome, the committee decides to recount the vote, removing Dimitry from consideration. This reduces the preference schedule to:

	5	5	6	4
1st choice	A	A	C	B
2nd choice	C	C	B	A
3rd choice	B	B	A	C

A vs B: 10 votes to 10 votes   A gets ½ point, B gets ½ point
A vs C: 14 votes to 6 votes   A gets 1 point
B vs C: 4 votes to 16 votes   C gets 1 point

Totaling:
A has 1 ½ points   B has ½ point
C has 1 point

Suddenly Anna is the winner! This leads us to another fairness criterion.

> **The Independence of Irrelevant Alternatives (IIA) Criterion**
> If a non-winning choice is removed from the ballot, it should not change the winner of the election.
>
> Equivalently, if choice A is preferred over choice B, introducing or removing a choice C should not cause B to be preferred over A.

In the election from Example 11, the IIA Criterion was violated.

This anecdote illustrating the IIA issue is attributed to Sidney Morgenbesser:

> After finishing dinner, Sidney Morgenbesser decides to order dessert. The waitress tells him he has two choices: apple pie and blueberry pie. Sidney orders the apple pie. After a few minutes the waitress returns and says that they also have cherry pie at which point Morgenbesser says "In that case I'll have the blueberry pie."

Another disadvantage of Copeland's Method is that it is fairly easy for the election to end in a tie. For this reason, Copeland's method is usually the first part of a more advanced method that uses more sophisticated methods for breaking ties and determining the winner when there is not a Condorcet Candidate.

So Where's the Fair Method?

At this point, you're probably asking why we keep looking at method after method just to point out that they are not fully fair. We must be holding out on the perfect method, right?

Unfortunately, no. A mathematical economist, Kenneth Arrow, was able to prove in 1949 that there is <u>no</u> voting method that will satisfy all the fairness criteria we have discussed.

> **Arrow's Impossibility Theorem**
> **Arrow's Impossibility Theorem** states, roughly, that it is not possible for a voting method to satisfy every fairness criteria that we've discussed.

To see a very simple example of how difficult voting can be, consider the election below:

	5	5	5
1st choice	A	C	B
2nd choice	B	A	C
3rd choice	C	B	A

Notice that in this election:
10 people prefer A to B
10 people prefer B to C
10 people prefer C to A

No matter whom we choose as the winner, 2/3 of voters would prefer someone else! This scenario is dubbed **Condorcet's Voting Paradox**, and demonstrates how voting preferences are not transitive (just because A is preferred over B, and B over C, does not mean A is preferred over C). In this election, there is no fair resolution.

It is because of this impossibility of a totally fair method that Plurality, IRV, Borda Count, Copeland's Method, and dozens of variants are all still used. Usually the decision of which method to use is based on what seems most fair for the situation in which it is being applied.

Approval Voting

Up until now, we've been considering voting methods that require ranking of candidates on a preference ballot. There is another method of voting that can be more appropriate in some decision making scenarios. With Approval Voting, the ballot asks you to mark all choices that you find acceptable. The results are tallied, and the option with the most approval is the winner.

Example 12

A group of friends is trying to decide upon a movie to watch. Three choices are provided, and each person is asked to mark with an "X" which movies they are willing to watch. The results are:

	Bob	Ann	Marv	Alice	Eve	Omar	Lupe	Dave	Tish	Jim
Titanic		X	X			X		X		X
Scream	X		X	X		X	X		X	
The Matrix	X	X	X	X	X		X			X

Totaling the results, we find
Titanic received 5 approvals
Scream received 6 approvals
The Matrix received 7 approvals.

In this vote, The Matrix would be the winner.

Try it Now 6
Our mathematicians deciding on a conference location from earlier decide to use Approval voting. Their votes are tallied below. Find the winner using Approval voting.

	30	10	15	20	15	5	5
Seattle	X	X	X			X	
Tacoma	X		X	X	X	X	
Puyallup		X		X	X	X	
Olympia			X		X		X

50

What's Wrong with Approval Voting?

Approval voting can very easily violate the Majority Criterion.

Consider the voting schedule:

	80	15	5
1st choice	A	B	C
2nd choice	B	C	B
3rd choice	C	A	A

Clearly A is the majority winner. Now suppose that this election was held using Approval Voting, and every voter marked approval of their top two candidates.

A would receive approval from 80 voters
B would receive approval from 100 voters
C would receive approval from 20 voters

B would be the winner. Some argue that Approval Voting tends to vote the least disliked choice, rather than the most liked candidate.

Additionally, Approval Voting is susceptible to strategic insincere voting, in which a voter does not vote their true preference to try to increase the chances of their choice winning. For example, in the movie example above, suppose Bob and Alice would much rather watch Scream. They remove The Matrix from their approval list, resulting in a different result.

	Bob	Ann	Marv	Alice	Eve	Omar	Lupe	Dave	Tish	Jim
Titanic		X	X			X		X		X
Scream	X		X	X		X	X		X	
The Matrix		X	X		X		X			X

Totaling the results, we find Titanic received 5 approvals, Scream received 6 approvals, and The Matrix received 5 approvals. By voting insincerely, Bob and Alice were able to sway the result in favor of their preference.

Voting in America

In American politics, there is a lot more to selecting our representatives than simply casting and counting ballots. The process of selecting the president is even more complicated, so we'll save that for the next chapter. Instead, let's look at the process by which state congressional representatives and local politicians get elected.

For most offices, a sequence of two public votes is held: a primary election and the general election. For non-partisan offices like sheriff and judge, in which political party affiliation is not declared, the primary election is usually used to narrow the field of candidates.

Typically, the two candidates receiving the most votes in the primary will then move forward to the general election. While somewhat similar to instant runoff voting, this is actually an example of **sequential voting** - a process in which voters cast totally new ballots after each round of eliminations. Sequential voting has become quite common in television, where it is used in reality competition shows like American Idol.

Congressional, county, and city representatives are partisan offices, in which candidates usually declare themselves a member of a political party, like the Democrats, Republicans, the Green Party, or one of the many other smaller parties. As with non-partisan offices, a primary election is usually held to narrow down the field prior to the general election. Prior to the primary election, the candidate would have met with the political party leaders and gotten their approval to run under that party's affiliation.

In some states a **closed primary** is used, in which only voters who are members of the Democrat party can vote on the Democratic candidates, and similar for Republican voters. In other states, an **open primary** is used, in which any voter can pick the party whose primary they want to vote in. In other states, **caucuses** are used, which are basically meetings of the political parties, only open to party members. Closed primaries are often disliked by independent voters, who like the flexibility to change which party they are voting in. Open primaries do have the disadvantage that they allow raiding, in which a voter will vote in their non-preferred party's primary with the intent of selecting a weaker opponent for their preferred party's candidate.

Washington State currently uses a different method, called a **top 2 primary**, in which voters select from the candidates from all political parties on the primary, and the top two candidates, regardless of party affiliation, move on to the general election. While this method is liked by independent voters, it gives the political parties incentive to select a top candidate internally before the primary, so that two candidates will not split the party's vote.

Regardless of the primary type, the general election is the main election, open to all voters. Except in the case of the top 2 primary, the top candidate from each major political party would be included in the general election. While rules vary state-to-state, for an independent or minor party candidate to get listed on the ballot, they typically have to gather a certain number of signatures to petition for inclusion.

Try it Now Answers
1. Using plurality method:
G gets 44+14+20 = 78 first-choice votes
M gets 70+22 = 92 first-choice votes
B gets 80+39 = 119 first-choice votes
Bunney (B) wins under plurality method.

2. Determining the Condorcet Winner:

G vs M: 44+14+20 = 78 prefer G, 70+22+80=172 prefer M: M preferred

G vs B: 44+14+20+70=148 prefer G, 22+80+39 = 141 prefer B: G preferred

M vs B: 44+70+22=136 prefer M, 14+80+39=133 prefer B: M preferred

M is the Condorcet winner, based on the information we have.

3. Using IRV:

G has the fewest first-choice votes, so is eliminated first. The 20 voters who did not list a second choice do not get transferred - they simply get eliminated

	136	133
1st choice	M	B
2nd choice	B	M

McCarthy (M) now has a majority, and is declared the winner.

4. Using Borda Count:

We give 1 point for 3rd place, 2 points for 2nd place, and 3 points for 1st place.

	44	14	20	70	22	80	39
1st choice	G 132 pt	G 42 pt	G 60 pt	M 210 pt	M 66 pt	B 240 pt	B 117 pt
2nd choice	M 88 pt	B 28 pt		G 140 pt	B 44 pt	M 160 pt	
3rd choice	B 44 pt	M 14 pt	M 20 pt B 20 pt	B 70 pt	G 22 pt	G 80 pt	M 39 pt G 39 pt

G: 132+42+60+140+22+80+39 = 515 pts

M: 88+14+20+210+66+160+39 = 597 pts

B: 44+28+20+70+44+240+117 = 563 pts

McCarthy (M) would be the winner using Borda Count.

5. Using Copeland's Method:

Looking back at our work from Try it Now #2, we see

G vs M: 44+14+20 = 78 prefer G, 70+22+80=172 prefer M: M preferred – 1 point

G vs B: 44+14+20+70=148 prefer G, 22+80+39 = 141 prefer B: G preferred – 1 point

M vs B: 44+70+22=136 prefer M, 14+80+39=133 prefer B: M preferred – 1 point

M earns 2 points; G earns 1 point. M wins under Copeland's method.

6. Using Approval voting:

Seattle has 30+10+15+5 = 60 approval votes

Tacoma has 30+15+20+15+5 = 85 approval votes

Puyallup has 10+20+25+5 = 50 approval votes

Olympia has 15+15+5 = 35 approval votes

Tacoma wins under this approval voting

Exercises

Skills

1. To decide on a new website design, the designer asks people to rank three designs that have been created (labeled A, B, and C). The individual ballots are shown below. Create a preference table.

   ABC, ABC, ACB, BAC, BCA, BCA, ACB, CAB, CAB, BCA, ACB, ABC

2. To decide on a movie to watch, a group of friends all vote for one of the choices (labeled A, B, and C). The individual ballots are shown below. Create a preference table.

   CAB, CBA, BAC, BCA, CBA, ABC, ABC, CBA, BCA, CAB, CAB, BAC

3. The planning committee for a renewable energy trade show is trying to decide what city to hold their next show in. The votes are shown below.

Number of voters	9	19	11	8
1st choice	Buffalo	Atlanta	Chicago	Buffalo
2nd choice	Atlanta	Buffalo	Buffalo	Chicago
3rd choice	Chicago	Chicago	Atlanta	Atlanta

   a. How many voters voted in this election?
   b. How many votes are needed for a majority? A plurality?
   c. Find the winner under the plurality method.
   d. Find the winner under the Borda Count Method.
   e. Find the winner under the Instant Runoff Voting method.
   f. Find the winner under Copeland's method.

4. A non-profit agency is electing a new chair of the board. The votes are shown below.

Number of voters	11	5	10	3
1st choice	Atkins	Cortez	Burke	Atkins
2nd choice	Cortez	Burke	Cortez	Burke
3rd choice	Burke	Atkins	Atkins	Cortez

   a. How many voters voted in this election?
   b. How many votes are needed for a majority? A plurality?
   c. Find the winner under the plurality method.
   d. Find the winner under the Borda Count Method.
   e. Find the winner under the Instant Runoff Voting method.
   f. Find the winner under Copeland's method.

5. The student government is holding elections for president. There are four candidates (labeled A, B, C, and D for convenience). The preference schedule for the election is:

Number of voters	120	50	40	90	60	100
1st choice	C	B	D	A	A	D
2nd choice	D	C	A	C	D	B
3rd choice	B	A	B	B	C	A
4th choice	A	D	C	D	B	C

    a. How many voters voted in this election?
    b. How many votes are needed for a majority? A plurality?
    c. Find the winner under the plurality method.
    d. Find the winner under the Borda Count Method.
    e. Find the winner under the Instant Runoff Voting method.
    f. Find the winner under Copeland's method.

6. The homeowners association is deciding a new set of neighborhood standards for architecture, yard maintenance, etc. Four options have been proposed. The votes are:

Number of voters	8	9	11	7	7	5
1st choice	B	A	D	A	B	C
2nd choice	C	D	B	B	A	D
3rd choice	A	C	C	D	C	A
4th choice	D	B	A	C	D	B

    a. How many voters voted in this election?
    b. How many votes are needed for a majority? A plurality?
    c. Find the winner under the plurality method.
    d. Find the winner under the Borda Count Method.
    e. Find the winner under the Instant Runoff Voting method.
    f. Find the winner under Copeland's method.

7. Consider an election with 129 votes.
    a. If there are 4 candidates, what is the smallest number of votes that a plurality candidate could have?
    b. If there are 8 candidates, what is the smallest number of votes that a plurality candidate could have?

8. Consider an election with 953 votes.
    a. If there are 7 candidates, what is the smallest number of votes that a plurality candidate could have?
    b. If there are 8 candidates, what is the smallest number of votes that a plurality candidate could have?

9. Does this voting system having a Condorcet Candidate? If so, find it.

Number of voters	14	15	2
1st choice	A	C	B
2nd choice	B	B	C
3rd choice	C	A	A

10. Does this voting system having a Condorcet Candidate? If so, find it.

Number of voters	8	7	6
1st choice	A	C	B
2nd choice	B	B	C
3rd choice	C	A	A

11. The marketing committee at a company decides to vote on a new company logo. They decide to use approval voting. Their results are tallied below. Each column shows the number of voters with the particular approval vote. Which logo wins under approval voting?

Number of voters	8	7	6	3
A	X	X		
B	X		X	X
C		X	X	X

12. The downtown business association is electing a new chairperson, and decides to use approval voting. The tally is below, where each column shows the number of voters with the particular approval vote. Which candidate wins under approval voting?

Number of voters	8	7	6	3	4	2	5
A	X	X			X		X
B	X		X	X			X
C		X	X	X		X	
D	X		X		X	X	

Concepts

13. An election resulted in Candidate A winning, with Candidate B coming in a close second, and candidate C being a distant third. If for some reason the election had to be held again and C decided to drop out of the election, which caused B to become the winner, which is the primary fairness criterion violated in this election?

14. An election resulted in Candidate A winning, with Candidate B coming in a close second, and candidate C being a distant third. If for some reason the election had to be held again and many people who had voted for C switched their preferences to favor A, which caused B to become the winner, which is the primary fairness criterion violated in this election?

15. An election resulted in Candidate A winning, with Candidate B coming in a close second, and candidate C being a distant third. If in a head-to-head comparison a majority of people prefer B to A or C, which is the primary fairness criterion violated in this election?

16. An election resulted in Candidate A winning, with Candidate B coming in a close second, and candidate C being a distant third. If B had received a majority of first place votes, which is the primary fairness criterion violated in this election?

Exploration

17. In the election shown below under the Plurality method, explain why voters in the third column might be inclined to vote insincerely. How could it affect the outcome of the election?

Number of voters	96	90	10
1st choice	A	B	C
2nd choice	B	A	B
3rd choice	C	C	A

18. In the election shown below under the Borda Count method, explain why voters in the second column might be inclined to vote insincerely. How could it affect the outcome of the election?

Number of voters	20	18
1st choice	A	B
2nd choice	B	A
3rd choice	C	C

19. Compare and contrast the motives of the insincere voters in the two questions above.

20. Consider a two party election with preferences shown below. Suppose a third candidate, C, entered the race, and a segment of voters sincerely voted for that third candidate, producing the preference schedule from #17 above. Explain how other voters might perceive candidate C.

Number of voters	96	100
1st choice	A	B
2nd choice	B	A

21. In question 18, we showed that the outcome of Borda Count can be manipulated if a group of individuals change their vote.
    a. Show that it is possible for a single voter to change the outcome under Borda Count if there are four candidates.
    b. Show that it is not possible for a single voter to change the outcome under Borda Count if there are three candidates.

22. Show that when there is a Condorcet winner in an election, it is impossible for a single voter to manipulate the vote to help a different candidate become a Condorcet winner.

23. The Pareto criterion is another fairness criterion that states: *If every voter prefers choice A to choice B, then B should not be the winner.* Explain why plurality, instant runoff, Borda count, and Copeland's method all satisfy the Pareto condition.

24. Sequential Pairwise voting is a method not commonly used for political elections, but sometimes used for shopping and games of pool. In this method, the choices are assigned an order of comparison, called an agenda. The first two choices are compared. The winner is then compared to the next choice on the agenda, and this continues until all choices have been compared against the winner of the previous comparison.
    a. Using the preference schedule below, apply Sequential Pairwise voting to determine the winner, using the agenda: A, B, C, D.

Number of voters	10	15	12
1st choice	C	A	B
2nd choice	A	B	D
3rd choice	B	D	C
4th choice	D	C	A

    b. Show that Sequential Pairwise voting can violate the Pareto criterion.
    c. Show that Sequential Pairwise voting can violate the Majority criterion.

25. The Coombs method is a variation of instant runoff voting. In Coombs method, the choice with the most last place votes is eliminated. Apply Coombs method to the preference schedules from questions 5 and 6.

26. Copeland's Method is designed to identify a Condorcet Candidate if there is one, and is considered a Condorcet Method. There are many Condorcet Methods, which vary primarily in how they deal with ties, which are very common when a Condorcet winner does not exist. Copeland's method does not have a tie-breaking procedure built-in. Research the Schulze method, another Condorcet method that is used by the Wikimedia foundation that runs Wikipedia, and give some examples of how it works.

27. The plurality method is used in most U.S. elections. Some people feel that Ross Perot in 1992 and Ralph Nader in 2000 changed what the outcome of the election would have been if they had not run. Research the outcomes of these elections and explain how each candidate could have affected the outcome of the elections (for the 2000 election, you may wish to focus on the count in Florida). Describe how an alternative voting method could have avoided this issue.

28. Instant Runoff Voting and Approval voting have supporters advocating that they be adopted in the United States and elsewhere to decide elections. Research comparisons between the two methods describing the advantages and disadvantages of each in practice. Summarize the comparisons, and form your own opinion about whether either method should be adopted.

29. In a primary system, a first vote is held with multiple candidates. In some states, each political party has its own primary. In Washington State, there is a "top two" primary, where all candidates are on the ballot and the top two candidates advance to the general election, regardless of party. Compare and contrast the top two primary with general election system to instant runoff voting, considering both differences in the methods, and practical differences like cost, campaigning, fairness, etc.

30. In a primary system, a first vote is held with multiple candidates. In some many states, where voters must declare a party to vote in the primary election, and they are only able to choose between candidates for their declared party. The top candidate from each party then advances to the general election. Compare and contrast this primary with general election system to instant runoff voting, considering both differences in the methods, and practical differences like cost, campaigning, fairness, etc.

31. Sometimes in a voting scenario it is desirable to rank the candidates, either to establish preference order between a set of choices, or because the election requires multiple winners. For example, a hiring committee may have 30 candidates apply, and need to select 6 to interview, so the voting by the committee would need to produce the top 6 candidates. Describe how Plurality, Instant Runoff Voting, Borda Count, and Copeland's Method could be extended to produce a ranked list of candidates.

Weighted Voting

In a corporate shareholders meeting, each shareholders' vote counts proportional to the amount of shares they own. An individual with one share gets the equivalent of one vote, while someone with 100 shares gets the equivalent of 100 votes. This is called **weighted voting**, where each vote has some weight attached to it. Weighted voting is sometimes used to vote on candidates, but more commonly to decide "yes" or "no" on a proposal, sometimes called a motion. Weighted voting is applicable in corporate settings, as well as decision making in parliamentary governments and voting in the United Nations Security Council.

In weighted voting, we are most often interested in the power each voter has in influencing the outcome.

Beginnings

We'll begin with some basic vocabulary for weighted voting systems.

Vocabulary for Weighted Voting

Each individual or entity casting a vote is called a **player** in the election. They're often notated as $P_1, P_2, P_3, \ldots, P_N$, where N is the total number of voters.

Each player is given a **weight**, which usually represents how many votes they get.

The **quota** is the minimum weight needed for the votes or weight needed for the proposal to be approved.

A weighted voting system will often be represented in a shorthand form:

$$[q: w_1, w_2, w_3, \ldots, w_n]$$

In this form, q is the quota, w_1 is the weight for player 1, and so on.

Example 1

In a small company, there are 4 shareholders. Mr. Smith has a 30% ownership stake in the company, Mr. Garcia has a 25% stake, Mrs. Hughes has a 25% stake, and Mrs. Lee has a 20% stake. They are trying to decide whether to open a new location. The company by-laws state that more than 50% of the ownership has to approve any decision like this. This could be represented by the weighted voting system:

[51: 30, 25, 25, 20]

Here we have treated the percentage ownership as votes, so Mr. Smith gets the equivalent of 30 votes, having a 30% ownership stake. Since more than 50% is required to approve the decision, the quota is 51, the smallest whole number over 50.

In order to have a meaningful weighted voting system, it is necessary to put some limits on the quota.

> **Limits on the Quota**
> The quota must be more than ½ the total number of votes.
> The quota can't be larger than the total number of votes.

Why? Consider the voting system $[q: 3, 2, 1]$

Here there are 6 total votes. If the quota was set at only 3, then player 1 could vote yes, players 2 and 3 could vote no, and both would reach quota, which doesn't lead to a decision being made. In order for only one decision to reach quota at a time, the quota must be at least half the total number of votes. If the quota was set to 7, then no group of voters could ever reach quota, and no decision can be made, so it doesn't make sense for the quota to be larger than the total number of voters.

Try it Now 1

In a committee there are four representatives from the management and three representatives from the workers' union. For a proposal to pass, four of the members must support it, including at least one member of the union. Find a voting system that can represent this situation.

A Look at Power

Consider the voting system [10: 11, 3, 2]. Notice that in this system, player 1 can reach quota without the support of any other player. When this happens, we say that player 1 is a **dictator**.

> **Dictator**
> A player will be a dictator if their weight is equal to or greater than the quota. The dictator can also block any proposal from passing; the other players cannot reach quota without the dictator.

In the voting system [8: 6, 3, 2], no player is a dictator. However, in this system, the quota can only be reached if player 1 is in support of the proposal; player 2 and 3 cannot reach quota without player 1's support. In this case, player 1 is said to have **veto power**. Notice that player 1 is not a dictator, since player 1 would still need player 2 or 3's support to reach quota.

> **Veto Power**
> A player has veto power if their support is necessary for the quota to be reached. It is possible for more than one player to have veto power, or for no player to have veto power.

With the system [10: 7, 6, 2], player 3 is said to be a **dummy**, meaning they have no influence in the outcome. The only way the quota can be met is with the support of both players 1 and 2 (both of which would have veto power here); the vote of player 3 cannot affect the outcome.

> **Dummy**
> A player is a dummy if their vote is never essential for a group to reach quota.

Example 2

> In the voting system [16: 7, 6, 3, 3, 2], are any players dictators? Do any have veto power? Are any dummies?
>
> No player can reach quota alone, so there are no dictators.
>
> Without player 1, the rest of the players' weights add to 14, which doesn't reach quota, so player 1 has veto power. Likewise, without player 2, the rest of the players' weights add to 15, which doesn't reach quota, so player 2 also has veto power.
>
> Since player 1 and 2 can reach quota with either player 3 or player 4's support, neither player 3 or player 4 have veto power. However they cannot reach quota with player 5's support alone, so player 5 has no influence on the outcome and is a dummy.

Try it Now 2
In the voting system [q: 10, 5, 3], which players are dictators, have veto power, and are dummies if the quota is 10? 12? 16?

To better define power, we need to introduce the idea of a **coalition**. A coalition is a group of players voting the same way. In the example above, {P_1, P_2, P_4} would represent the coalition of players 1, 2 and 4. This coalition has a combined weight of 7+6+3 = 16, which meets quota, so this would be a winning coalition.

A player is said to be **critical** in a coalition if them leaving the coalition would change it from a winning coalition to a losing coalition. In the coalition {P_1, P_2, P_4}, every player is critical. In the coalition {P_3, P_4, P_5}, no player is critical, since it wasn't a winning coalition to begin with. In the coalition {P_1, P_2, P_3, P_4, P_5}, only players 1 and 2 are critical; any other player could leave the coalition and it would still meet quota.

> **Coalitions and Critical Players**
> A coalition is any group of players voting the same way.
>
> A coalition is a **winning coalition** if the coalition has enough weight to meet quota.
>
> A player is **critical** in a coalition if them leaving the coalition would change it from a winning coalition to a losing coalition.

Example 3

In the Scottish Parliament in 2009 there were 5 political parties: 47 representatives for the Scottish National Party, 46 for the Labour Party, 17 for the Conservative Party, 16 for the Liberal Democrats, and 2 for the Scottish Green Party. Typically all representatives from a party vote as a block, so the parliament can be treated like the weighted voting system:

[65: 47, 46, 17, 16, 2]

Consider the coalition $\{P_1, P_3, P_4\}$. No two players alone could meet the quota, so all three players are critical in this coalition.

In the coalition $\{P_1, P_3, P_4, P_5\}$, any player except P_1 could leave the coalition and it would still meet quota, so only P_1 is critical in this coalition.

Notice that a player with veto power will be critical in every winning coalition, since removing their support would prevent a proposal from passing.

Likewise, a dummy will never be critical, since their support will never change a losing coalition to a winning one.

Dictators, Veto, and Dummies and Critical Players

A player is a **dictator** if the single-player coalition containing them is a winning coalition.

A player has **veto power** if they are critical in every winning coalition.

A player is a **dummy** if they are not critical in any winning coalition.

Calculating Power: Banzhaf Power Index

The **Banzhaf power index** was originally created in 1946 by Lionel Penrose, but was reintroduced by John Banzhaf in 1965. The power index is a numerical way of looking at power in a weighted voting situation.

Calculating Banzhaf Power Index

To calculate the Banzhaf power index:

1. List all winning coalitions
2. In each coalition, identify the players who are critical
3. Count up how many times each player is critical
4. Convert these counts to fractions or decimals by dividing by the total times any player is critical

---

Example 4

Find the Banzhaf power index for the voting system [8: 6, 3, 2].

We start by listing all winning coalitions. If you aren't sure how to do this, you can list all coalitions, then eliminate the non-winning coalitions. No player is a dictator, so we'll only consider two and three player coalitions.

$\{P_1, P_2\}$	Total weight: 9. Meets quota.
$\{P_1, P_3\}$	Total weight: 8. Meets quota.
$\{P_2, P_3\}$	Total weight: 5. Does not meet quota.
$\{P_1, P_2, P_3\}$	Total weight: 11. Meets quota.

Next we determine which players are critical in each winning coalition. In the winning two-player coalitions, both players are critical since no player can meet quota alone. Underlining the critical players to make it easier to count:
$\{\underline{P_1}, \underline{P_2}\}$
$\{\underline{P_1}, \underline{P_3}\}$

In the three-person coalition, either P_2 or P_3 could leave the coalition and the remaining players could still meet quota, so neither is critical. If P_1 were to leave, the remaining players could not reach quota, so P_1 is critical.
$\{\underline{P_1}, P_2, P_3\}$

Altogether, P_1 is critical 3 times, P_2 is critical 1 time, and P_3 is critical 1 time. Converting to percents:
$P_1 = 3/5 = 60\%$
$P_2 = 1/5 = 20\%$
$P_3 = 1/5 = 20\%$

Example 5

Consider the voting system [16: 7, 6, 3, 3, 2]. Find the Banzhaf power index.

The winning coalitions are listed below, with the critical players underlined.
$\{\underline{P_1}, \underline{P_2}, \underline{P_3}\}$
$\{\underline{P_1}, \underline{P_2}, \underline{P_4}\}$
$\{\underline{P_1}, \underline{P_2}, P_3, P_4\}$
$\{\underline{P_1}, \underline{P_2}, \underline{P_3}, P_5\}$
$\{\underline{P_1}, \underline{P_2}, \underline{P_4}, P_5\}$
$\{\underline{P_1}, \underline{P_2}, P_3, P_4, P_5\}$

Counting up times that each player is critical:
$P_1 = 6$
$P_2 = 6$
$P_3 = 2$
$P_4 = 2$
$P_5 = 0$
Total of all: 16

Divide each player's count by 16 to convert to fractions or percents:

$P_1 = 6/16 = 3/8 = 37.5\%$
$P_2 = 6/16 = 3/8 = 37.5\%$
$P_3 = 2/16 = 1/8 = 12.5\%$
$P_4 = 2/16 = 1/8 = 12.5\%$
$P_5 = 0/16 = 0 = 0\%$

The Banzhaf power index measures a player's ability to influence the outcome of the vote. Notice that player 5 has a power index of 0, indicating that there is no coalition in which they would be critical power and could influence the outcome. This means player 5 is a dummy, as we noted earlier.

Example 6

Revisiting the Scottish Parliament, with voting system [65: 47, 46, 17, 16, 2], the winning coalitions are listed, with the critical players underlined.

$\{\underline{P_1}, \underline{P_2}\}$
$\{\underline{P_1}, \underline{P_2}, P_3\}$ $\{\underline{P_1}, \underline{P_2}, P_4\}$
$\{\underline{P_1}, \underline{P_2}, P_5\}$ $\{\underline{P_1}, \underline{P_3}, \underline{P_4}\}$
$\{\underline{P_1}, \underline{P_3}, \underline{P_5}\}$ $\{\underline{P_1}, \underline{P_4}, \underline{P_5}\}$
$\{\underline{P_2}, \underline{P_3}, \underline{P_4}\}$ $\{\underline{P_2}, \underline{P_3}, \underline{P_5}\}$
$\{P_1, P_2, \underline{P_3}, \underline{P_4}\}$ $\{\underline{P_1}, P_2, P_3, \underline{P_5}\}$
$\{\underline{P_1}, P_2, P_4, P_5\}$ $\{\underline{P_1}, P_3, P_4, P_5\}$
$\{\underline{P_2}, \underline{P_3}, P_4, P_5\}$
$\{P_1, P_2, P_3, P_4, P_5\}$

Counting up times that each player is critical:

District	Times critical	Power index
P_1 (Scottish National Party)	9	$9/27 = 33.3\%$
P_2 (Labour Party)	7	$7/27 = 25.9\%$
P_3 (Conservative Party)	5	$5/27 = 18.5\%$
P_4 (Liberal Democrats Party)	3	$3/27 = 11.1\%$
P_5 (Scottish Green Party)	3	$3/27 = 11.1\%$

Interestingly, even though the Liberal Democrats party has only one less representative than the Conservative Party, and 14 more than the Scottish Green Party, their Banzhaf power index is the same as the Scottish Green Party's. In parliamentary governments, forming coalitions is an essential part of getting results, and a party's ability to help a coalition reach quota defines its influence.

Try it Now 3
Find the Banzhaf power index for the weighted voting system [36: 20, 17, 16, 3].

Example 7

Banzhaf used this index to argue that the weighted voting system used in the Nassau County Board of Supervisors in New York was unfair. The county was divided up into 6 districts, each getting voting weight proportional to the population in the district, as shown below. Calculate the power index for each district.

District	Weight
Hempstead #1	31
Hempstead #2	31
Oyster Bay	28
North Hempstead	21
Long Beach	2
Glen Cove	2

Translated into a weighted voting system, assuming a simple majority is needed for a proposal to pass:

$$[58: 31, 31, 28, 21, 2, 2]$$

Listing the winning coalitions and marking critical players:

{H1, H2}	{H1, OB, NH}	{H2, OB, NH, LB}
{H1, OB}	{H1, OB, LB}	{H2, OB, NH, GC}
{H2, OB}	{H1, OB, GC}	{H2, OB, LB, GC}
{H1, H2, NH}	{H1, OB, NH, LB}	{H2, OB, NH, LB, GC}
{H1, H2, LB}	{H1, OB, NH, GC}	{H1, H2, OB}
{H1, H2, GC}	{H1, OB, LB, GC}	{H1, H2, OB, NH}
{H1, H2, NH, LB}	{H1, OB, NH, LB. GC}	{H1, H2, OB, LB}
{H1, H2, NH, GC}	{H2, OB, NH}	{H1, H2, OB, GC}
{H1, H2, LB, GC}	{H2, OB, LB}	{H1, H2, OB, NH, LB}
{H1, H2, NH, LB. GC}	{H2, OB, GC}	{H1, H2, OB, NH, GC}
		{H1, H2, OB, NH, LB, GC}

There are a lot of them! Counting up how many times each player is critical,

District	Times critical	Power index
Hempstead #1	16	16/48 = 1/3 = 33%
Hempstead #2	16	16/48 = 1/3 = 33%
Oyster Bay	16	16/48 = 1/3 = 33%
North Hempstead	0	0/48 = 0%
Long Beach	0	0/48 = 0%
Glen Cove	0	0/48 = 0%

It turns out that the three smaller districts are dummies. Any winning coalition requires two of the larger districts.

The weighted voting system that Americans are most familiar with is the Electoral College system used to elect the President. In the Electoral College, states are given a number of votes equal to the number of their congressional representatives (house + senate). Most states give all their electoral votes to the candidate that wins a majority in their state, turning the Electoral College into a weighted voting system, in which the states are the players. As I'm sure you can imagine, there are billions of possible winning coalitions, so the power index for the Electoral College has to be computed by a computer using approximation techniques.

Calculating Power: Shapley-Shubik Power Index

The **Shapley-Shubik** power index was introduced in 1954 by economists Lloyd Shapley and Martin Shubik, and provides a different approach for calculating power.

In situations like political alliances, the order in which players join an alliance could be considered the most important consideration. In particular, if a proposal is introduced, the player that joins the coalition and allows it to reach quota might be considered the most essential. The Shapley-Shubik power index counts how likely a player is to be **pivotal**. What does it mean for a player to be pivotal?

First, we need to change our approach to coalitions. Previously, the coalition $\{P_1, P_2\}$ and $\{P_2, P_1\}$ would be considered equivalent, since they contain the same players. We now need to consider the *order* in which players join the coalition. For that, we will consider **sequential coalitions** – coalitions that contain all the players in which the order players are listed reflect the order they joined the coalition. For example, the sequential coalition $<P_2, P_1, P_3>$ would mean that P_2 joined the coalition first, then P_1, and finally P_3. The angle brackets $<\,>$ are used instead of curly brackets to distinguish sequential coalitions.

Pivotal Player

A sequential coalition lists the players in the order in which they joined the coalition.

A **pivotal player** is the player in a sequential coalition that changes a coalition from a losing coalition to a winning one. Notice there can only be one pivotal player in any sequential coalition.

Example 8

In the weighted voting system [8: 6, 4, 3, 2], which player is pivotal in the sequential coalition $<P_3, P_2, P_4, P_1>$?

The sequential coalition shows the order in which players joined the coalition. Consider the running totals as each player joins:

P_3	Total weight: 3	Not winning
P_3, P_2	Total weight: 3+4 = 7	Not winning
P_3, P_2, P_4	Total weight: 3+4+2 = 9	Winning
P_3, P_2, P_4, P_1	Total weight: 3+4+2+6 = 15	Winning

Since the coalition *becomes* winning when P_4 joins, P_4 is the pivotal player in this coalition.

> **Calculating Shapley-Shubik Power Index**
> To calculate the Shapley-Shubik Power Index:
> 1. List all sequential coalitions
> 2. In each sequential coalition, determine the pivotal player
> 3. Count up how many times each player is pivotal
> 4. Convert these counts to fractions or decimals by dividing by the total number of sequential coalitions

How many sequential coalitions should we expect to have? If there are N players in the voting system, then there are N possibilities for the first player in the coalition, $N - 1$ possibilities for the second player in the coalition, and so on. Combining these possibilities, the total number of coalitions would be: $N(N-1)(N-2)(N-3)\cdots(3)(2)(1)$. This calculation is called a **factorial**, and is notated $N!$ The number of sequential coalitions with N players is $N!$

Example 9

How many sequential coalitions will there be in a voting system with 7 players?

There will be $7!$ sequential coalitions. $7! = 7 \cdot 6 \cdot 5 \cdot 4 \cdot 3 \cdot 2 \cdot 1 = 5040$

As you can see, computing the Shapley-Shubik power index by hand would be very difficult for voting systems that are not very small.

Example 10

Consider the weighted voting system [6: 4, 3, 2]. We will list all the sequential coalitions and identify the pivotal player. We will have $3! = 6$ sequential coalitions. The coalitions are listed, and the pivotal player is underlined.

$< P_1, \underline{P_2}, P_3 >$ $< P_1, \underline{P_3}, P_2 >$ $< P_2, \underline{P_1}, P_3 >$

$< P_2, P_3, \underline{P_1} >$ $< P_3, P_2, \underline{P_1} >$ $< P_3, \underline{P_1}, P_2 >$

P_1 is pivotal 4 times, P_2 is pivotal 1 time, and P_3 is pivotal 1 time.

Player	Times pivotal	Power index
P_1	4	$4/6 = 66.7\%$
P_2	1	$1/6 = 16.7\%$
P_3	1	$1/6 = 16.7\%$

For comparison, the Banzhaf power index for the same weighted voting system would be P_1: 60%, P_2: 20%, P_3: 20%. While the Banzhaf power index and Shapley-Shubik power index are usually not terribly different, the two different approaches usually produce somewhat different results.

Find the Shapley-Shubik power index for the weighted voting system [36: 20, 17, 15].

Try it Now Answers

1. If we represent the players as $M_1, M_2, M_3, M_4, U_1, U_2, U_3$, then we may be tempted to set up a system like [4: 1, 1, 1, 1, 1, 1, 1]. While this system would meet the first requirement that four members must support a proposal for it to pass, this does not satisfy the requirement that at least one member of the union must support it.

To accomplish that, we might try increasing the voting weight of the union members: [5: 1, 1, 1, 1, 2, 2, 2]. The quota was set at 5 so that the four management members alone would not be able to reach quota without one of the union members. Unfortunately, now the three union members can reach quota alone. To fix this, three management members need to have more weight than two union members.

After trying several other guesses, we land on the system [13: 3, 3, 3, 3, 4, 4, 4]. Here, the four management members have combined weight of 12, so cannot reach quota. Likewise, the three union members have combined weight of 12, so cannot reach quota alone. But, as required, any group of 4 members that includes at least one union member will reach the quota of 13. For example, three management members and one union member have combined weight of 3+3+3+4=13, and reach quota.

2. In the voting system [q: 10, 5, 3], if the quota is 10, then player 1 is a dictator since they can reach quota without the support of the other players. This makes the other two players automatically dummies.

If the quota is 12, then player 1 is necessary to reach quota, so has veto power. Since at this point either player 2 or player 3 would allow player 1 to reach quota, neither player is a dummy, so they are regular players (not dictators, no veto power, and not a dummy).

If the quota is 16, then both players 1 and 2 are necessary to reach quota, so they both have veto power. Player 3's support is not necessary, so player 3 is a dummy.

3. The voting system tells us that the quota is 36, that Player 1 has 20 votes (or equivalently, has a weight of 20), Player 2 has 17 votes, Player 3 has 16 votes, and Player 4 has 3 votes.

A coalition is any group of one or more players. What we're looking for is winning coalitions - coalitions whose combined votes (weights) add to up to the quota or more. So the coalition {P3, P4} is not a winning coalition because the combined weight is 16+3=19, which is below the quota.

So we look at each possible combination of players and identify the winning ones:

{P1, P2} (weight: 37)   {P1, P3} (weight: 36)
{P1, P2, P3} (weight: 53)   {P1, P2, P4} (weight: 40)
{P1, P3, P4} (weight: 39)   {P1, P2, P3, P4} (weight: 56)
{P2, P3, P4} (weight: 36)

Now, in each coalition, we need to identify which players are critical. A player is critical if the coalition would no longer reach quota without that person. So, in the coalition {P1, P2}, both players are necessary to reach quota, so both are critical. However in the coalition {P1, P2, P3}, we can see from the earlier two coalitions that either P2 or P3 could leave the coalition and it would still reach quota. But if P1 left, it would not reach quota, so P1 is the only player critical in this coalition. We evaluate the rest of the coalitions similarly, giving us this (underlining the critical players)

{<u>P1</u>, <u>P2</u>}   {<u>P1</u>, <u>P3</u>}
{<u>P1</u>, P2, P3}   {<u>P1</u>, <u>P2</u>, P4}
{<u>P1</u>, <u>P3</u>, P4}   {P1, P2, P3, P4}
{<u>P2</u>, <u>P3</u>, <u>P4</u>}

Next we count how many times each player is critical:
P1: 5 times
P2: 3 times
P3: 3 times
P4: 1 time

In total, there were 5+3+3+1 = 12 times anyone was critical, so we take our counts and turn them into fractions, giving us our Banzhaf power:
P1: 5/12
P2: 3/12 = 1/4
P3: 3/12 = 1/4
P4: 1/12

[36: 20, 17, 15].

4. Listing all sequential coalitions and identifying the pivotal player:

$< P_1, \underline{P_2}, P_3 >$ $< P_1, P_3, \underline{P_2} >$ $< P_2, \underline{P_1}, P_3 >$
$< P_2, P_3, \underline{P_1} >$ $< P_3, P_2, \underline{P_1} >$ $< P_3, \underline{P_1}, P_2 >$

P_1 is pivotal 4 times, P_2 is pivotal 1 time, and P_3 is pivotal 1 time.

Player	Times pivotal	Power index
P_1	3	3/6 = 50%
P_2	3	3/6 = 50%
P_3	0	0/6 = 0%

Exercises

Skills

1. Consider the weighted voting system [47: 10,9,9,5,4,4,3,2,2]
   a. How many players are there?
   b. What is the total number (weight) of votes?
   c. What is the quota in this system?

2. Consider the weighted voting system [31: 10,10,8,7,6,4,1,1]
   a. How many players are there?
   b. What is the total number (weight) of votes?
   c. What is the quota in this system?

3. Consider the weighted voting system [q: 7,5,3,1,1]
   a. What is the smallest value that the quota q can take?
   b. What is the largest value that the quota q can take?
   c. What is the value of the quota if at least two-thirds of the votes are required to pass a motion?

4. Consider the weighted voting system [q: 10,9,8,8,8,6]
   a. What is the smallest value that the quota q can take?
   b. What is the largest value that the quota q can take?
   c. What is the value of the quota if at least two-thirds of the votes are required to pass a motion?

5. Consider the weighted voting system [13: 13, 6, 4, 2]
   a. Identify the dictators, if any.
   b. Identify players with veto power, if any
   c. Identify dummies, if any.

6. Consider the weighted voting system [11: 9, 6, 3, 1]
   a. Identify the dictators, if any.
   b. Identify players with veto power, if any
   c. Identify dummies, if any.

7. Consider the weighted voting system [19: 13, 6, 4, 2]
   a. Identify the dictators, if any.
   b. Identify players with veto power, if any
   c. Identify dummies, if any.

8. Consider the weighted voting system [17: 9, 6, 3, 1]
   a. Identify the dictators, if any.
   b. Identify players with veto power, if any
   c. Identify dummies, if any.

9. Consider the weighted voting system [15: 11, 7, 5, 2]
   a. What is the weight of the coalition $\{P_1,P_2,P_4\}$
   b. In the coalition $\{P_1,P_2,P_4\}$ which players are critical?

10. Consider the weighted voting system [17: 13, 9, 5, 2]
    a. What is the weight of the coalition $\{P_1,P_2,P_3\}$
    b. In the coalition $\{P_1,P_2,P_3\}$ which players are critical?

11. Find the Banzhaf power distribution of the weighted voting system
    [27: 16, 12, 11, 3]

12. Find the Banzhaf power distribution of the weighted voting system
    [33: 18, 16, 15, 2]

13. Consider the weighted voting system [q: 15, 8, 3, 1] Find the Banzhaf power
    distribution of this weighted voting system,
    a. When the quota is 15
    b. When the quota is 16
    c. When the quota is 18

14. Consider the weighted voting system [q: 15, 8, 3, 1] Find the Banzhaf power
    distribution of this weighted voting system,
    a. When the quota is 19
    b. When the quota is 23
    c. When the quota is 26

15. Consider the weighted voting system [17: 13, 9, 5, 2]. In the sequential coalition
    $<P_3,P_2,P_1,P_4>$ which player is pivotal?

16. Consider the weighted voting system [15: 13, 9, 5, 2]. In the sequential coalition
    $<P_1,P_4,P_2,P_3>$ which player is pivotal?

17. Find the Shapley-Shubik power distribution for the system [24: 17, 13, 11]

18. Find the Shapley-Shubik power distribution for the system [25: 17, 13, 11]

Concepts

19. Consider the weighted voting system [q: 7, 3, 1]
    a. Which values of q result in a dictator (list all possible values)
    b. What is the smallest value for q that results in exactly one player with veto
       power but no dictators?
    c. What is the smallest value for q that results in exactly two players with veto
       power?

20. Consider the weighted voting system [q: 9, 4, 2]
    a. Which values of q result in a dictator (list all possible values)
    b. What is the smallest value for q that results in exactly one player with veto power?
    c. What is the smallest value for q that results in exactly two players with veto power?

21. Using the Shapley-Shubik method, is it possible for a dummy to be pivotal?

22. If a specific weighted voting system requires a unanimous vote for a motion to pass:
    a. Which player will be pivotal in any sequential coalition?
    b. How many winning coalitions will there be?

23. Consider a weighted voting system with three players. If Player 1 is the only player with veto power, there are no dictators, and there are no dummies:
    a. Find the Banzhof power distribution.
    b. Find the Shapley-Shubik power distribution

24. Consider a weighted voting system with three players. If Players 1 and 2 have veto power but are not dictators, and Player 3 is a dummy:
    a. Find the Banzhof power distribution.
    b. Find the Shapley-Shubik power distribution

25. An executive board consists of a president (P) and three vice-presidents (V_1, V_2, V_3). For a motion to pass it must have three yes votes, one of which must be the president's. Find a weighted voting system to represent this situation.

26. On a college's basketball team, the decision of whether a student is allowed to play is made by four people: the head coach and the three assistant coaches. To be allowed to play, the student needs approval from the head coach and at least one assistant coach. Find a weighted voting system to represent this situation.

27. In a corporation, the shareholders receive 1 vote for each share of stock they hold, which is usually based on the amount of money the invested in the company. Suppose a small corporation has two people who invested $30,000 each, two people who invested $20,000 each, and one person who invested $10,000. If they receive one share of stock for each $1000 invested, and any decisions require a majority vote, set up a weighted voting system to represent this corporation's shareholder votes.

28. A contract negotiations group consists of 4 workers and 3 managers. For a proposal to be accepted, a majority of workers and a majority of managers must approve of it. Calculate the Banzhaf power distribution for this situation. Who has more power: a worker or a manager?

29. The United Nations Security Council consists of 15 members, 10 of which are elected, and 5 of which are permanent members. For a resolution to pass, 9 members must support it, which must include all 5 of the permanent members. Set up a weighted voting system to represent the UN Security Council and calculate the Banzhaf power distribution.

Exploration

30. In the U.S., the Electoral College is used in presidential elections. Each state is awarded a number of electors equal to the number of representatives (based on population) and senators (2 per state) they have in congress. Since most states award the winner of the popular vote in their state all their state's electoral votes, the Electoral College acts as a weighted voting system. To explore how the Electoral College works, we'll look at a mini-country with only 4 states. Here is the outcome of a hypothetical election:

State	Smalota	Medigan	Bigonia	Hugodo
Population	50,000	70,000	100,000	240,000
Votes for A	40,000	50,000	80,000	50,000
Votes for B	10,000	20,000	20,000	190,000

a. If this country did not use an Electoral College, which candidate would win the election?

b. Suppose that each state gets 1 electoral vote for every 10,000 people. Set up a weighted voting system for this scenario, calculate the Banzhaf power index for each state, then calculate the winner if each state awards all their electoral votes to the winner of the election in their state.

c. Suppose that each state gets 1 electoral vote for every 10,000 people, plus an additional 2 votes. Set up a weighted voting system for this scenario, calculate the Banzhaf power index for each state, then calculate the winner if each state awards all their electoral votes to the winner of the election in their state.

d. Suppose that each state gets 1 electoral vote for every 10,000 people, and awards them based on the number of people who voted for each candidate. Additionally, they get 2 votes that are awarded to the majority winner in the state. Calculate the winner under these conditions.

e. Does it seem like an individual state has more power in the Electoral College under the vote distribution from part c or from part d?

f. Research the history behind the Electoral College to explore why the system was introduced instead of using a popular vote. Based on your research and experiences, state and defend your opinion on whether the Electoral College system is or is not fair.

31. The value of the Electoral College (see previous problem for an overview) in modern elections is often debated. Find an article or paper providing an argument for or against the Electoral College. Evaluate the source and summarize the article, then give your opinion of why you agree or disagree with the writer's point of view. If done in class, form groups and hold a debate.

Apportionment

Apportionment is the problem of dividing up a fixed number of things among groups of different sizes. In politics, this takes the form of allocating a limited number of representatives amongst voters. This problem, presumably, is older than the United States, but the best-known ways to solve it have their origins in the problem of assigning each state an appropriate number of representatives in the new Congress when the country was formed. States also face this apportionment problem in defining how to draw districts for state representatives. The apportionment problem comes up in a variety of non-political areas too, though. We face several restrictions in this process:

Apportionment rules
1. The things being divided up can exist only in whole numbers.
2. We must use all of the things being divided up, and we cannot use any more.
3. Each group must get at least one of the things being divided up.
4. The number of things assigned to each group should be at least approximately proportional to the population of the group. (Exact proportionality isn't possible because of the whole number requirement, but we should try to be close, and in any case, if Group A is larger than Group B, then Group B shouldn't get more of the things than Group A does.)

In terms of the apportionment of the United States House of Representatives, these rules imply:
1. We can only have whole representatives (a state can't have 3.4 representatives)
2. We can only use the (currently) 435 representatives available. If one state gets another representative, another state has to lose one.
3. Every state gets at least one representative
4. The number of representatives each state gets should be approximately proportional to the state population. This way, the number of constituents each representative has should be approximately equal.

We will look at four ways of solving the apportionment problem. Three of them (Lowndes's method is the exception) have been used at various times to apportion the U.S. Congress, although the method currently in use (the Huntington-Hill method) is significantly more complicated.

Hamilton's Method

Alexander Hamilton proposed the method that now bears his name. His method was approved by Congress in 1791, but was vetoed by President Washington. It was later adopted in 1852 and used through 1911. He begins by determining, to several decimal places, how many things each group should get. Since he was interested in the question of Congressional representation, we'll use the language of states and representatives, so he determines how many representatives each state should get. He follows these steps:

> **Hamilton's Method**
> 1. Determine how many people each representative should represent. Do this by dividing the total population of all the states by the total number of representatives. This answer is called the **divisor**.
>
> 2. Divide each state's population by the divisor to determine how many representatives it should have. Record this answer to several decimal places. This answer is called the **quota**.
>
> Since we can only allocate whole representatives, Hamilton resolves the whole number problem, as follows:
>
> 3. Cut off all the decimal parts of all the quotas (but don't forget what the decimals were). These are called the **lower quotas**. Add up the remaining whole numbers. This answer will always be less than or equal to the total number of representatives (and the "or equal to" part happens only in very specific circumstances that are incredibly unlikely to turn up).
>
> 4. Assuming that the total from Step 3 was less than the total number of representatives, assign the remaining representatives, one each, to the states whose decimal parts of the quota were largest, until the desired total is reached.
>
> Make sure that each state ends up with at least one representative!

Note on rounding: Today we have technological advantages that Hamilton (and the others) couldn't even have imagined. Take advantage of them, and keep several decimal places.

Example 1

The state of Delaware has three counties: Kent, New Castle, and Sussex. The Delaware state House of Representatives has 41 members. If Delaware wants to divide this representation along county lines (which is *not* required, but let's pretend they do), let's use Hamilton's method to apportion them. The populations of the counties are as follows (from the 2010 Census):

County	Population
Kent	162,310
New Castle	538,479
Sussex	197,145
Total	**897,934**

1. First, we determine the divisor: $897,934/41 = 21,900.82927$

2. Now we determine each county's quota by dividing the county's population by the divisor:

County	Population	Quota
Kent	162,310	7.4111
New Castle	538,479	24.5872
Sussex	197,145	9.0017
Total	**897,934**	

3. Removing the decimal parts of the quotas gives:

County	Population	Quota	Initial
Kent	162,310	7.4111	7
New Castle	538,479	24.5872	24
Sussex	197,145	9.0017	9
Total	**897,934**		**40**

4. We need 41 representatives and this only gives 40. The remaining one goes to the county with the largest decimal part, which is New Castle:

County	Population	Quota	Initial	Final
Kent	162,310	7.4111	7	7
New Castle	538,479	24.5872	24	25
Sussex	197,145	9.0017	9	9
Total	**897,934**		**40**	

Example 2

Use Hamilton's method to apportion the 75 seats of Rhode Island's House of Representatives among its five counties.

County	Population
Bristol	49,875
Kent	166,158
Newport	82,888
Providence	626,667
Washington	126,979
Total	**1,052,567**

1. The divisor is $1,052,567/75 = 14,034.22667$

2. Determine each county's quota by dividing its population by the divisor:

County	Population	Quota
Bristol	49,875	3.5538
Kent	166,158	11.8395
Newport	82,888	5.9061
Providence	626,667	44.6528
Washington	126,979	9.0478
Total	**1,052,567**	

3. Remove the decimal part of each quota:

County	Population	Quota	Initial
Bristol	49,875	3.5538	3
Kent	166,158	11.8395	11
Newport	82,888	5.9061	5
Providence	626,667	44.6528	44
Washington	126,979	9.0478	9
Total	**1,052,567**		**72**

4. We need 75 representatives and we only have 72, so we assign the remaining three, one each, to the three counties with the largest decimal parts, which are Newport, Kent, and Providence:

County	Population	Quota	Initial	Final
Bristol	49,875	3.5538	3	3
Kent	166,158	11.8395	11	12
Newport	82,888	5.9061	5	6
Providence	626,667	44.6528	44	45
Washington	126,979	9.0478	9	9
Total	**1,052,567**		**72**	**75**

Note that even though Bristol County's decimal part is greater than .5, it isn't big enough to get an additional representative, because three other counties have greater decimal parts.

Hamilton's method obeys something called the Quota Rule. The Quota Rule isn't a law of any sort, but just an idea that some people, including Hamilton, think is a good one.

> **Quota Rule**
> The Quota Rule says that the final number of representatives a state gets should be within one of that state's quota. Since we're dealing with whole numbers for our final answers, that means that each state should either go up to the next whole number above its quota, or down to the next whole number below its quota.

Controversy

After seeing Hamilton's method, many people find that it makes sense, it's not that difficult to use (or, at least, the difficulty comes from the numbers that are involved and the amount of computation that's needed, not from the method), and they wonder why anyone would want another method. The problem is that Hamilton's method is subject to several paradoxes. Three of them happened, on separate occasions, when Hamilton's method was used to apportion the United States House of Representatives.

The **Alabama Paradox** is named for an incident that happened during the apportionment that took place after the 1880 census. (A similar incident happened ten years earlier involving the state of Rhode Island, but the paradox is named after Alabama.) The post-1880 apportionment had been completed, using Hamilton's method and the new population numbers from the census. Then it was decided that because of the country's growing population, the House of Representatives should be made larger. That meant that the apportionment would need to be done again, still using Hamilton's method and the same 1880 census numbers, but with more representatives. The assumption was that some states would gain another representative and others would stay with the same number they already had (since there weren't enough new representatives being added to give one more to every state). The paradox is that Alabama ended up *losing* a representative in the process, even though no populations were changed and the total number of representatives increased.

The **New States Paradox** happened when Oklahoma became a state in 1907. Oklahoma had enough population to qualify for five representatives in Congress. Those five representatives would need to come from somewhere, though, so five states, presumably, would lose one representative each. That happened, but another thing also happened: Maine gained a representative (from New York).

The **Population Paradox** happened between the apportionments after the census of 1900 and of 1910. In those ten years, Virginia's population grew at an average annual rate of 1.07%, while Maine's grew at an average annual rate of 0.67%. Virginia started with more people, grew at a faster rate, grew by more people, and ended up with more people than Maine. By itself, that doesn't mean that Virginia should gain representatives or Maine shouldn't, because there are lots of other states involved. But Virginia ended up losing a representative *to Maine*.

Jefferson's Method

Thomas Jefferson proposed a different method for apportionment. After Washington vetoed Hamilton's method, Jefferson's method was adopted, and used in Congress from 1791 through 1842. Jefferson, of course, had political reasons for wanting his method to be used rather than Hamilton's. Primarily, his method favors larger states, and his own home state of Virginia was the largest in the country at the time. He would also argue that it's the ratio of people to representatives that is the critical thing, and apportionment methods should be based on that. But the paradoxes we saw also provide mathematical reasons for concluding that Hamilton's method isn't so good, and while Jefferson's method might or might not be the best one to replace it, at least we should look for other possibilities.

The first steps of Jefferson's method are the same as Hamilton's method. He finds the same divisor and the same quota, and cuts off the decimal parts in the same way, giving a total number of representatives that is less than the required total. The difference is in how Jefferson resolves that difference. He says that since we ended up with an answer that is too small, our divisor must have been too big. He changes the divisor by making it smaller, finding new quotas with the new divisor, cutting off the decimal parts, and looking at the new total, until we find a divisor that produces the required total.

> **Jefferson's Method**
> 1. Determine how many people each representative should represent. Do this by dividing the total population of all the states by the total number of representatives. This answer is called the **standard divisor**.
>
> 2. Divide each state's population by the divisor to determine how many representatives it should have. Record this answer to several decimal places. This answer is called the **quota**.
>
> 3. Cut off all the decimal parts of all the quotas (but don't forget what the decimals were). These are the **lower quotas**. Add up the remaining whole numbers. This answer will always be less than or equal to the total number of representatives.
>
> 4. If the total from Step 3 was less than the total number of representatives, reduce the divisor and recalculate the quota and allocation. Continue doing this until the total in Step 3 is equal to the total number of representatives. The divisor we end up using is called the **modified divisor** or **adjusted divisor**.

Example 3

We'll return to Delaware and apply Jefferson's method. We begin, as we did with Hamilton's method, by finding the quotas with the original divisor, 21,900.82927:

County	Population	Quota	Initial
Kent	162,310	7.4111	7
New Castle	538,479	24.5872	24
Sussex	197,145	9.0017	9
Total	**897,934**		**40**

We need 41 representatives, and this divisor gives only 40. We must reduce the divisor until we get 41 representatives. Let's try 21,500 as the divisor:

County	Population	Quota	Initial
Kent	162,310	7.5493	7
New Castle	538,479	25.0455	25
Sussex	197,145	9.1695	9
Total	**897,934**		**41**

This gives us the required 41 representatives, so we're done. If we had fewer than 41, we'd need to reduce the divisor more. If we had more than 41, we'd need to choose a divisor less than the original but greater than the second choice.

Notice that with the new, lower divisor, the quota for New Castle County (the largest county in the state) increased by much more than those of Kent County or Sussex County.

Example 4

We'll apply Jefferson's method for Rhode Island. The original divisor of 14,034.22667 gave these results:

County	Population	Quota	Initial
Bristol	49,875	3.5538	3
Kent	166,158	11.8395	11
Newport	82,888	5.9061	5
Providence	626,667	44.6528	44
Washington	126,979	9.0478	9
Total	**1,052,567**		**72**

We need 75 representatives and we only have 72, so we need to use a smaller divisor. Let's try 13,500:

County	Population	Quota	Initial
Bristol	49,875	3.6944	3
Kent	166,158	12.3080	12
Newport	82,888	6.1399	6
Providence	626,667	46.4198	46
Washington	126,979	9.4059	9
Total	**1,052,567**		**76**

We've gone too far. We need a divisor that's greater than 13,500 but less than 14,034.22667. Let's try 13,700:

County	Population	Quota	Initial
Bristol	49,875	3.6405	3
Kent	166,158	12.1283	12
Newport	82,888	6.0502	6
Providence	626,667	45.7421	45
Washington	126,979	9.2685	9
Total	**1,052,567**		**75**

This works.

Notice, in comparison to Hamilton's method, that although the results were the same, they came about in a different way, and the outcome was almost different. Providence County (the largest) almost went up to 46 representatives before Kent (which is much smaller) got to 12. Although that didn't happen here, it can. Divisor-adjusting methods like Jefferson's are not guaranteed to follow the quota rule!

Webster's Method

Daniel Webster (1782-1852) proposed a method similar to Jefferson's in 1832. It was adopted by Congress in 1842, but replaced by Hamilton's method in 1852. It was then adopted again in 1901. The difference is that Webster rounds the quotas to the nearest whole number rather than dropping the decimal parts. If that doesn't produce the desired results at the beginning, he says, like Jefferson, to adjust the divisor until it does. (In Jefferson's case, at least the first adjustment will always be to make the divisor smaller. That is not always the case with Webster's method.)

Webster's Method

1. Determine how many people each representative should represent. Do this by dividing the total population of all the states by the total number of representatives. This answer is called the **standard divisor**.

2. Divide each state's population by the divisor to determine how many representatives it should have. Record this answer to several decimal places. This answer is called the **quota**.

3. Round all the quotas to the nearest whole number (but don't forget what the decimals were). Add up the remaining whole numbers.

4. If the total from Step 3 was less than the total number of representatives, reduce the divisor and recalculate the quota and allocation. If the total from step 3 was larger than the total number of representatives, increase the divisor and recalculate the quota and allocation. Continue doing this until the total in Step 3 is equal to the total number of representatives. The divisor we end up using is called the **modified divisor** or **adjusted divisor**.

Example 5

Again, Delaware, with an initial divisor of 21,900.82927:

County	Population	Quota	Initial
Kent	162,310	7.4111	7
New Castle	538,479	24.5872	25
Sussex	197,145	9.0017	9
Total	**897,934**		**41**

This gives the required total, so we're done.

Example 6

Again, Rhode Island, with an initial divisor of 14,034.22667:

County	Population	Quota	Initial
Bristol	49,875	3.5538	4
Kent	166,158	11.8395	12
Newport	82,888	5.9061	6
Providence	626,667	44.6528	45
Washington	126,979	9.0478	9
Total	**1,052,567**		**76**

This is too many, so we need to increase the divisor. Let's try 14,100:

County	Population	Quota	Initial
Bristol	49,875	3.5372	4
Kent	166,158	11.7843	12
Newport	82,888	5.8786	6
Providence	626,667	44.4445	44
Washington	126,979	9.0056	9
Total	**1,052,567**		**75**

This works, so we're done.

Like Jefferson's method, Webster's method carries a bias in favor of states with large populations, but rounding the quotas to the nearest whole number greatly reduces this bias. (Notice that Providence County, the largest, is the one that gets a representative trimmed because of the increased quota.) Also like Jefferson's method, Webster's method does not always follow the quota rule, but it follows the quota rule much more often than Jefferson's method does. (In fact, if Webster's method had been applied to every apportionment of Congress in all of American history, it would have followed the quota rule every single time.)

In 1980, two mathematicians, Peyton Young and Mike Balinski, proved what we now call the Balinski-Young Impossibility Theorem.

Balinski-Young Impossibility Theorem
The Balinski-Young Impossibility Theorem shows that any apportionment method which always follows the quota rule will be subject to the possibility of paradoxes like the Alabama, New States, or Population paradoxes. In other words, we can choose a method that avoids those paradoxes, but only if we are willing to give up the guarantee of following the quota rule.

Huntington-Hill Method

In 1920, no new apportionment was done, because Congress couldn't agree on the method to be used. They appointed a committee of mathematicians to investigate, and they recommended the Huntington-Hill Method. They continued to use Webster's method in 1931, but after a second report recommending Huntington-Hill, it was adopted in 1941 and is the current method of apportionment used in Congress.

The Huntington-Hill Method is similar to Webster's method, but attempts to minimize the percent differences of how many people each representative will represent.

Huntington-Hill Method

1. Determine how many people each representative should represent. Do this by dividing the total population of all the states by the total number of representatives. This answer is called the **standard divisor**.

2. Divide each state's population by the divisor to determine how many representatives it should have. Record this answer to several decimal places. This answer is called the **quota**.

3. Cut off the decimal part of the quota to obtain the lower quota, which we'll call n. Compute $\sqrt{n(n+1)}$, which is the **geometric mean** of the lower quota and one value higher.

4. If the quota is larger than the geometric mean, round up the quota; if the quota is smaller than the geometric mean, round down the quota. Add up the resulting whole numbers to get the **initial allocation**.

5. If the total from Step 4 was less than the total number of representatives, reduce the divisor and recalculate the quota and allocation. If the total from step 4 was larger than the total number of representatives, increase the divisor and recalculate the quota and allocation. Continue doing this until the total in Step 4 is equal to the total number of representatives. The divisor we end up using is called the **modified divisor** or **adjusted divisor**.

Example 7

Again, Delaware, with an initial divisor of 21,900.82927:

County	Population	Quota	Lower Quota	Geom Mean	Initial
Kent	162,310	7.4111	7	7.48	7
New Castle	538,479	24.5872	24	24.49	25
Sussex	197,145	9.0017	9	9.49	9
Total	**897,934**				**41**

This gives the required total, so we're done.

Example 8

Again, Rhode Island, with an initial divisor of 14,034.22667:

County	Population	Quota	Lower Quota	Geom Mean	Initial
Bristol	49,875	3.5538	3	3.46	4
Kent	166,158	11.8395	11	11.49	12
Newport	82,888	5.9061	5	5.48	6
Providence	626,667	44.6528	44	44.50	45
Washington	126,979	9.0478	9	9.49	9
Total	**1,052,567**				**76**

This is too many, so we need to increase the divisor. Let's try 14,100:

County	Population	Quota	Lower Quota	Geom Mean	Initial
Bristol	49,875	3.5372	3	3.46	4
Kent	166,158	11.7843	11	11.49	12
Newport	82,888	5.8786	5	5.48	6
Providence	626,667	44.4445	44	44.50	44
Washington	126,979	9.0056	9	9.49	9
Total	**1,052,567**				**75**

This works, so we're done.

In both these cases, the apportionment produced by the Huntington-Hill method was the same as those from Webster's method.

Example 9

Consider a small country with 5 states, two of which are much larger than the others. We need to apportion 70 representatives. We will apportion using both Webster's method and the Huntington-Hill method.

State	Population
A	300,500
B	200,000
C	50,000
D	38,000
E	21,500

1. The total population is 610,000. Dividing this by the 70 representatives gives the divisor: 8714.286

2. Dividing each state's population by the divisor gives the quotas

State	Population	Quota
A	300,500	34.48361
B	200,000	22.95082
C	50,000	5.737705
D	38,000	4.360656
E	21,500	2.467213

Webster's Method

3. Using Webster's method, we round each quota to the nearest whole number

State	Population	Quota	Initial
A	300,500	34.48361	34
B	200,000	22.95082	23
C	50,000	5.737705	6
D	38,000	4.360656	4
E	21,500	2.467213	2

4. Adding these up, they only total 69 representatives, so we adjust the divisor down. Adjusting the divisor down to 8700 gives an updated allocation totaling 70 representatives

State	Population	Quota	Initial
A	300,500	34.54023	35
B	200,000	22.98851	23
C	50,000	5.747126	6
D	38,000	4.367816	4
E	21,500	2.471264	2

Huntington-Hill Method

3. Using the Huntington-Hill method, we round down to find the lower quota, then calculate the geometric mean based on each lower quota. If the quota is less than the geometric mean, we round down; if the quota is more than the geometric mean, we round up.

State	Population	Quota	Lower Quota	Geometric Mean	Initial
A	300,500	34.48361	34	34.49638	34
B	200,000	22.95082	22	22.49444	23
C	50,000	5.737705	5	5.477226	6
D	38,000	4.360656	4	4.472136	4
E	21,500	2.467213	2	2.44949	3

These allocations add up to 70, so we're done.

Notice that this allocation is different than that produced by Webster's method. In this case, state E got the extra seat instead of state A.

Lowndes' Method

William Lowndes (1782-1822) was a Congressman from South Carolina (a small state) who proposed a method of apportionment that was more favorable to smaller states. Unlike the methods of Hamilton, Jefferson, and Webster, Lowndes's method has never been used to apportion Congress.

Lowndes believed that an additional representative was much more valuable to a small state than to a large one. If a state already has 20 or 30 representatives, getting one more doesn't matter very much. But if it only has 2 or 3, one more is a big deal, and he felt that the additional representatives should go where they could make the most difference.

Like Hamilton's method, Lowndes's method follows the quota rule. In fact, it arrives at the same quotas as Hamilton and the rest, and like Hamilton and Jefferson, it drops the decimal parts. But in deciding where the remaining representatives should go, we divide the decimal part of each state's quota by the whole number part (so that the same decimal part with a smaller whole number is worth more, because it matters more to that state).

Lowndes's Method

1. Determine how many people each representative should represent. Do this by dividing the total population of all the states by the total number of representatives. This answer is called the **divisor**.

2. Divide each state's population by the divisor to determine how many representatives it should have. Record this answer to several decimal places. This answer is called the **quota**.

3. Cut off all the decimal parts of all the quotas (but don't forget what the decimals were). Add up the remaining whole numbers.

4. Assuming that the total from Step 3 was less than the total number of representatives, divide the decimal part of each state's quota by the whole number part. Assign the remaining representatives, one each, to the states whose **ratio** of decimal part to whole part were largest, until the desired total is reached.

Example 10

We'll do Delaware again. We begin in the same way as with Hamilton's method:

County	Population	Quota	Initial
Kent	162,310	7.4111	7
New Castle	538,479	24.5872	24
Sussex	197,145	9.0017	9
Total	**897,934**		**40**

We need one more representative. To find out which county should get it, Lowndes says to divide each county's decimal part by its whole number part, with the largest result getting the extra representative:

Kent: $0.4111/7 \approx 0.0587$
New Castle: $0.5872/24 \approx 0.0245$
Sussex: $0.0017/9 \approx 0.0002$

The largest of these is Kent's, so Kent gets the 41st representative:

County	Population	Quota	Initial	Ratio	Final
Kent	162,310	7.4111	7	0.0587	8
New Castle	538,479	24.5872	24	0.0245	24
Sussex	197,145	9.0017	9	0.0002	9
Total	**897,934**		**40**		**41**

Example 11

Rhode Island, again beginning in the same way as Hamilton:

County	Population	Quota	Initial
Bristol	49,875	3.5538	3
Kent	166,158	11.8395	11
Newport	82,888	5.9061	5
Providence	626,667	44.6528	44
Washington	126,979	9.0478	9
Total	**1,052,567**		**72**

We divide each county's quota's decimal part by its whole number part to determine which three should get the remaining representatives:

Bristol: $0.5538/3 \approx$ 0.1846
Kent: $0.8395/11 \approx$ 0.0763
Newport: $0.9061/5 \approx$ 0.1812
Providence: $0.6528/44 \approx$ 0.0148
Washington: $0.0478/9 \approx$ 0.0053

The three largest of these are Bristol, Newport, and Kent, so they get the remaining three representatives:

County	Population	Quota	Initial	Ratio	Final
Bristol	49,875	3.5538	3	0.1846	4
Kent	166,158	11.8395	11	0.0763	12
Newport	82,888	5.9061	5	0.1812	6
Providence	626,667	44.6528	44	0.0148	44
Washington	126,979	9.0478	9	0.0053	9
Total	**1,052,567**		**72**		**75**

As you can see, there is no "right answer" when it comes to choosing a method for apportionment. Each method has its virtues, and favors different sized states.

Apportionment of Legislative Districts

In most states, there are a fixed number of representatives to the state legislature. Rather than apportioning each county a number of representatives, legislative districts are drawn so that each legislator represents a district. The apportionment process, then, comes in the drawing of the legislative districts, with the goal of having each district include approximately the same number of constituents. Because of this goal, a geographically small city may have several representatives, while a large rural region may be represented by one legislator.

When populations change, it becomes necessary to redistrict the regions each legislator represents (Incidentally, this also occurs for the regions that federal legislators represent). The process of redistricting is typically done by the legislature itself, so not surprisingly it is common to see **gerrymandering**.

> **Gerrymandering**
> Gerrymandering is when districts are drawn based on the political affiliation of the constituents to the advantage of those drawing the boundary.

Example 12

Consider three districts, simplified to the three boxes below. On the left there is a college area that typically votes Democratic. On the right is a rural area that typically votes Republican. The rest of the people are more evenly split. The middle district has been voting 50% Democratic and 50% Republican.

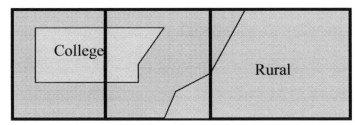

As part of a redistricting, a Democratic led committee could redraw the boundaries so that the middle district includes less of the typically Republican voters, thereby making it more likely that their party will win in that district, while increasing the Republican majority in the third district.

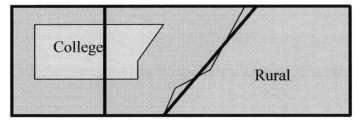

90

Example 13

The map to the right shows the 38th congressional district in California in 2004[1]. This district was created through a bi-partisan committee of incumbent legislators. This gerrymandering leads to districts that are not competitive; the prevailing party almost always wins with a large margin.

Congressional District 38

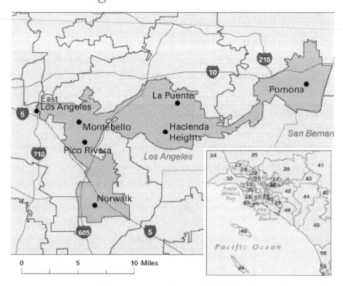

The map to the right shows the 4th congressional district in Illinois in 2004.[2] This district was drawn to contain the two predominantly Hispanic areas of Chicago. The largely Puerto Rican area to the north and the southern Mexican areas are only connected in this districting by a piece of the highway to the west.

Congressional District 4

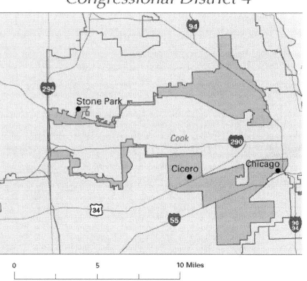

[1] http://en.wikipedia.org/wiki/File:California_District_38_2004.png
[2] http://en.wikipedia.org/wiki/File:Illinois_District_4_2004.png

Exercises

In exercises 1-8, determine the apportionment using
  a. Hamilton's Method
  b. Jefferson's Method
  c. Webster's Method
  d. Huntington-Hill Method
  e. Lowndes' method

1. A college offers tutoring in Math, English, Chemistry, and Biology. The number of students enrolled in each subject is listed below. If the college can only afford to hire 15 tutors, determine how many tutors should be assigned to each subject.
    Math: 330   English: 265 Chemistry: 130   Biology: 70

2. Reapportion the previous problem if the college can hire 20 tutors.

3. The number of salespeople assigned to work during a shift is apportioned based on the average number of customers during that shift. Apportion 20 salespeople given the information below.

Shift	Morning	Midday	Afternoon	Evening
Average number of customers	95	305	435	515

4. Reapportion the previous problem if the store has 25 salespeople.

5. Three people invest in a treasure dive, each investing the amount listed below. The dive results in 36 gold coins. Apportion those coins to the investors.
    Alice: $7,600   Ben: $5,900   Carlos: $1,400

6. Reapportion the previous problem if 37 gold coins are recovered.

7. A small country consists of five states, whose populations are listed below. If the legislature has 119 seats, apportion the seats.
    A: 810,000   B: 473,000   C: 292,000   D: 594,000   E: 211,000

8. A small country consists of six states, whose populations are listed below. If the legislature has 200 seats, apportion the seats.
    A: 3,411   B: 2,421   C: 11,586   D: 4,494   E: 3,126   F: 4,962

9. A small country consists of three states, whose populations are listed below.
            A: 6,000   B: 6,000   C: 2,000
    a. If the legislature has 10 seats, use Hamilton's method to apportion the seats.
    b. If the legislature grows to 11 seats, use Hamilton's method to apportion the seats.
    c. Which apportionment paradox does this illustrate?

10. A state with five counties has 50 seats in their legislature. Using Hamilton's method, apportion the seats based on the 2000 census, then again using the 2010 census. Which apportionment paradox does this illustrate?

County	2000 Population	2010 Population
Jefferson	60,000	60,000
Clay	31,200	31,200
Madison	69,200	72,400
Jackson	81,600	81,600
Franklin	118,000	118,400

11. A school district has two high schools: Lowell, serving 1715 students, and Fairview, serving 7364. The district could only afford to hire 13 guidance counselors.
    a. Determine how many counselors should be assigned to each school using Hamilton's method.
    b. The following year, the district expands to include a third school, serving 2989 students. Based on the divisor from above, how many additional counselors should be hired for the new school?
    c. After hiring that many new counselors, the district recalculates the reapportion using Hamilton's method. Determine the outcome.
    d. Does this situation illustrate any apportionment issues?

12. A small country consists of four states, whose populations are listed below. If the legislature has 116 seats, apportion the seats using Hamilton's method. Does this illustrate any apportionment issues?
    A: 33,700   B: 559,500   C: 141,300   D: 89,100

Exploration
13. Explore and describe the similarities, differences, and interplay between weighted voting, fair division (if you've studied it yet), and apportionment.

14. In the methods discussed in the text, it was assumed that the number of seats being apportioned was fixed. Suppose instead that the number of seats could be adjusted slightly, perhaps 10% up or down. Create a method for apportioning that incorporates this additional freedom, and describe why you feel it is the best approach. Apply your method to the apportionment in Exercise 7.

15. Lowndes felt that small states deserved additional seats more than larger states. Suppose you were a legislator from a larger state, and write an argument refuting Lowndes.

16. Research how apportionment of legislative seats is done in other countries around the world. What are the similarities and differences compared to how the United States apportions congress?

17. Adams's method is similar to Jefferson's method, but rounds quotas *up* rather than *down*. This means we usually need a modified divisor that is smaller than the standard divisor. Rework problems 1-8 using Adam's method. Which other method are the results most similar to?

Fair Division

Whether it is two kids sharing a candy bar or a couple splitting assets during a divorce, there are times in life where items of value need to be divided between two or more parties. While some cases can be handled through mutual agreement or mediation, in others the parties are adversarial or cannot reach a decision all feel is fair. In these cases, fair division methods can be utilized.

Fair Division Method

A fair division method is a procedure that can be followed that will result in a division of items in a way so that each party feels they have received their fair share. For these methods to work, we have to make a few assumptions:

- The parties are non-cooperative, so the method must operate without communication between the parties.
- The parties have no knowledge of what the other players like (their valuations).
- The parties act rationally, meaning they act in their best interest, and do not make emotional decisions.
- The method should allow the parties to make a fair division without requiring an outside arbitrator or other intervention.

With these methods, each party will be entitled to some **fair share**. When there are N parties equally dividing something, that fair share would be $1/N$.   For example, if there were 4 parties, each would be entitled to a fair share of $¼ = 25\%$ of the whole. More specifically, they are entitled to a share that *they* value as 25% of the whole.

Fair Share

When N parties divide something equally, each party's **fair share** is the amount they entitled to. As a fraction, it will be $1/N$

It should be noted that a fair division method simply needs to guarantee that each party will receive a share they view as fair. A basic fair division does not need to be envy free; an **envy-free** division is one in which no party would prefer another party's share over their own. A basic fair division also does not need to be Pareto optimal; a **Pareto optimal** division is one in which no other division would make a participant better off without making someone else worse off. Nor does fair division have to be **equitable**; an equitable division is one in which the proportion of the whole each party receives, judged by their own valuation, is the same. Basically, a simple fair division doesn't have to be the best possible division – it just has to give each party their fair share.

Example 1

Suppose that 4 classmates are splitting equally a $12 pizza that is half pepperoni, half veggie that someone else bought them. What is each person's fair share?

Since they all are splitting the pizza equally, each person's fair share is $3, or pieces they value as 25% of the pizza.

It is important to keep in mind that each party might value portions of the whole differently. For example, a vegetarian would probably put zero value on the pepperoni half of the pizza.

Example 2

Suppose that 4 classmates are splitting equally a $12 pizza that is half pepperoni, half veggie. Steve likes pepperoni twice as much as veggie. Describe a fair share for Steve.

He would value the veggie half as being worth $4 and the pepperoni half as $8, twice as much. If the pizza was divided up into 4 pepperoni slices and 4 veggie slices, he would value a pepperoni slice as being worth $2, and a veggie slice as being worth $1.

If we weren't able to guess the values, we could take a more algebraic approach. If Steve values a veggie slice as worth x dollars, then he'd value a pepperoni slice as worth $2x$ dollars – twice as much. Four veggie slices would be worth $4 \cdot x = 4x$ dollars, and 4 pepperoni slices would be worth $4 \cdot 2x = 8x$ dollars. Altogether, the eight slices would be worth $4x + 8x = 12x$ dollars. Since the total value of the pizza was $12, then $12x = \$12$. Solving we get $x = \$1$; the value of a veggie slice is $1, and the value of a pepperoni slice is $2x = \$2$.

A fair share for Steve would be one pepperoni slice and one veggie slice ($2 + $1 = $3 value), 1½ pepperoni slices (1½ · $2 = $3 value), 3 veggie slices (3 · $1 = $3 value), or a variety of more complicated possibilities.

Try it Now 1

Suppose Kim is another classmate splitting the pizza, but Kim is vegetarian, so won't eat pepperoni. Describe a fair share for Kim.

You will find that many examples and exercises in this topic involve dividing food – dividing candy, cutting cakes, sharing pizza, etc. This may make this topic seem somewhat trivial, but instead of cutting a cake, we might be drawing borders dividing Germany after WWII. Instead of splitting a bag of candy, siblings might be dividing belongings from an inheritance. Mathematicians often characterize very important and contentious issues in terms of simple items like cake to separate any emotional influences from the mathematical method.

Because of this, our requirement that the players not communicate about their preferences can seem silly. After all, why wouldn't four classmates talk about what kind of pizza they like if they're splitting a pizza? Just remember that in issues of politics, business, finance, divorce settlements, etc. the players are usually less cooperative and more concerned about the other players trying to get more than their fair share.

There are two broad classifications of fair division methods: those that apply to **continuously divisible** items, and those that apply to **discretely divisible** items. Continuously divisible items are things that can be divided into pieces of any size, like dividing a candy bar into two pieces or drawing borders to split a piece of land into smaller plots. Discretely divisible items are when you are dividing several items that cannot be broken apart easily, such as assets in a divorce (house, car, furniture, etc).

Divider-Chooser

The first method we will look at is a method for continuously divisible items. This method will be familiar to many parents - it is the "You cut, I choose" method. In this method, one party is designated the **divider** and the other the **chooser**, perhaps with a coin toss. The method works as follows:

> **Divider-Chooser Method**
> 1. The divider cuts the item into two pieces that are, in his eyes, equal in value.
> 2. The chooser selects either of the two pieces
> 3. The divider receives the remaining piece

Notice that the divider-chooser method is specific to a two-party division. Examine why this method guarantees a fair division: since the divider doesn't know which piece he will receive, the rational action for him to take would be to divide the whole into two pieces he values equally. There is no incentive for the divider to attempt to "cheat" since he doesn't know which piece he will receive. Since the chooser can pick either piece, she is guaranteed that one of them is worth at least 50% of the whole in her eyes. The chooser is guaranteed a piece she values as at least 50%, and the divider is guaranteed a piece he values at 50%.

Example 3

> Two retirees, Fred and Martha, buy a vacation beach house in Florida together, with the agreement that they will split the year into two parts.
>
> Fred is chosen to be the divider, and splits the year into two pieces: November – February and March – October. Even though the first piece is 4 months and the second is 8 months, Fred places equal value on both pieces since he really likes to be in Florida during the winter. Martha gets to pick whichever piece she values more. Suppose she values all months equally. In this case, she would choose the March – October time, resulting in a piece that she values as 8/12 = 66.7% of the whole. Fred is left with the November – February slot which he values as 50% of the whole.

Of course, in this example, Fred and Martha probably could have discussed their preferences and reached a mutually agreeable decision. The divider-chooser method is more necessary in cases where the parties are suspicious of each other's motives, or are unable to communicate effectively, such as two countries drawing a border, or two children splitting a candy bar.

Dustin and Quinn were given an apple pie and a chocolate cake, and need to divide them. Dustin values the apple pie at $6 and the chocolate cake at $4. Quinn values the apple pie as $4 and the chocolate cake at $10. Describe a fair division if Quinn is dividing, and specify which "half" Dustin will choose.

Things quickly become more complicated when we have more than two parties involved. We will look at three different approaches. But first, let us look at one that doesn't work.

How not to divide with 3 parties

When first approaching the question of 3-party fair division, it is very tempting to propose this method: Randomly designate one participant to be the divider, and designate the rest choosers. Proceed as follows:

1) Have the divider divide the item into 3 pieces
2) Have the first chooser select any of the three pieces they feel is worth a fair share
3) Have the second chooser select either of the remaining pieces
4) The divider gets the piece left.

Example 4. Don't do this – it is bad!

Suppose we have three people splitting a cake. We can immediately see that the divider will receive a fair share as long as they cut the cake fairly at the beginning. The first chooser certainly will also receive a fair share. What about the second chooser? Suppose each person values the three pieces like this:

	Piece 1	Piece 2	Piece 3
Chooser 1	40%	30%	30%
Chooser 2	45%	30%	25%
Divider	33.3%	33.3%	33.3%

Since the first chooser will clearly select Piece 1, the second chooser is left to select between Piece 2 and Piece 3, neither of which she values as a fair share (1/3 or about 33.3%). This example shows that this method does *not* guarantee a fair division.

To handle division with 3 or more parties, we'll have to take a more clever approach.

Lone Divider

The lone divider method works for any number of parties – we will use N for the number of parties. One participant is randomly designated the divider, and the rest of the participants are designated as choosers.

Lone Divider Method

The Lone Divider method proceeds as follows:

1) The divider divides the item into N pieces, which we'll label $s_1, s_2, ..., s_N$.
2) Each of the choosers will separately list which pieces they consider to be a fair share. This is called their **declaration**, or **bid**.
3) The lists are examined. There are two possibilities:
   a. If it is possible to give each party a piece they declared then do so, and the divider gets the remaining piece.
   b. If two or more parties both want the same pieces and no others, then give a non-contested piece to the divider. The rest of the pieces are combined and repeat the entire procedure with the remaining parties. If there are only two parties left, they can use divider-chooser.

Example 5

Consider the example from earlier, in which the pieces were valued as:

	Piece 1	Piece 2	Piece 3
Chooser 1	40%	30%	30%
Chooser 2	45%	30%	25%
Divider	33.3%	33.3%	33.3%

Each chooser makes a declaration of which pieces they value as a fair share. In this case,

Chooser 1 would make the declaration: Piece 1
Chooser 2 would make the declaration: Piece 1

Since both choosers want the same piece, we cannot immediately allocate the pieces. The lone divider method specifies that we give a non-contested piece to the divider. Both pieces 2 and 3 are uncontested, so we flip a coin and give Piece 2 to the divider. Piece 1 and 3 are then recombined to make a piece worth 70% to Chooser 1, and 70% to Chooser 2. Since there are only two players left, they can divide the recombined pieces using divider-chooser. Each is guaranteed a piece they value as at least 35%, which is a fair share.

Try it Now 3

Use the Lone Divider method to complete the fair division given the values below.

	Piece 1	Piece 2	Piece 3
Chooser 1	40%	30%	30%
Chooser 2	40%	35%	25%
Divider	33.3%	33.3%	33.3%

Example 6

Suppose that Abby, Brian, Chris, and Dorian are dividing a plot of land. Dorian was selected to be the divider through a coin toss. Each person's valuation of each piece is shown below.

	Piece 1	Piece 2	Piece 3	Piece 4
Abby	15%	30%	20%	35%
Brian	30%	35%	10%	25%
Chris	20%	45%	20%	15%
Dorian	25%	25%	25%	25%

Based on this, their declarations should be:
Abby: Piece 2, Piece 4
Brian: Piece 1, Piece 2, Piece 4
Chris: Piece 2

This case can be settled simply – by awarding Piece 2 to Chris, Piece 4 to Abby, Piece 1 to Brian, and Piece 3 to Dorian. Each person receives a piece that they value as at least a fair share (25% value).

Example 7

Suppose the valuations in the previous problem were:

	Piece 1	Piece 2	Piece 3	Piece 4
Abby	15%	30%	20%	35%
Brian	20%	35%	10%	35%
Chris	20%	45%	20%	15%
Dorian	25%	25%	25%	25%

The declarations would be:
Abby: Piece 2, Piece 4
Brian: Piece 2, Piece 4
Chris: Piece 2

Notice in this case that there is no simple settlement. So, the piece no one else declared, Piece 3, is awarded to the original divider Dorian, and the procedure is repeated with the remaining three players.

Suppose that on the second round of this method Brian is selected to be the divider, three new pieces are cut, and the valuations are as follows:

	Piece 1	Piece 2	Piece 3
Abby	40%	30%	30%
Brian	33.3%	33.3%	33.3%
Chris	50%	20%	30%

The declarations here would be:
Abby: Piece 1
Chris: Piece 1

Once again we have a standoff. Brian can be awarded either of Piece 2 or Piece 3, and the remaining pieces can be recombined. Since there are only two players left, they can divide the remaining land using the basic divider-chooser method.

Try it Now 4
Four investors are dividing a piece of land valued at $320,000. One was chosen as the divider, and their values of the division (in thousands) are shown below. Who was the divider? Describe the outcome of the division.

	Piece 1	Piece 2	Piece 3	Piece 4
Sonya	$90	$70	$80	$80
Cesar	$80	$80	$80	$80
Adrianna	$60	$70	$100	$90
Raquel	$70	$50	$90	$110

Last Diminisher

The Last Diminisher method is another approach to division among 3 or more parties.

Last Diminisher Method
In this method, the parties are randomly assigned an order, perhaps by pulling names out of a hat. The method then proceeds as follows:

1) The first person cuts a slice they value as a fair share.
2) The second person examines the piece
   a. If they think it is worth less than a fair share, they then pass on the piece unchanged.
   b. If they think the piece is worth more than a fair share, they trim off the excess and lay claim to the piece. The trimmings are added back into the to-be-divided pile.
3) Each remaining person, in turn, can either pass or trim the piece
4) After the last person has made their decision, the last person to trim the slice receives it. If no one has modified the slice, then the person who cut it receives it.
5) Whoever receives the piece leaves with their piece and the process repeats with the remaining people. Continue until only 2 people remain; they can divide what is left with the divider-chooser method.

Example 8

Suppose that four salespeople are dividing up Washington State into sales regions; each will get one region to work in. They pull names from a hat to decide play order.

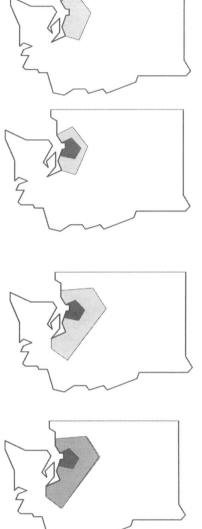

Round 1. The first salesman, Bob, draws a region around Seattle, the most populous area of the state. The piece Bob cuts and automatically lays claim to is shown in yellow.

The second salesman, Henry, felt that this region was worth more than 25%, each player's fair share. Because of this, Henry opts to trim this piece. The new piece is shown in pink. The trimmings (in yellow) return to the to-be-divided portion of the state. Henry automatically lays claim to this smaller piece since he trimmed it.

The third saleswoman, Marjo, feels this piece is worth less than 25% and passes, as does the fourth saleswoman, Beth. Since both pass, the last person to trim it, Henry, receives the piece.

Round 2. The second round begins with Bob laying claim to a piece, shown again in yellow. Henry already has a piece, so is out of the process now. Marjo passes on this piece, feeling it is worth less than a fair share.

Beth, on the other hand, feels the piece as currently drawn is worth 35%. Beth is in an advantageous position, being the last to make a decision. Even though Beth values this piece at 35%, she can cut a very small amount and still lay claim to it. So Beth barely cuts the piece, resulting in a piece (blue) that is essentially worth 35% to her. Since she is the last to trim, she receives the piece.

Round 3. At this point, Bob and Marjo are the only players without a piece. Since there are two of them, they can finish the division using the divider-chooser method. They flip a coin, and determine that Marjo will be the divider. Marjo draws a line dividing the remainder of the state into two pieces. Bob chooses the Eastern piece, leaving Marjo with the Western half.

Notice that in this division, Henry and Marjo ended up with pieces that they feel are worth exactly 25% - a fair share. Beth was able to receive a piece she values as more than a fair share, and Bob may feel the piece he received is worth more than 25%.

Example 9

Marcus, Abby, Julian, and Ben are splitting a pizza that is 4 slices of cheese and 4 slices of veggie with total value $12. Marcus and Ben like both flavors equally, Abby only likes cheese, and Julian likes veggie twice as much as cheese. They divide the pizza using last diminisher method, playing in the order Marcus, Abby, Ben, then Julian.

Notice Ben and Marcus both value any slice of pizza at $1.50.
Abby values each slice of cheese at $3, and veggie at $0.
Julian values each slice of cheese at $1, and each slice of veggie at $2. (see Example 2)
A fair share for any player is $3.

In the first round, suppose Marcus cuts out 2 slices of cheese, which he values at $3.
Abby only likes cheese, so will value this cut at $6. She will trim it to 1 slice of cheese, which she values as her fair share of $3.
Ben will view this piece as less than a fair share, and will pass.
Julian will view this piece as less than a fair share, and will pass.
Abby receives the piece.

In the second round, suppose Marcus cuts a slice that is 2 slices of veggie.
Abby already received a slice so is out.
Ben will view this piece as having value $3. He can barely trim it and lay claim to it.
Julian will value this piece as having value $4, so will barely trim it and claim it.

Marcus and Ben can then split the remaining 3 slices of cheese and 2 slices of veggie using the divider-chooser method.

In the second round, both Ben and Julian will make tiny trims (pulling off a small crumb) in order to lay claim to the piece without practically reducing the value. The piece Julian receives is still essentially worth $4 to him; we don't worry about the value of that crumb.

Try it Now 5

Five players are dividing a $20 cake. In the first round, Player 1 makes the initial cut and claims the piece. For each of the remaining players, the value of the current piece (which may have been trimmed) at the time it is their turn is shown below. Describe the outcome of the first round.

	P_2	P_3	P_4	P_5
Value of the current piece	$3	$5	$3.50	$3

In the second round, Player 1 again makes the initial cut and claims the piece, and the current values are shown again. Describe the outcome of the second round.

	P_2	P_4	P_5
Value of the current piece	$7	$3	$5

Moving Knife

A somewhat different approach to continuous fair division is called the moving knife procedure.

Moving Knife Method

In this method, applied to a cake,

1) A referee starts moving a knife from left to right across a cake.
2) As soon as any player feels the piece to the left of the knife is worth a fair share, they shout "STOP." The referee then cuts the cake at the current knife position and the player who called stop gets the piece to the left of the knife.
3) This procedure continues until there is only one player left. The player left gets the remaining cake.

Example 10

Suppose that our four salespeople from above decided to use this approach to divide Washington. Rather than move the "knife" from left to right, they decide to move it from top to bottom.

The referee starts moving a line down a map of the state. Henry is the first to call STOP when the knife is at the position shown, giving him the portion of the state above the line.

Marjo is the next to call STOP when the knife is at the position shown, giving her the second portion of the state.

Bob is the next to call STOP, leaving Beth with the southernmost portion of the state

While this method guarantees a fair division, it clearly results in some potentially silly divisions in a case like this. The method is probably better suited to situations like dividing an actual cake.

Sealed Bids Method

The Sealed Bids method provides a method for *discrete* fair division, allowing for the division of items that cannot be split into smaller pieces, like a house or a car. Because of this, the method requires that all parties have a large amount of cash at their disposal to balance out the difference in item values.

Sealed Bids Method

The method begins by compiling a list of items to be divided. Then:

1) Each party involved lists, in secret, a dollar amount they value each item to be worth. This is their sealed bid.
2) The bids are collected. For each party, the value of all the items is totaled, and divided by the number of parties. This defines their fair share.
3) Each item is awarded to the highest bidder.
4) For each party, the value of all items received is totaled. If the value is more than that party's fair share, they pay the difference into a holding pile. If the value is less than that party's fair share, they receive the difference from the holding pile. This ends the initial allocation.
5) In most cases, there will be a surplus, or leftover money, in the holding pile. The surplus is divided evenly between all the players. This produces the final allocation.

While the assumptions we made for fair division methods specified that an arbitrator should not be necessary, it is common for an independent third party to collect the bids and announce the outcome. While not technically necessary, since the method can be executed without a third party involved, this protects the secrecy of the bids, which can sometimes help avoid resentment or bad feelings between the players.

Example 11

Sam and Omar have cohabitated for the last 3 years, during which time they shared the expense of purchasing several items for their home. Sam has accepted a job in another city, and now they find themselves needing to divide their shared assets.

Each records their value of each item, as shown below.

	Sam	Omar
Couch	$150	$100
TV	$200	$250
Video game system	$250	$150
Surround sound system	$50	$100

Sam's total valuation of the items is $150+$200+$250+$50 = $650, making a fair share for Sam $650/2 = $325.

Omar's total valuation of the items is $100+$250+$150+$100 = $600, making a fair share for Omar $600/2 = $300.

Each item is now awarded to the highest bidder. Sam will receive the couch and video game system, providing $150+$250 = $400 of value to Sam. Since this exceeds his fair share, he has to pay the difference, $75, into a holding pile.

Omar will receive the TV and surround sound system, providing $250+$100=$350 in value. This is more than his fair share, so he has to pay the difference, $50, into the holding pile.

Thus, in the initial allocation, Sam receives the couch and video game system and pays $75 into the holding pile. Omar receives the TV and surround sound system and pays $50 into the holding pile. At this point, both players would feel they have received a fair share.

There is now $125 remaining in the holding pile. This is the surplus from the division. This is now split evenly, and both Sam and Omar are given back $62.50. Since Sam had paid in $75, the net effect is that he paid $12.50. Since Omar had originally paid in $50, the net effect is that he receives $12.50.

Thus, in the final allocation, Sam receives the couch and video game system and pays $12.50 to Omar. Omar receives the TV and surround sound system and receives $12.50. At this point, both players feel they have received more than a fair share.

Example 12

Four small graphic design companies are merging operations to become one larger corporation. In this merger, there are a number of issues that need to be settled. Each company is asked to place a monetary value (in thousands of dollars) on each issue:

	Super Designs	DesignByMe	LayoutPros	Graphix
Company name	$5	$3	$3	$6
Company location	$8	$9	$7	$6
CEO	$10	$5	$6	$7
Chair of the board	$7	$6	$6	$8

We can then calculate for each company:

	Super Designs	DesignByMe	LayoutPros	Graphix
Total value of all issues	$30	$23	$22	$27
Fair share	$7.50	$5.75	$5.50	$6.75

The items would then be allocated to the company that bid the most for each.
Company name would be awarded to Graphix
Company location would be awarded to DesignByMe
CEO would be awarded to Super Designs
Chair of the board would be awarded to Graphix

For each company, we calculate the total value of the items they receive, and how much they get or pay in the initial allocation

	Super Designs	DesignByMe	LayoutPros	Graphix
Total value of issues awarded	$10	$9	$0	$14
Amount they pay/get	$7.50-$10 Pay $2.50	$5.75 - $9 Pay $3.25	$5.50 - $0 Get $5.50	$6.75 - $14 Pay $7.25

After the initial allocation, there is a total of $2.50+$3.25-$5.50+$7.25 = $7.50 in surplus. Dividing that evenly, each company gets $1.875 (approximately)

	Super Designs	DesignByMe	LayoutPros	Graphix
Amount they pay/get	Pay $2.50 Get $1.875	Pay $3.25 Get $1.875	Get $5.50 Get $1.875	Pay $7.25 Get $1.875
After surplus	Pay $0.625	Pay $1.375	Get $7.375	Pay $5.375

So in the final allocation,
Super Designs wins the CEO, and pays $625 ($0.625 thousand)
DesignByMe wins the company location and pays $1,375 ($1.375 thousand)
LayoutPros wins no issues, but receives $7,375 in compensation
Graphix wins the company name and chair of the board, and pays $5,375.

Try it Now 6

Jamal, Maggie, and Kendra are dividing an estate consisting of a house, a vacation home, and a small business. Their valuations (in thousands) are shown below. Determine the final allocation.

	Jamal	Maggie	Kendra
House	$250	$300	$280
Vacation home	$170	$180	$200
Small business	$300	$255	$270

Example 13

Fair division does not always have to be used for items of value. It can also be used to divide undesirable items. Suppose Chelsea and Mariah are sharing an apartment, and need to split the chores for the household. They list the chores, assigning a *negative* dollar value to each item; in other words, the amount they would pay for someone else to do the chore (a per week amount). We will assume, however, that they are committed to doing all the chores themselves and not hiring a maid.

	Chelsea	Mariah
Vacuuming	-$10	-$8
Cleaning bathroom	-$14	-$20
Doing dishes	-$4	-$6
Dusting	-$6	-$4

We can then calculate fair share:

	Chelsea	Mariah
Total value	-$34	-$38
Fair Share	-$17	-$19

We award to the person with the largest bid. For example, we award vacuuming to Mariah since she dislikes it less (remember -8 > -10).
Chelsea gets cleaning the bathroom and doing dishes. Value: -$18
Mariah gets vacuuming and dusting. Value: -$12

Notice that Chelsea's fair share is -$17 but she is doing chores she values at -$18. She should get $1 to bring her to a fair share. Mariah is doing chores valued at -$12, but her fair share is -$19. She needs to pay $7 to bring her to a fair share.

This creates a surplus of $6, which will be divided between the two. In the final allocation:
Chelsea gets cleaning the bathroom and doing dishes, and receives $1 + $3 = $4/week.
Mariah gets vacuuming and dusting, and pays $7 - $3 = $4/week.

Try it Now Answers
1. Kim will value the veggie half of the pizza at the full value, $12, and the pepperoni half as worth $0. Since a fair share is 25%, a fair share for Kim is one slice of veggie, which she'll value at $3. Of course, Kim only getting one slice doesn't really seem very fair, but if every player had the same valuation as Kim, this would be the only fair outcome. Luckily, if the classmates splitting the pizza are friends, they are probably cooperative, and will talk about what kind of pizza they like.

2. There are a lot of possible fair divisions Quinn could make. Since she values the two desserts at $14 together, a fair share in her eyes is $7. Notice since Dustin values the desserts at $10 together, a fair share in his eyes is $5 of value. A couple possible divisions:

Piece 1	Piece 2	Dustin would choose
7/10 chocolate ($7) No pie ($0)	3/10 chocolate ($3) All pie ($4)	Piece 2. Value in his eyes: $6 + 3/10·$4 = $7.20
½ chocolate ($5) ½ pie ($2)	½ chocolate ($5) ½ pie ($2)	Either. Value in his eyes: ½·$6 + ½·$4 = $5

3. Chooser 1 would make the declaration: Piece 1
Chooser 2 would make the declaration: Piece 1, Piece 2
We can immediately allocate the pieces, giving Piece 2 to Chooser 2, Piece 1 to Chooser 1, and Piece 3 to the Divider. All players receive a piece they value as a fair share.

Try it Now Answers Continued

4. A fair share would be $80,000. Cesar was the divider.
Their declarations would be:
Sonya: Pierce 1, Piece 3, Piece 4.
Adrianna: Piece 3, Piece 4
Raquel: Piece 3, Piece 4

Sonya would receive Piece 1 and Cesar would receive the uncontested Piece 2.
Since Adrianna and Raquel both want Piece 3 or Piece 4, a coin would be flipped to allocate those pieces between them.

5. In the first round, Player 1 will cut a piece he values as a fair share of $4. Player 2 values the piece as $3, so will pass. Player 3 values the piece as $5, so will claim it and trim it to something she values as $4. Player 4 receives that piece and values it as $3.50 so will pass. Player 5 values the piece at $3 and will also pass. Player 3 receives the trimmed piece she values at $4.

In the second round, Player 1 will again cut a piece he values as a fair share of $4. Player 2 values the piece as $7, so will claim it and trim it to something he values as $4. Player 4 values the trimmed piece at $3 and passes. Player 5 values the piece at $5, so will claim it. Since Player 5 is the last player, she has an advantage and can claim then barely trim the piece.   Player 5 receives a piece she values at $5.

6. Jamal's total value is $250 + $170 + $300 = $720. His fair share is $240 thousand.
Maggie's total value is $300 + $180 + $255 = $735. Her fair share is $245 thousand.
Kendra's total value is$280 + $200 + $270 = $750. Her fair share is $250 thousand.

In the initial allocation,
Jamal receives the business, and pays $300 - $240 = $60 thousand into holding.
Maggie receives the house, and pays $300 - $245 = $55 thousand into holding.
Kendra receives the vacation home, and gets $250 - $200 = $50 thousand from holding.

There is a surplus of $60 + $55 - $50 = $65 thousand in holding, so each person will receive $21,667 from surplus. In the final allocation,
Jamal receives the business, and pays $38,333.
Maggie receives the house, and pays $33,333.
Kendra receives the vacation home, and gets $71,667.

Exercises

Skills

1. Chance and Brianna buy a pizza for $10 that is half pepperoni and half veggie. They cut the pizza into 8 slices.

   If Chance likes veggie three times as much as pepperoni, what is the value of a slice that is half pepperoni, half veggie?

2. Ahmed and Tiana buy a cake for $14 that is half chocolate and half vanilla. They cut the cake into 8 slices.

   If Ahmed likes chocolate four times as much as vanilla, what is the value of a slice that is half chocolate, half vanilla?

3. Erin, Catherine, and Shannon are dividing a large bag of candy. They randomly split the bag into three bowls. The values of the entire bag and each of the three bowls in the eyes of each of the players are shown below. For each player, identify which bowls they value as a fair share.

	Whole Bag	Bowl 1	Bowl 2	Bowl 3
Erin	$5	$2.75	$1.25	$1.00
Catherine	$4	$0.75	$2.50	$0.75
Shannon	$8	$1.75	$2.25	$4.00

4. Jenna, Tatiana, and Nina are dividing a large bag of candy. They randomly split the bag into three bowls. The values of the entire bag and each of the three bowls in the eyes of each of the players are shown below. For each player, identify which bowls they value as a fair share.

	Whole Bag	Bowl 1	Bowl 2	Bowl 3
Jenna	$8	$4.50	$0.75	$2.75
Tatiana	$4	$1.00	$1.00	$2.00
Nina	$6	$1.75	$2.50	$1.75

5. Dustin and Kendra want to split a bag of fun-sized candy, and decide to use the divider-chooser method. The bag contains 100 Snickers, 100 Milky Ways, and 100 Reese's, which Dustin values at $1, $5, and $2 respectively. (This means Dustin values the 100 Snickers together at $1, or $0.01 for 1 Snickers).

   If Kendra is the divider, and in one half puts:
   25 Snickers, 20 Milky Ways, and 60 Reese's
   a. What is the value of this half in Dustin's eyes?
   b. Does Dustin consider this a fair share?
   c. If Dustin was a divider, find a possible division that is consistent with his value system.

6. Dustin and Kendra want to split a bag of fun-sized candy, and decide to use the divider-chooser method. The bag contains 100 Snickers, 100 Milky Ways, and 100 Reese's, which Dustin values at $1, $3, and $5 respectively. (This means Dustin values the 100 Snickers together at $1, or $0.01 for 1 Snickers).

    If Kendra is the divider, and in one half puts:
    30 Snickers, 40 Milky Ways, and 66 Reese's
    a. What is the value of this half in Dustin's eyes?
    b. Does Dustin consider this a fair share?
    c. If Dustin was a divider, find a possible division that is consistent with his value system.

7. Maggie, Meredith, Holly, and Zoe are dividing a piece of land using the lone-divider method. The values of the four pieces of land in the eyes of the each player are shown below.

	Piece 1	Piece 2	Piece 3	Piece 4
Maggie	21%	27%	32%	20%
Meredith	27%	29%	22%	22%
Holly	23%	14%	41%	22%
Zoe	25%	25%	25%	25%

    a. Who was the divider?
    b. If playing honestly, what will each player's declaration be?
    c. Find the final division.

8. Cody, Justin, Ahmed, and Mark are going to share a vacation property. The year will be divided into 4 time slots using the lone-divider method. The values of each time slot in the eyes of the each player are shown below.

	Time 1	Time 2	Time 3	Time 4
Cody	10%	35%	34%	21%
Justin	25%	25%	25%	25%
Ahmed	19%	24%	30%	27%
Mark	23%	31%	22%	24%

    a. Who was the divider?
    b. If playing honestly, what will each player's declaration be?
    c. Find the final division.

9. A 6-foot sub valued at $30 is divided among five players (P_1, P_2, P_3, P_4, P_5) using the last-diminisher method. The players play in a fixed order, with P_1 first, P_2 second, and so on. In round 1, P_1 makes the first cut and makes a claim on a piece. For each of the remaining players, the value of the *current* claimed piece at the time it is their turn is given in the following table:

	P_2	P_3	P_4	P_5
Value of the current claimed piece	$6.00	$8.00	$7.00	$6.50

   a. Which player gets his or her share at the end of round 1?
   b. What is the value of the share to the player receiving it?
   c. How would your answer change if the values were:

	P_2	P_3	P_4	P_5
Value of the current claimed piece	$6.00	$8.00	$7.00	$4.50

10. A huge collection of low-value baseball cards appraised at $100 is being divided by 5 kids (P_1, P_2, P_3, P_4, P_5) using the last-diminisher method. The players play in a fixed order, with P_1 first, P_2 second, and so on. In round 1, P_1 makes the first selection and makes a claim on a pile of cards. For each of the remaining players, the value of the *current* pile of cards at the time it is their turn is given in the following table:

	P_2	P_3	P_4	P_5
Value of the current pile of cards	$15.00	$22.00	$18.00	$19.00

   a. Which player gets his or her share at the end of round 1?
   b. What is the value of the share to the player receiving it?
   c. How would your answer change if the values were:

	P_2	P_3	P_4	P_5
Value of the current pile of cards	$15.00	$22.00	$18.00	$21.00

11. Four heirs (A, B, C, and D) must fairly divide an estate consisting of two items - a desk and a vanity - using the method of sealed bids. The players' bids (in dollars) are:

	A	B	C	D
Desk	320	240	300	260
Vanity	220	140	200	180

   a. What is A's fair share?
   b. Find the initial allocation.
   c. Find the final allocation.

12. Three heirs (A, B, C) must fairly divide an estate consisting of three items - a house, a car, and a coin collection - using the method of sealed bids. The players' bids (in dollars) are:

	A	B	C
House	180,000	210,000	·220,000
Car	,12,000	10,000	8,000
Coins	3,000	6,000	2,000

    a. What is A's fair share?
    b. Find the initial allocation.
    c. Find the final allocation.

13. As part of an inheritance, four children, Abby, Ben and Carla, are dividing four vehicles using Sealed Bids. Their bids (in thousands of dollars) for each item is shown below. Find the final allocation.

	Abby	Ben	Carla
Motorcycle	,10	9	8
Car	10	,11	9
Tractor	4	1	2
Boat	·7	6	4

14. As part of an inheritance, four children, Abby, Ben, Carla, and Dan, are dividing four vehicles using Sealed Bids. Their bids (in thousands of dollars) for each item is shown below. Find the final allocation.

	Abby	Ben	Carla	Dan
Motorcycle	6	7	11	8
Car	8	13	10	11
Tractor	3	1	5	4
Boat	7	6	3	8

15. After living together for a year, Sasha and Megan have decided to go their separate ways. They have several items they bought together to divide, as well as some moving-out chores. The values each place are shown below. Find the final allocation.

	Sasha	Megan
Couch	120	80
TV	200	250
Stereo	40	50
Detail cleaning	-40	-60
Cleaning carpets	-50	-40

16. After living together for a year, Emily, Kayla, and Kendra have decided to go their separate ways. They have several items they bought together to divide, as well as some moving-out chores. The values each place are shown below. Find the final allocation.

	Emily	Kayla	Kendra
Dishes	20	30	40
Vacuum cleaner	100	120	80
Dining table	100	80	130
Detail cleaning	-70	-40	-50
Cleaning carpets	-30	-60	-50

Exploration

17. This question explores how bidding dishonestly can end up hurting the cheater. Four partners are dividing a million-dollar property using the lone-divider method. Using a map, Danny divides the property into four parcels s_1, s_2, s_3, and s_4. The following table shows the value of the four parcels in the eyes of each partner (in thousands of dollars):

	s_1	s_2	s_3	s_4
Danny	$250	$250	$250	$250
Brianna	$460	$180	$200	$160
Carlos	$260	$310	$220	$210
Greedy	$330	$300	$270	$100

a. Assuming all players bid honestly, which piece will Greedy receive?

b. Assume Brianna and Carlos bid honestly, but Greedy decides to bid only for s1, figuring that doing so will get him s1. In this case there is a standoff between Brianna and Greedy. Since Danny and Carlos are not part of the standoff, they can receive their fair shares. Suppose Danny gets s3 and Carlos gets s2, and the remaining pieces are put back together and Brianna and Greedy will split them using the basic divider-chooser method. If Greedy gets selected to be the divider, what will be the value of the piece he receives?

c. Extension: Create a Sealed Bids scenario that shows that sometimes a player can successfully cheat and increase the value they receive by increasing their bid on an item, but if they increase it too much, they could end up receiving less than their fair share.

18. Explain why divider-chooser method with two players will always result in an envy-free division.

19. Will the lone divider method always result in an envy-free division? If not, will it ever result in an envy-free division?

20. The Selfridge-Conway method is an envy-free division method for three players. Research how the method works and prepare a demonstration for the class.

21. Suppose that two people are dividing a $12 pizza that is half pepperoni, half cheese. Steve likes both equally, but Maria likes cheese twice as much as pepperoni. As divider, Steve divides the pizza so that one piece is 1/3 cheese and 2/3 pepperoni, and the second piece is 1/3 pepperoni and 2/3 cheese.
    a.   Describe the value of each piece to each player
    b.   Since the value to each player is not the same, this division is not equitable. Find a division that would be equitable. Is it still envy-free?
    c.   The original division is not Pareto optimal. To show this, find another division that would increase the value to one player without decreasing the value to the other player. Is this division still envy-free?
    d.   Would it be possible with this set of preferences to find a division that is both equitable and Pareto optimal? If so, find it. If not, explain why.

22. Is the Sealed Bids method Pareto optimal when used with two players? If not, can you adjust the method to be so?

23. Is the Sealed Bids method envy-free when used with two players? If not, can you adjust the method to be so?

24. Is the Sealed Bids method equitable when used with two players? If not, can you adjust the method to be so?

25. All the problems we have looked at in this chapter have assumed that all participants receive an equal share of what is being divided. Often, this does not occur in real life. Suppose Fred and Maria are going to divide a cake using the divider-chooser method. However, Fred is only entitled to 30% of the cake, and Maria is entitled to 70% of the cake (maybe it was a $10 cake, and Fred put in $3 and Maria put in $7). Adapt the divider-choose method to allow them to divide the cake fairly.

Assume (as we have throughout this chapter) that different parts of the cake may have different values to Fred and Maria, and that they don't communicate their preferences/values with each other. You goal is to come up with a method of fair division, meaning that although the participants may not receive equal shares, they should be *guaranteed* their fair share. Your method needs to be designed so that each person will always be guaranteed a share that they value as being worth at least as much as they're entitled to.

The last few questions will be based on the Adjusted Winner method, described here:
For discrete division between two players, there is a method called Adjusted Winner that produces an outcome that is always equitable, envy-free, and Pareto optimal. It does, however, require that items can be split or shared. The method works like this:

1) Each player gets 100 points that they assign to the items to be divided based on their relative worth.
2) In the initial allocation, each item is given to the party that assigned it more points. If there were any items with both parties assigned the same number of points, they'd go to the person with the fewest points so far.
3) If the assigned point values are not equal, then begin transferring items from the person with more points to the person with fewer points. Start with the items that have the smallest point ratio, calculated as (higher valuation/lower valuation).
4) If transferring an entire item would move too many points, only transfer a portion of the item.

Example: A couple is attempting to settle a contentious divorce[1]. They assign their 100 points to the issues in contention:

	Mike	Carol
Custody of children	25	65
Alimony	60	25
House	15	10

In the initial allocation, Mike gets his way on alimony and house, and Carol gets custody of the children. In the initial allocation, Mike has 75 points and Carol has 65 points. To decide what to transfer, we calculate the point ratios.

	Mike	Carol	Point ratio
Custody of children	25	65	65/25 = 2.6
Alimony	60	25	60/25 = 2.4
House	15	10	15/10 = 1.5

Since the house has the smallest point ratio, the house will be the item we work with first. Since transferring the entire house would give Carol too many points, we instead need to transfer some fraction, p, of the house to that Carol and Mike will end up with the same point values. If Carol receives a fraction p of the house, then Mike will give up $(1-p)$ of the house. The value Carol will receive is $10p$: the fraction p of the 10 points Carol values the house at. The value Mike will get is $15(1-p)$. We set their point totals equal to solve for p:

$65 + 10p = 60 + 15(1-p)$
$65 + 10p = 60 + 15 - 15p$
$25p = 10$
$p = 10/25 = 0.4 = 40\%$. So Carol should receive 40% of the house.

---

[1] From *Negotiating to Settlement in Divorce*, 1987

26. Apply the Adjusted Winner method to settle a divorce where the couples have assigned the point values below

	Sandra	Kenny
Home	20	30
Summer home	15	10
Retirement account	50	40
Investments	10	10
Other	5	10

27. In 1974, the United States and Panama negotiated over US involvement and interests in the Panama Canal. Suppose that these were the issues and point values assigned by each side[2]. Apply the Adjusted Winner method.

	United States	Panama
US defense rights	22	9
Use rights	22	15
Land and water	15	15
Expansion rights	14	3
Duration	11	15
Expansion routes	6	5
Jurisdiction	2	7
US military rights	2	7
Defense role of Panama	2	13

[2] Taken from *The Art and Science of Negotiation*, 1982

Graph Theory and Network Flows

In the modern world, planning efficient routes is essential for business and industry, with applications as varied as product distribution, laying new fiber optic lines for broadband internet, and suggesting new friends within social network websites like Facebook.

This field of mathematics started nearly 300 years ago as a look into a mathematical puzzle (we'll look at it in a bit). The field has exploded in importance in the last century, both because of the growing complexity of business in a global economy and because of the computational power that computers have provided us.

Graphs

Drawing Graphs

Example 1

Here is a portion of a housing development from Missoula, Montana[1]. As part of her job, the development's lawn inspector has to walk down every street in the development making sure homeowners' landscaping conforms to the community requirements.

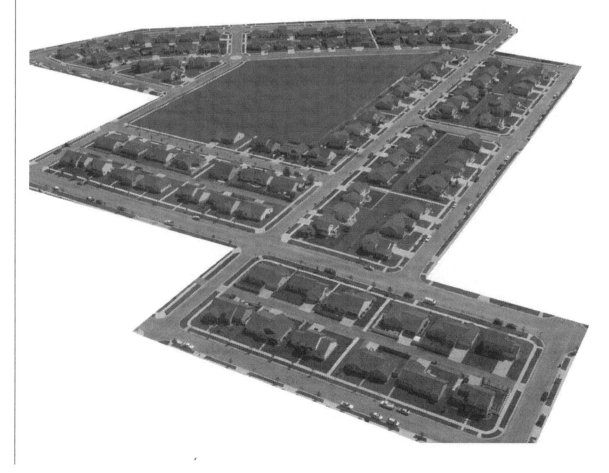

---

[1] Sam Beebe. http://www.flickr.com/photos/sbeebe/2850476641/

118

Naturally, she wants to minimize the amount of walking she has to do. Is it possible for her to walk down every street in this development without having to do any backtracking? While you might be able to answer that question just by looking at the picture for a while, it would be ideal to be able to answer the question for any picture regardless of its complexity.

To do that, we first need to simplify the picture into a form that is easier to work with. We can do that by drawing a simple line for each street. Where streets intersect, we will place a dot.

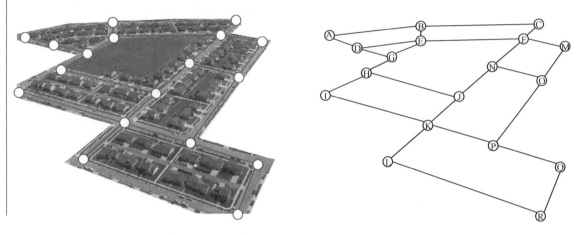

This type of simplified picture is called a **graph**.

Graphs, Vertices, and Edges
A **graph** consists of a set of dots, called **vertices**, and a set of **edges** connecting pairs of vertices.

While we drew our original graph to correspond with the picture we had, there is nothing particularly important about the layout when we analyze a graph. Both of the graphs below are equivalent to the one drawn above since they show the same edge connections between the same vertices as the original graph.

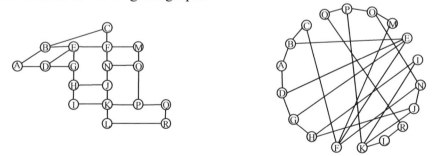

You probably already noticed that we are using the term *graph* differently than you may have used the term in the past to describe the graph of a mathematical function.

Example 2

Back in the 18th century in the Prussian city of Königsberg, a river ran through the city and seven bridges crossed the forks of the river. The river and the bridges are highlighted in the picture to the right[2].

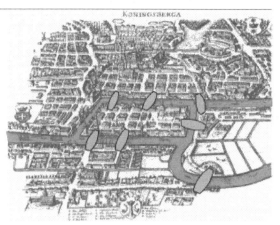

As a weekend amusement, townsfolk would see if they could find a route that would take them across every bridge once and return them to where they started.

Leonard Euler (pronounced OY-lur), one of the most prolific mathematicians ever, looked at this problem in 1735, laying the foundation for graph theory as a field in mathematics. To analyze this problem, Euler introduced edges representing the bridges:

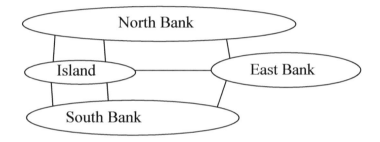

Since the size of each land mass it is not relevant to the question of bridge crossings, each can be shrunk down to a vertex representing the location:

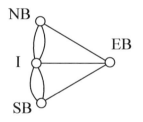

Notice that in this graph there are *two* edges connecting the north bank and island, corresponding to the two bridges in the original drawing. Depending upon the interpretation of edges and vertices appropriate to a scenario, it is entirely possible and reasonable to have more than one edge connecting two vertices.

While we haven't answered the actual question yet of whether or not there is a route which crosses every bridge once and returns to the starting location, the graph provides the foundation for exploring this question.

---

[2] Bogdan Giuşcă. http://en.wikipedia.org/wiki/File:Konigsberg_bridges.png

120

Definitions

While we loosely defined some terminology earlier, we now will try to be more specific.

Vertex

A vertex is a dot in the graph that could represent an intersection of streets, a land mass, or a general location, like "work" or "school". Vertices are often connected by edges. Note that vertices only occur when a dot is explicitly placed, not whenever two edges cross. Imagine a freeway overpass – the freeway and side street cross, but it is not possible to change from the side street to the freeway at that point, so there is no intersection and no vertex would be placed.

Edges

Edges connect pairs of vertices. An edge can represent a physical connection between locations, like a street, or simply that a route connecting the two locations exists, like an airline flight.

Loop

A loop is a special type of edge that connects a vertex to itself. Loops are not used much in street network graphs.

Degree of a vertex

The degree of a vertex is the number of edges meeting at that vertex. It is possible for a vertex to have a degree of zero or larger.

Degree 0	Degree 1	Degree2	Degree 3	Degree 4
○	⌒	⌄	Y	X

Path

A path is a sequence of vertices using the edges. Usually we are interested in a path between two vertices. For example, a path from vertex A to vertex M is shown below. It is one of many possible paths in this graph.

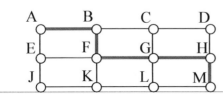

Circuit

A circuit is a path that begins and ends at the same vertex. A circuit starting and ending at vertex A is shown below.

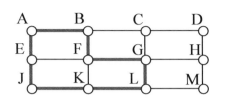

Connected

A graph is connected if there is a path from any vertex to any other vertex. Every graph drawn so far has been connected. The graph below is **disconnected**; there is no way to get from the vertices on the left to the vertices on the right.

Weights

Depending upon the problem being solved, sometimes weights are assigned to the edges. The weights could represent the distance between two locations, the travel time, or the travel cost. It is important to note that the distance between vertices in a graph does not necessarily correspond to the weight of an edge.

Try it Now 1

1. The graph below shows 5 cities. The weights on the edges represent the airfare for a one-way flight between the cities.
   a. How many vertices and edges does the graph have?
   b. Is the graph connected?
   c. What is the degree of the vertex representing LA?
   d. If you fly from Seattle to Dallas to Atlanta, is that a path or a circuit?
   e. If you fly from LA to Chicago to Dallas to LA, is that a path or a circuit?

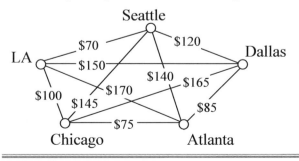

122

Shortest Path

When you visit a website like Google Maps or use your Smartphone to ask for directions from home to your Aunt's house in Pasadena, you are usually looking for a shortest path between the two locations. These computer applications use representations of the street maps as graphs, with estimated driving times as edge weights.

While often it is possible to find a shortest path on a small graph by guess-and-check, our goal in this chapter is to develop methods to solve complex problems in a systematic way by following **algorithms**. An algorithm is a step-by-step procedure for solving a problem. Dijkstra's (pronounced dike-stra) algorithm will find the shortest path between two vertices.

Dijkstra's Algorithm
1. Mark the ending vertex with a distance of zero. Designate this vertex as current.
2. Find all vertices leading to the current vertex. Calculate their distances to the end. Since we already know the distance the current vertex is from the end, this will just require adding the most recent edge. Don't record this distance if it is longer than a previously recorded distance.
3. Mark the current vertex as visited. We will never look at this vertex again.
4. Mark the vertex with the smallest distance as current, and repeat from step 2.

Example 3

Suppose you need to travel from Tacoma, WA (vertex T) to Yakima, WA (vertex Y). Looking at a map, it looks like driving through Auburn (A) then Mount Rainier (MR) might be shortest, but it's not totally clear since that road is probably slower than taking the major highway through North Bend (NB). A graph with travel times in minutes is shown below. An alternate route through Eatonville (E) and Packwood (P) is also shown.

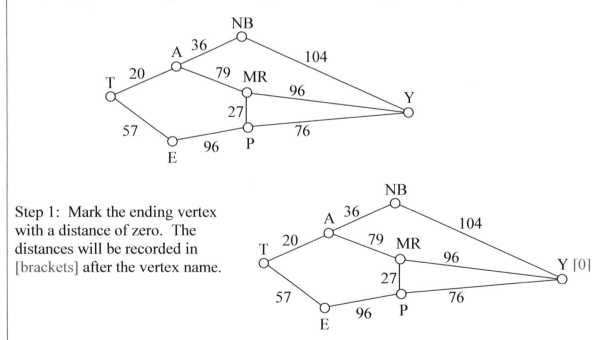

Step 1: Mark the ending vertex with a distance of zero. The distances will be recorded in [brackets] after the vertex name.

Step 2: For each vertex leading to Y, we calculate the distance to the end. For example, NB is a distance of 104 from the end, and MR is 96 from the end. Remember that distances in this case refer to the travel time in minutes.

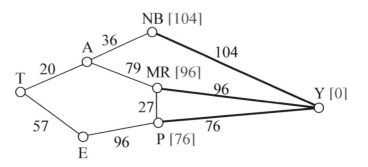

Step 3 & 4: We mark Y as visited, and mark the vertex with the smallest recorded distance as current. At this point, P will be designated current. Back to step 2.

Step 2 (#2): For each vertex leading to P (and not leading to a visited vertex) we find the distance from the end. Since E is 96 minutes from P, and we've already calculated P is 76 minutes from Y, we can compute that E is 96+76 = 172 minutes from Y.

If we make the same computation for MR, we'd calculate 76+27 = 103. Since this is larger than the previously recorded distance from Y to MR, we will *not* replace it.

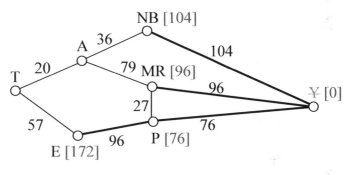

Step 3 & 4 (#2): We mark P as visited, and designate the vertex with the smallest recorded distance as current: MR. Back to step 2.

Step 2 (#3): For each vertex leading to MR (and not leading to a visited vertex) we find the distance to the end. The only vertex to be considered is A, since we've already visited Y and P. Adding MR's distance 96 to the length from A to MR gives the distance 96+79 = 175 minutes from A to Y.

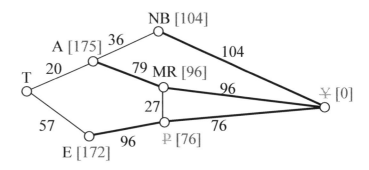

Step 3 & 4 (#3): We mark MR as visited, and designate the vertex with smallest recorded distance as current: NB. Back to step 2.

Step 2 (#4): For each vertex leading to NB, we find the distance to the end. We know the shortest distance from NB to Y is 104 and the distance from A to NB is 36, so the distance from A to Y through NB is 104+36 = 140. Since this distance *is* shorter than the previously calculated distance from Y to A through MR, we replace it.

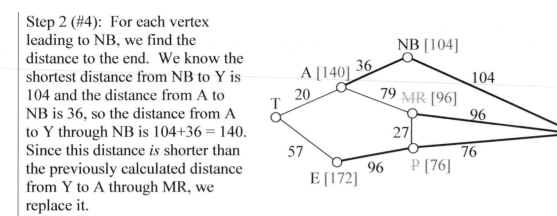

Step 3 & 4 (#4): We mark NB as visited, and designate A as current, since it now has the shortest distance.

Step 2 (#5): T is the only non-visited vertex leading to A, so we calculate the distance from T to Y through A: 20+140 = 160 minutes.

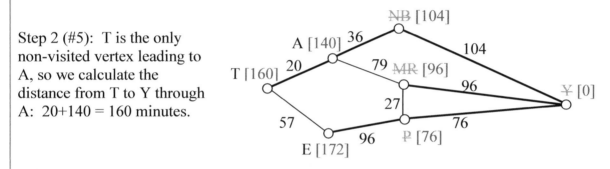

Step 3 & 4 (#5): We mark A as visited, and designate E as current.

Step 2 (#6): The only non-visited vertex leading to E is T. Calculating the distance from T to Y through E, we compute 172+57 = 229 minutes. Since this is longer than the existing marked time, we do not replace it.

Step 3 (#6): We mark E as visited. Since all vertices have been visited, we are done.

From this, we know that the shortest path from Tacoma to Yakima will take 160 minutes. Tracking which sequence of edges yielded 160 minutes, we see the shortest path is T-A-NB-Y.

Dijkstra's algorithm is an **optimal algorithm**, meaning that it always produces the actual shortest path, not just a path that is pretty short, provided one exists. This algorithm is also **efficient**, meaning that it can be implemented in a reasonable amount of time. Dijkstra's algorithm takes around V^2 calculations, where V is the number of vertices in a graph[3]. A graph with 100 vertices would take around 10,000 calculations. While that would be a lot to do by hand, it is not a lot for computer to handle. It is because of this efficiency that your car's GPS unit can compute driving directions in only a few seconds.

---

[3] It can be made to run faster through various optimizations to the implementation.

In contrast, an **inefficient** algorithm might try to list all possible paths then compute the length of each path. Trying to list all possible paths could easily take 10^{25} calculations to compute the shortest path with only 25 vertices; that's a 1 with 25 zeros after it! To put that in perspective, the fastest computer in the world would still spend over 1000 years analyzing all those paths.

Example 4

A shipping company needs to route a package from Washington, D.C. to San Diego, CA. To minimize costs, the package will first be sent to their processing center in Baltimore, MD then sent as part of mass shipments between their various processing centers, ending up in their processing center in Bakersfield, CA. From there it will be delivered in a small truck to San Diego.

The travel times, in hours, between their processing centers are shown in the table below. Three hours has been added to each travel time for processing. Find the shortest path from Baltimore to Bakersfield.

	Baltimore	Denver	Dallas	Chicago	Atlanta	Bakersfield
Baltimore	*			15	14	
Denver		*		18	24	19
Dallas			*	18	15	25
Chicago	15	18	18	*	14	
Atlanta	14	24	15	14	*	
Bakersfield		19	25			*

While we could draw a graph, we can also work directly from the table.

Step 1: The ending vertex, Bakersfield, is marked as current.

Step 2: All cities connected to Bakersfield, in this case Denver and Dallas, have their distances calculated; we'll mark those distances in the column headers.

Step 3 & 4: Mark Bakersfield as visited. Here, we are doing it by shading the corresponding row and column of the table. We mark Denver as current, shown in bold, since it is the vertex with the shortest distance.

	Baltimore	**Denver** [19]	Dallas [25]	Chicago	Atlanta	Bakersfield [0]
Baltimore	*			15	14	
Denver		*		18	24	19
Dallas			*	18	15	25
Chicago	15	18	18	*	14	
Atlanta	14	24	15	14	*	
Bakersfield		19	25			*

Step 2 (#2): For cities connected to Denver, calculate distance to the end. For example, Chicago is 18 hours from Denver, and Denver is 19 hours from the end, the distance for Chicago to the end is 18+19 = 37 (Chicago to Denver to Bakersfield). Atlanta is 24 hours from Denver, so the distance to the end is 24+19 = 43 (Atlanta to Denver to Bakersfield).

Step 3 & 4 (#2): We mark Denver as visited and mark Dallas as current.

	Baltimore	Denver [19]	**Dallas** [25]	Chicago [37]	Atlanta [43]	Bakersfield [0]
Baltimore	*			15	14	
Denver		*		18	24	19
Dallas			*	18	15	25
Chicago	15	18	18	*	14	
Atlanta	14	24	15	14	*	
Bakersfield		19	25			*

Step 2 (#3): For cities connected to Dallas, calculate the distance to the end. For Chicago, the distance from Chicago to Dallas is 18 and from Dallas to the end is 25, so the distance from Chicago to the end through Dallas would be 18+25 = 43. Since this is longer than the currently marked distance for Chicago, we do not replace it. For Atlanta, we calculate 15+25 = 40. Since this is shorter than the currently marked distance for Atlanta, we replace the existing distance.

Step 3 & 4 (#3): We mark Dallas as visited, and mark Chicago as current.

	Baltimore	Denver [19]	Dallas [25]	**Chicago** [37]	Atlanta [40]	Bakersfield [0]
Baltimore	*			15	14	
Denver		*		18	24	19
Dallas			*	18	15	25
Chicago	15	18	18	*	14	
Atlanta	14	24	15	14	*	
Bakersfield		19	25			*

Step 2 (#4): Baltimore and Atlanta are the only non-visited cities connected to Chicago. For Baltimore, we calculate 15+37 = 52 and mark that distance. For Atlanta, we calculate 14+37 = 51. Since this is longer than the existing distance of 40 for Atlanta, we do not replace that distance.

Step 3 & 4 (#4): Mark Chicago as visited and Atlanta as current.

	Baltimore [52]	Denver [19]	Dallas [25]	Chicago [37]	**Atlanta** [40]	Bakersfield [0]
Baltimore	*			15	14	
Denver		*		18	24	19
Dallas			*	18	15	25
Chicago	15	18	18	*	14	
Atlanta	14	24	15	14	*	
Bakersfield		19	25			*

Step 2 (#5): The distance from Atlanta to Baltimore is 14. Adding that to the distance already calculated for Atlanta gives a total distance of 14+40 = 54 hours from Baltimore to Bakersfield through Atlanta. Since this is larger than the currently calculated distance, we do not replace the distance for Baltimore.

Step 3 & 4 (#5): We mark Atlanta as visited. All cities have been visited and we are done.

The shortest route from Baltimore to Bakersfield will take 52 hours, and will route through Chicago and Denver.

Try it Now 2
Find the shortest path between vertices A and G in the graph below.

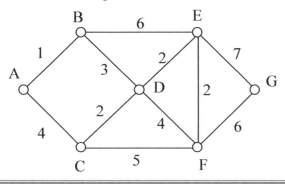

Euler Circuits and the Chinese Postman Problem

In the first section, we created a graph of the Königsberg bridges and asked whether it was possible to walk across every bridge once. Because Euler first studied this question, these types of paths are named after him.

Euler Path
An **Euler path** is a path that uses every edge in a graph with no repeats. Being a path, it does not have to return to the starting vertex.

Example 5

In the graph shown below, there are several Euler paths. One such path is CABDCB. The path is shown in arrows to the right, with the order of edges numbered.

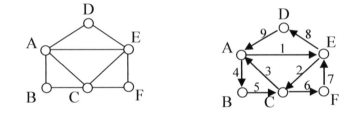

Euler Circuit
An **Euler circuit** is a circuit that uses every edge in a graph with no repeats. Being a circuit, it must start and end at the same vertex.

Example 6

The graph below has several possible Euler circuits. Here's a couple, starting and ending at vertex A: ADEACEFCBA and AECABCFEDA. The second is shown in arrows.

Look back at the example used for Euler paths – does that graph have an Euler circuit? A few tries will tell you no; that graph does not have an Euler circuit. When we were working with shortest paths, we were interested in the optimal path. With Euler paths and circuits, we're primarily interested in whether an Euler path or circuit *exists*.

Why do we care if an Euler circuit exists? Think back to our housing development lawn inspector from the beginning of the chapter. The lawn inspector is interested in walking as little as possible. The ideal situation would be a circuit that covers every street with no repeats. That's an Euler circuit! Luckily, Euler solved the question of whether or not an Euler path or circuit will exist.

Euler's Path and Circuit Theorems
A graph will contain an Euler path if it contains at most two vertices of odd degree.

A graph will contain an Euler circuit if all vertices have even degree

Example 7

In the graph below, vertices A and C have degree 4, since there are 4 edges leading into each vertex. B is degree 2, D is degree 3, and E is degree 1. This graph contains two vertices with odd degree (D and E) and three vertices with even degree (A, B, and C), so Euler's theorems tell us this graph has an Euler path, but not an Euler circuit.

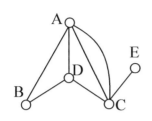

Example 8

Is there an Euler circuit on the housing development lawn inspector graph we created earlier in the chapter? All the highlighted vertices have odd degree. Since there are more than two vertices with odd degree, there are no Euler paths or Euler circuits on this graph. Unfortunately our lawn inspector will need to do some backtracking.

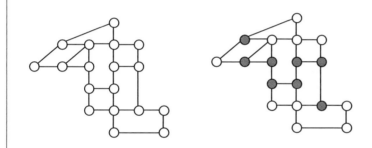

Example 9

When it snows in the same housing development, the snowplow has to plow both sides of every street. For simplicity, we'll assume the plow is out early enough that it can ignore traffic laws and drive down either side of the street in either direction. This can be visualized in the graph by drawing two edges for each street, representing the two sides of the street.

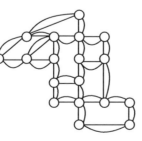

Notice that every vertex in this graph has even degree, so this graph does have an Euler circuit.

Now we know how to determine if a graph has an Euler circuit, but if it does, how do we find one? While it usually is possible to find an Euler circuit just by pulling out your pencil and trying to find one, the more formal method is **Fleury's algorithm.**

Fleury's Algorithm
1. Start at any vertex if finding an Euler circuit. If finding an Euler path, start at one of the two vertices with odd degree.
2. Choose any edge leaving your current vertex, provided deleting that edge will not separate the graph into two disconnected sets of edges.
3. Add that edge to your circuit, and delete it from the graph.
4. Continue until you're done.

Example 10

Let's find an Euler Circuit on this graph using Fleury's algorithm, starting at vertex A.

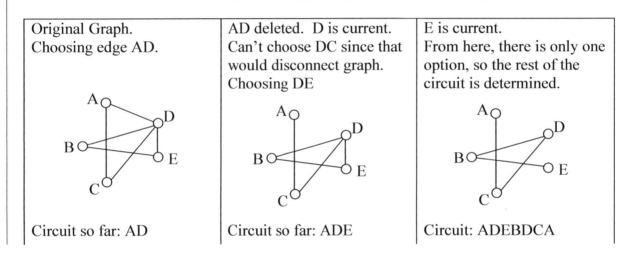

Original Graph. Choosing edge AD.	AD deleted. D is current. Can't choose DC since that would disconnect graph. Choosing DE	E is current. From here, there is only one option, so the rest of the circuit is determined.
Circuit so far: AD	Circuit so far: ADE	Circuit: ADEBDCA

Try it Now 3

Does the graph below have an Euler Circuit? If so, find one.

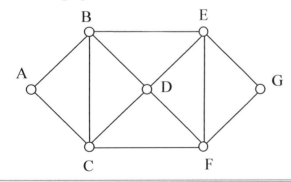

Eulerization and the Chinese Postman Problem

Not every graph has an Euler path or circuit, yet our lawn inspector still needs to do her inspections. Her goal is to minimize the amount of walking she has to do. In order to do that, she will have to duplicate some edges in the graph until an Euler circuit exists.

Eulerization
Eulerization is the process of adding edges to a graph to create an Euler circuit on a graph. To eulerize a graph, edges are duplicated to connect pairs of vertices with odd degree. Connecting two odd degree vertices increases the degree of each, giving them both even degree. When two odd degree vertices are not directly connected, we can duplicate all edges in a path connecting the two.

Note that we can only duplicate edges, not create edges where there wasn't one before. Duplicating edges would mean walking or driving down a road twice, while creating an edge where there wasn't one before is akin to installing a new road!

Example 11

For the rectangular graph shown, three possible eulerizations are shown. Notice in each of these cases the vertices that started with odd degrees have even degrees after eulerization, allowing for an Euler circuit.

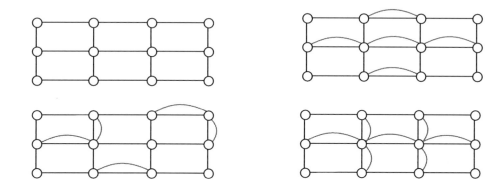

In the example above, you'll notice that the last eulerization required duplicating seven edges, while the first two only required duplicating five edges. If we were eulerizing the graph to find a walking path, we would want the eulerization with minimal duplications. If the edges had weights representing distances or costs, then we would want to select the eulerization with the minimal total added weight.

Try it Now 4
Eulerize the graph shown, then find an Euler circuit on the eulerized graph.

132

Example 12

Looking again at the graph for our lawn inspector from Examples 1 and 8, the vertices with odd degree are shown highlighted. With eight vertices, we will always have to duplicate at least four edges. In this case, we need to duplicate five edges since two odd degree vertices are not directly connected. Without weights we can't be certain this is the eulerization that minimizes walking distance, but it looks pretty good.

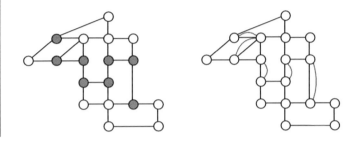

The problem of finding the optimal eulerization is called the Chinese Postman Problem, a name given by an American in honor of the Chinese mathematician Mei-Ko Kwan who first studied the problem in 1962 while trying to find optimal delivery routes for postal carriers. This problem is important in determining efficient routes for garbage trucks, school buses, parking meter checkers, street sweepers, and more.

Unfortunately, algorithms to solve this problem are fairly complex. Some simpler cases are considered in the exercises.

Hamiltonian Circuits and the Traveling Salesman Problem

In the last section, we considered optimizing a walking route for a postal carrier. How is this different than the requirements of a package delivery driver? While the postal carrier needed to walk down every street (edge) to deliver the mail, the package delivery driver instead needs to visit every one of a set of delivery locations. Instead of looking for a circuit that covers every edge once, the package deliverer is interested in a circuit that visits every vertex once.

> **Hamiltonian Circuits and Paths**
> A **Hamiltonian circuit** is a circuit that visits every vertex once with no repeats. Being a circuit, it must start and end at the same vertex. A **Hamiltonian path** also visits every vertex once with no repeats, but does not have to start and end at the same vertex.

Hamiltonian circuits are named for William Rowan Hamilton who studied them in the 1800's.

Example 13

One Hamiltonian circuit is shown on the graph below. There are several other Hamiltonian circuits possible on this graph. Notice that the circuit only has to visit every vertex once; it does not need to use every edge.

This circuit could be notated by the sequence of vertices visited, starting and ending at the same vertex: ABFGCDHMLKJEA. Notice that the same circuit could be written in reverse order, or starting and ending at a different vertex.

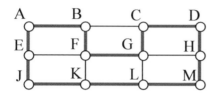

Unlike with Euler circuits, there is no nice theorem that allows us to instantly determine whether or not a Hamiltonian circuit exists for all graphs.[4]

Example 14

Does a Hamiltonian path or circuit exist on the graph below?

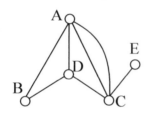

We can see that once we travel to vertex E there is no way to leave without returning to C, so there is no possibility of a Hamiltonian circuit. If we start at vertex E we can find several Hamiltonian paths, such as ECDAB and ECABD.

With Hamiltonian circuits, our focus will not be on existence, but on the question of optimization; given a graph where the edges have weights, can we find the optimal Hamiltonian circuit; the one with lowest total weight.

This problem is called the **Traveling salesman problem** (TSP) because the question can be framed like this: Suppose a salesman needs to give sales pitches in four cities. He looks up the airfares between each city, and puts the costs in a graph. In what order should he travel to visit each city once then return home with the lowest cost?

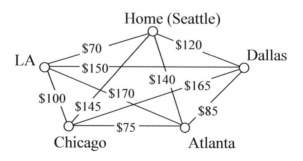

---

[4] There are some theorems that can be used in specific circumstances, such as Dirac's theorem, which says that a Hamiltonian circuit must exist on a graph with n vertices if each vertex has degree $n/2$ or greater.

To answer this question of how to find the lowest cost Hamiltonian circuit, we will consider some possible approaches. The first option that might come to mind is to just try all different possible circuits.

Brute Force Algorithm (a.k.a. exhaustive search)
1. List all possible Hamiltonian circuits
2. Find the length of each circuit by adding the edge weights
3. Select the circuit with minimal total weight.

Example 15

Apply the Brute force algorithm to find the minimum cost Hamiltonian circuit on the graph below.

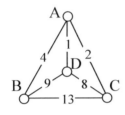

To apply the Brute force algorithm, we list all possible Hamiltonian circuits and calculate their weight:

Circuit	Weight
ABCDA	4+13+8+1 = 26
ABDCA	4+9+8+2 = 23
ACBDA	2+13+9+1 = 25

Note: These are the unique circuits on this graph. All other possible circuits are the reverse of the listed ones or start at a different vertex, but result in the same weights.

From this we can see that the second circuit, ABDCA, is the optimal circuit.

The Brute force algorithm is optimal; it will always produce the Hamiltonian circuit with minimum weight. Is it efficient? To answer that question, we need to consider how many Hamiltonian circuits a graph could have. For simplicity, let's look at the worst-case possibility, where every vertex is connected to every other vertex. This is called a **complete graph.**

Suppose we had a complete graph with five vertices like the air travel graph above. From Seattle there are four cities we can visit first. From each of those, there are three choices. From each of those cities, there are two possible cities to visit next. There is then only one choice for the last city before returning home.

This can be shown visually:

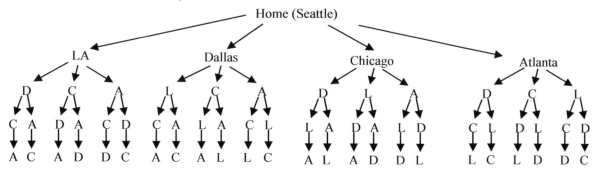

Counting the number of routes, we can see there are $4 \cdot 3 \cdot 2 \cdot 1 = 24$ routes. For six cities there would be $5 \cdot 4 \cdot 3 \cdot 2 \cdot 1 = 120$ routes.

> **Number of Possible Circuits**
> For N vertices in a complete graph, there will be $(n-1)! = (n-1)(n-2)(n-3) \cdots 3 \cdot 2 \cdot 1$
>
> routes. Half of these are duplicates in reverse order, so there are $\dfrac{(n-1)!}{2}$ unique
>
> circuits.
>
> The exclamation symbol, !, is read "factorial" and is shorthand for the product shown.

Example 16

How many circuits would a complete graph with 8 vertices have?

A complete graph with 8 vertices would have $(8-1)! = 7! = 7 \cdot 6 \cdot 5 \cdot 4 \cdot 3 \cdot 2 \cdot 1 = 5040$ possible Hamiltonian circuits. Half of the circuits are duplicates of other circuits but in reverse order, leaving 2520 unique routes.

While this is a lot, it doesn't seem unreasonably huge. But consider what happens as the number of cities increase:

Cities	Unique Hamiltonian Circuits
9	$8!/2 = 20,160$
10	$9!/2 = 181,440$
11	$10!/2 = 1,814,400$
15	$14!/2 = 43,589,145,600$
20	$19!/2 = 60,822,550,204,416,000$

As you can see the number of circuits is growing extremely quickly. If a computer looked at one billion circuits a second, it would still take almost two years to examine all the possible circuits with only 20 cities! Certainly Brute Force is **not** an efficient algorithm.

Unfortunately, no one has yet found an efficient *and* optimal algorithm to solve the TSP, and it is very unlikely anyone ever will. Since it is not practical to use brute force to solve the problem, we turn instead to **heuristic algorithms**; efficient algorithms that give approximate solutions. In other words, heuristic algorithms are fast, but may or may not produce the optimal circuit.

Nearest Neighbor Algorithm (NNA)
1. Select a starting point.
2. Move to the nearest unvisited vertex (the edge with smallest weight).
3. Repeat until the circuit is complete.

Example 17

Consider our earlier graph, shown to the right.
Starting at vertex A, the nearest neighbor is vertex D with a weight of 1.
From D, the nearest neighbor is C, with a weight of 8.
From C, our only option is to move to vertex B, the only unvisited vertex, with a cost of 13.
From B we return to A with a weight of 4.

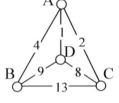

The resulting circuit is ADCBA with a total weight of 1+8+13+4 = 26.

We ended up finding the worst circuit in the graph! What happened? Unfortunately, while it is very easy to implement, the NNA is a **greedy** algorithm, meaning it only looks at the immediate decision without considering the consequences in the future. In this case, following the edge AD forced us to use the very expensive edge BC later.

Example 18

Consider again our salesman. Starting in Seattle, the nearest neighbor (cheapest flight) is to LA, at a cost of $70. From there:

LA to Chicago: $100
Chicago to Atlanta: $75
Atlanta to Dallas: $85
Dallas to Seattle: $120
Total cost: $450

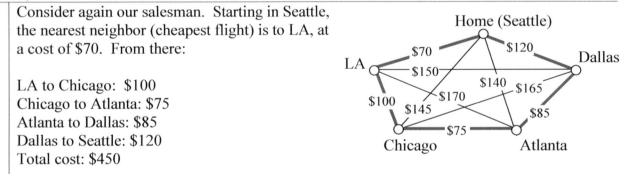

In this case, nearest neighbor did find the optimal circuit.

Going back to our first example, how could we improve the outcome? One option would be to redo the nearest neighbor algorithm with a different starting point to see if the result changed. Since nearest neighbor is so fast, doing it several times isn't a big deal.

> **Repeated Nearest Neighbor Algorithm (RNNA)**
> 1. Do the Nearest Neighbor Algorithm starting at each vertex
> 2. Choose the circuit produced with minimal total weight

Example 19

We will revisit the graph from Example 17.

Starting at vertex A resulted in a circuit with weight 26.

Starting at vertex B, the nearest neighbor circuit is BADCB with a weight of $4+1+8+13 = 26$. This is the same circuit we found starting at vertex A. No better.

Starting at vertex C, the nearest neighbor circuit is CADBC with a weight of $2+1+9+13 = 25$. Better!

Starting at vertex D, the nearest neighbor circuit is DACBA. Notice that this is actually the same circuit we found starting at C, just written with a different starting vertex.

The RNNA was able to produce a slightly better circuit with a weight of 25, but still not the optimal circuit in this case. Notice that even though we found the circuit by starting at vertex C, we could still write the circuit starting at A: ADBCA or ACBDA.

Try it Now 5

The table below shows the time, in milliseconds, it takes to send a packet of data between computers on a network. If data needed to be sent in sequence to each computer, then notification needed to come back to the original computer, we would be solving the TSP. The computers are labeled A-F for convenience.

	A	B	C	D	E	F
A	--	44	34	12	40	41
B	44	--	31	43	24	50
C	34	31	--	20	39	27
D	12	43	20	--	11	17
E	40	24	39	11	--	42
F	41	50	27	17	42	--

a. Find the circuit generated by the NNA starting at vertex B.
b. Find the circuit generated by the RNNA.

138

While certainly better than the basic NNA, unfortunately, the RNNA is still greedy and will produce very bad results for some graphs. As an alternative, our next approach will step back and look at the "big picture" – it will select first the edges that are shortest, and then fill in the gaps.

Sorted Edges Algorithm (a.k.a. Cheapest Link Algorithm)
1. Select the cheapest unused edge in the graph.
2. Repeat step 1, adding the cheapest unused edge to the circuit, unless:
   a. adding the edge would create a circuit that doesn't contain all vertices, or
   b. adding the edge would give a vertex degree 3.
3. Repeat until a circuit containing all vertices is formed.

Example 20

Using the four vertex graph from earlier, we can use the Sorted Edges algorithm.

The cheapest edge is AD, with a cost of 1. We highlight that edge to mark it selected. The next shortest edge is AC, with a weight of 2, so we highlight that edge.

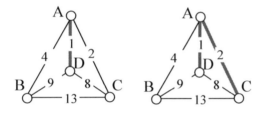

For the third edge, we'd like to add AB, but that would give vertex A degree 3, which is not allowed in a Hamiltonian circuit. The next shortest edge is CD, but that edge would create a circuit ACDA that does not include vertex B, so we reject that edge. The next shortest edge is BD, so we add that edge to the graph.

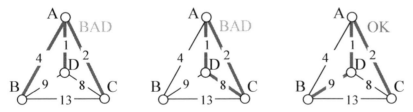

We then add the last edge to complete the circuit: ACBDA with weight 25.

Notice that the algorithm did not produce the optimal circuit in this case; the optimal circuit is ACDBA with weight 23.

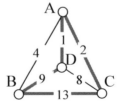

While the Sorted Edge algorithm overcomes some of the shortcomings of NNA, it is still only a heuristic algorithm, and does not guarantee the optimal circuit.

Example 21

Your teacher's band, *Derivative Work*, is doing a bar tour in Oregon. The driving distances are shown below. Plan an efficient route for your teacher to visit all the cities and return to the starting location. Use NNA starting at Portland, and then use Sorted Edges.

	Ashland	Astoria	Bend	Corvallis	Crater Lake	Eugene	Newport	Portland	Salem	Seaside
Ashland	-	374	200	223	108	178	252	285	240	356
Astoria	374	-	255	166	433	199	135	95	136	17
Bend	200	255	-	128	277	128	180	160	131	247
Corvallis	223	166	128	-	430	47	52	84	40	155
Crater Lake	108	433	277	430	-	453	478	344	389	423
Eugene	178	199	128	47	453	-	91	110	64	181
Newport	252	135	180	52	478	91	-	114	83	117
Portland	285	95	160	84	344	110	114	-	47	78
Salem	240	136	131	40	389	64	83	47	-	118
Seaside	356	17	247	155	423	181	117	78	118	-

Using NNA with a large number of cities, you might find it helpful to mark off the cities as they're visited to keep from accidently visiting them again. Looking in the row for Portland, the smallest distance is 47, to Salem. Following that idea, our circuit will be:

Portland to Salem	47
Salem to Corvallis	40
Corvallis to Eugene	47
Eugene to Newport	91
Newport to Seaside	117
Seaside to Astoria	17
Astoria to Bend	255
Bend to Ashland	200
Ashland to Crater Lake	108
Crater Lake to Portland	344
Total trip length:	1266 miles

Using Sorted Edges, you might find it helpful to draw an empty graph, perhaps by drawing vertices in a circular pattern. Adding edges to the graph as you select them will help you visualize any circuits or vertices with degree 3.

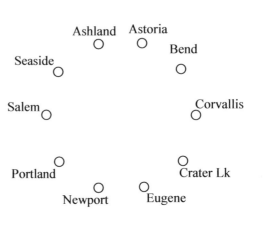

We start adding the shortest edges:

Seaside to Astoria   17 miles
Corvallis to Salem   40 miles
Portland to Salem   47 miles
Corvallis to Eugene   47 miles

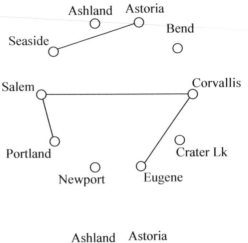

The graph after adding these edges is shown to the right.   The next shortest edge is from Corvallis to Newport at 52 miles, but adding that edge would give Corvallis degree 3.

Continuing on, we can skip over any edge pair that contains Salem or Corvallis, since they both already have degree 2.

Portland to Seaside   78 miles
Eugene to Newport   91 miles
Portland to Astoria   (reject – closes circuit)
Ashland to Crater Lk   108 miles

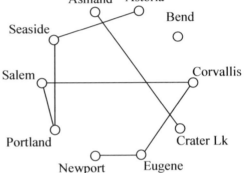

The graph after adding these edges is shown to the right. At this point, we can skip over any edge pair that contains Salem, Seaside, Eugene, Portland, or Corvallis since they already have degree 2.

Newport to Astoria   (reject – closes circuit)
Newport to Bend   180 miles
Bend to Ashland   200 miles

At this point the only way to complete the circuit is to add:
Crater Lk to Astoria   433 miles

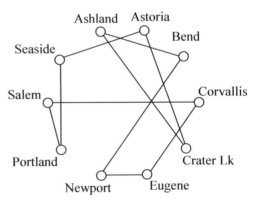

The final circuit, written to start at Portland, is:
Portland, Salem, Corvallis, Eugene, Newport, Bend, Ashland, Crater Lake, Astoria, Seaside, Portland.

Total trip length: 1241 miles.

While better than the NNA route, neither algorithm produced the optimal route. The following route can make the tour in 1069 miles:
Portland, Astoria, Seaside, Newport, Corvallis, Eugene, Ashland, Crater Lake, Bend, Salem, Portland

Try it Now 6

Find the circuit produced by the Sorted Edges algorithm using the graph below.

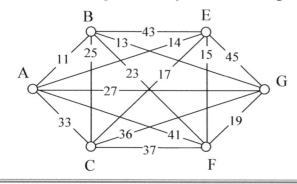

Spanning Trees

A company requires reliable internet and phone connectivity between their five offices (named A, B, C, D, and E for simplicity) in New York, so they decide to lease dedicated lines from the phone company. The phone company will charge for each link made. The costs, in thousands of dollars per year, are shown in the graph.

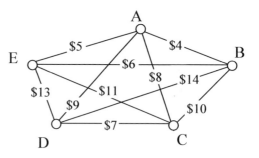

In this case, we don't need to find a circuit, or even a specific path; all we need to do is make sure we can make a call from any office to any other. In other words, we need to be sure there is a path from any vertex to any other vertex.

Spanning Tree
A **spanning tree** is a connected graph using all vertices in which there are no circuits. In other words, there is a path from any vertex to any other vertex, but no circuits.

Some examples of spanning trees are shown below. Notice there are no circuits in the trees, and it is fine to have vertices with degree higher than two.

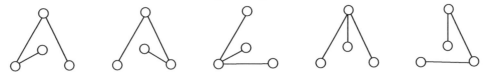

Usually we have a starting graph to work from, like in the phone example above. In this case, we form our spanning tree by finding a **subgraph** – a new graph formed using all the vertices but only some of the edges from the original graph. No edges will be created where they didn't already exist.

Of course, any random spanning tree isn't really what we want. We want the **minimum cost spanning tree (MCST)**.

> **Minimum Cost Spanning Tree (MCST)**
> The minimum cost spanning tree is the spanning tree with the smallest total edge weight.

A nearest neighbor style approach doesn't make as much sense here since we don't need a circuit, so instead we will take an approach similar to sorted edges.

> **Kruskal's Algorithm**
> 1. Select the cheapest unused edge in the graph.
> 2. Repeat step 1, adding the cheapest unused edge, <u>unless</u>:
>   a. adding the edge would create a circuit
> 3. Repeat until a spanning tree is formed

Example 22

Using our phone line graph from above, begin adding edges:

AB	$4	OK
AE	$5	OK
BE	$6	reject – closes circuit ABEA
DC	$7	OK
AC	$8	OK

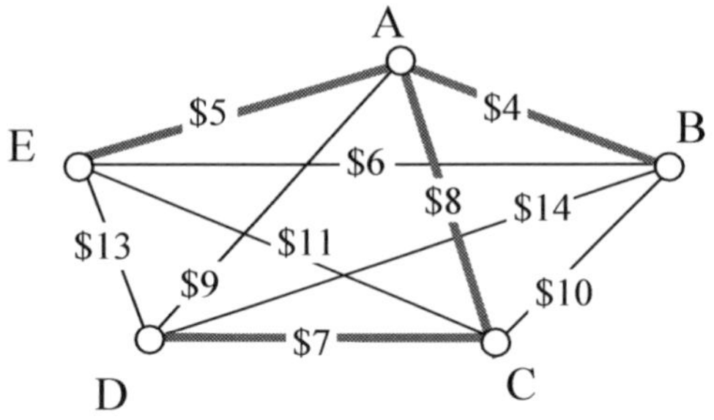

At this point we stop – every vertex is now connected, so we have formed a spanning tree with cost $24 thousand a year.

Remarkably, Kruskal's algorithm is both optimal and efficient; we are guaranteed to always produce the optimal MCST.

Example 23

The power company needs to lay updated distribution lines connecting the ten Oregon cities below to the power grid. How can they minimize the amount of new line to lay?

	Ashland	Astoria	Bend	Corvallis	Crater Lake	Eugene	Newport	Portland	Salem	Seaside
Ashland	-	374	200	223	108	178	252	285	240	356
Astoria	374	-	255	166	433	199	135	95	136	17
Bend	200	255	-	128	277	128	180	160	131	247
Corvallis	223	166	128	-	430	47	52	84	40	155
Crater Lake	108	433	277	430	-	453	478	344	389	423
Eugene	178	199	128	47	453	-	91	110	64	181
Newport	252	135	180	52	478	91	-	114	83	117
Portland	285	95	160	84	344	110	114	-	47	78
Salem	240	136	131	40	389	64	83	47	-	118
Seaside	356	17	247	155	423	181	117	78	118	-

Using Kruskal's algorithm, we add edges from cheapest to most expensive, rejecting any that close a circuit. We stop when the graph is connected.

Seaside to Astoria	17 miles
Corvallis to Salem	40 miles
Portland to Salem	47 miles
Corvallis to Eugene	47 miles
Corvallis to Newport	52 miles
Salem to Eugene	reject – closes circuit
Portland to Seaside	78 miles

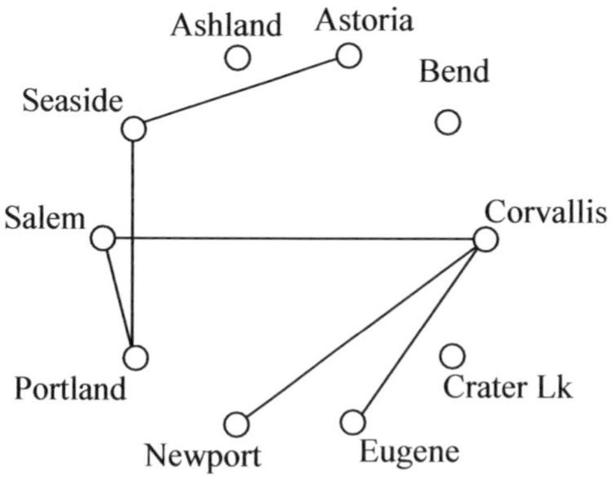

The graph up to this point is shown to the right. Continuing,

Newport to Salem	reject
Corvallis to Portland	reject
Eugene to Newport	reject
Portland to Astoria	reject
Ashland to Crater Lk	108 miles
Eugene to Portland	reject
Newport to Portland	reject
Newport to Seaside	reject
Salem to Seaside	reject
Bend to Eugene	128 miles
Bend to Salem	reject

Astoria to Newport	reject
Salem to Astoria	reject
Corvallis to Seaside	reject
Portland to Bend	reject
Astoria to Corvallis	reject
Eugene to Ashland	178 miles

This connects the graph. The total length of cable to lay would be 695 miles.

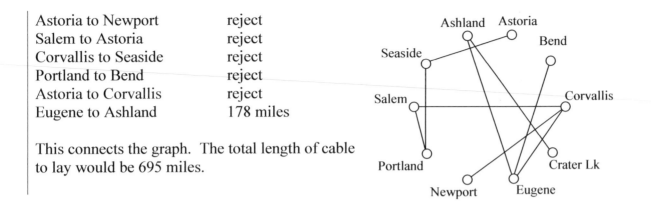

Try it Now 7
Find a minimum cost spanning tree on the graph below using Kruskal's algorithm.

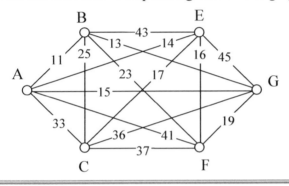

Try it Now Answers
1. a. 5 vertices, 10 edges
b. Yes, it is connected
c. The vertex is degree 4
d. A path
e. A circuit

2. The shortest path is ABDEG, with length 13

3. Yes, all vertices have even degree so this graph has an Euler Circuit. There are several possibilities. One is: ABEGFCDFEDBCA

4. This graph can be eulerized by duplicating the edge BC, as shown. One possible Euler circuit on the eulerized graph is ACDBCBA

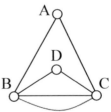

Try it Now Answers Continued

5. At each step, we look for the nearest location we haven't already visited.
From B the nearest computer is E with time 24.
From E, the nearest computer is D with time 11.
From D the nearest is A with time 12.
From A the nearest is C with time 34.
From C, the only computer we haven't visited is F with time 27
From F, we return back to B with time 50.

The NNA circuit from B is BEDACFB with time 158 milliseconds.

Using NNA again from other starting vertices:
Starting at A: ADEBCFA: time 146
Starting at C: CDEBAFC: time 167
Starting at D: DEBCFAD: time 146
Starting at E: EDACFBE: time 158
Starting at F: FDEBCAF: time 158

The RNNA found a circuit with time 146 milliseconds: ADEBCFA. We could also write this same circuit starting at B if we wanted: BCFADEB or BEDAFCB.

6.
AB: Add, cost 11
BG: Add, cost 13
AE: Add, cost 14
EF: Add, cost 15
EC: Skip (degree 3 at E)
FG: Skip (would create a circuit not including C)
BF, BC, AG, AC: Skip (would cause a vertex to have degree 3)
GC: Add, cost 36
CF: Add, cost 37, completes the circuit

Final circuit: ABGCFEA

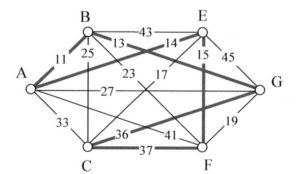

7.
AB: Add, cost 11
BG: Add, cost 13
AE: Add, cost 14
AG: Skip, would create circuit ABGA
EF: Add, cost 16
EC: Add, cost 17

This completes the spanning tree

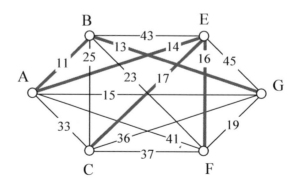

146

Exercises

Skills

1. To deliver mail in a particular neighborhood, the postal carrier needs to walk along each of the streets with houses (the dots). Create a graph with edges showing where the carrier must walk to deliver the mail.

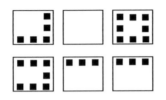

2. Suppose that a town has 7 bridges as pictured below. Create a graph that could be used to determine if there is a path that crosses all bridges once.

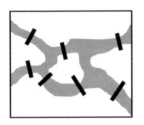

3. The table below shows approximate driving times (in minutes, without traffic) between five cities in the Dallas area. Create a weighted graph representing this data.

	Plano	Mesquite	Arlington	Denton
Fort Worth	54	52	19	42
Plano		38	53	41
Mesquite			43	56
Arlington				50

4. Shown in the table below are the one-way airfares between 5 cities[5]. Create a graph showing this data.

	Honolulu	London	Moscow	Cairo
Seattle	$159	$370	$654	$684
Honolulu		$830	$854	$801
London			$245	$323
Moscow				$329

5. Find the degree of each vertex in the graph below.

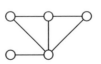

---

[5] Cheapest fares found when retrieved Sept 1, 2009 for travel Sept 22, 2009

6. Find the degree of each vertex in the graph below.

7. Which of these graphs are connected?

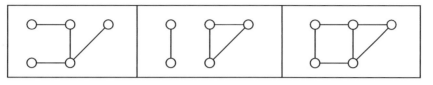

8. Which of these graphs are connected?

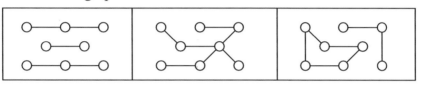

9. Travel times by rail for a segment of the Eurail system is shown below with travel times in hours and minutes[6]. Find path with shortest travel time from Bern to Berlin by applying Dijkstra's algorithm.

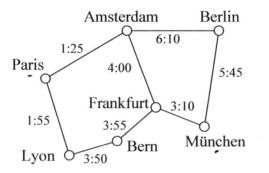

10. Using the graph from the previous problem, find the path with shortest travel time from Paris to München.

11. Does each of these graphs have an Euler circuit? If so, find it.

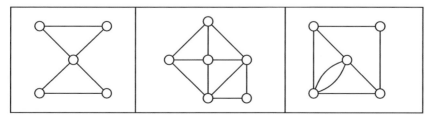

---

[6] From http://www.eurail.com/eurail-railway-map

148

12. Does each of these graphs have an Euler circuit? If so, find it.

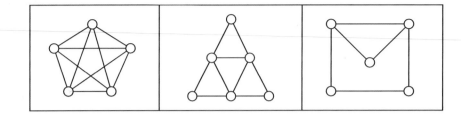

13. Eulerize this graph using as few edge duplications as possible. Then, find an Euler circuit.

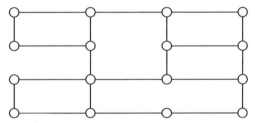

14. Eulerize this graph using as few edge duplications as possible. Then, find an Euler circuit.

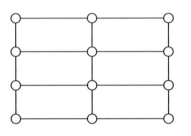

15. The maintenance staff at an amusement park need to patrol the major walkways, shown in the graph below, collecting litter. Find an efficient patrol route by finding an Euler circuit. If necessary, eulerize the graph in an efficient way.

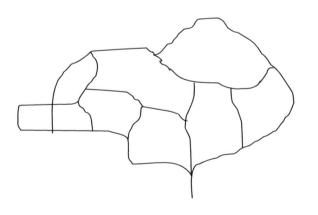

16. After a storm, the city crew inspects for trees or brush blocking the road. Find an efficient route for the neighborhood below by finding an Euler circuit. If necessary, eulerize the graph in an efficient way.

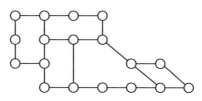

17. Does each of these graphs have at least one Hamiltonian circuit? If so, find one.

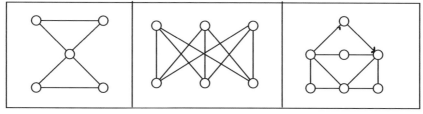

18. Does each of these graphs have at least one Hamiltonian circuit? If so, find one.

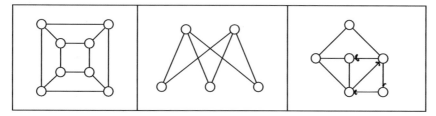

19. A company needs to deliver product to each of their 5 stores around the Dallas, TX area. Driving distances between the stores are shown below. Find a route for the driver to follow, returning to the distribution center in Fort Worth:
   a. Using Nearest Neighbor starting in Fort Worth
   b. Using Repeated Nearest Neighbor
   c. Using Sorted Edges

	Plano	Mesquite	Arlington	Denton
Fort Worth	54	52	19	42
Plano		38	53	41
Mesquite			43	56
Arlington				50

20. A salesperson needs to travel from Seattle to Honolulu, London, Moscow, and Cairo. Use the table of flight costs from problem #4 to find a route for this person to follow:
   a. Using Nearest Neighbor starting in Seattle
   b. Using Repeated Nearest Neighbor
   c. Using Sorted Edges

· 21. When installing fiber optics, some companies will install a sonet ring; a full loop of cable connecting multiple locations. This is used so that if any part of the cable is damaged it does not interrupt service, since there is a second connection to the hub. A company has 5 buildings. Costs (in thousands of dollars) to lay cables between pairs of buildings are shown below. Find the circuit that will minimize cost:
   a. Using Nearest Neighbor starting at building A
   b. Using Repeated Nearest Neighbor
   c. Using Sorted Edges

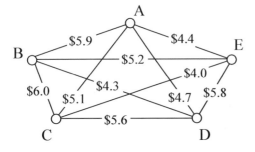

· 22. A tourist wants to visit 7 cities in Israel. Driving distances, in kilometers, between the cities are shown below[7]. Find a route for the person to follow, returning to the starting city:
   a. Using Nearest Neighbor starting in Jerusalem
   b. Using Repeated Nearest Neighbor
   c. Using Sorted Edges

	Jerusalem	Tel Aviv	Haifa	Tiberias	Beer Sheba	Eilat
Jerusalem	--					
Tel Aviv	58	--				
Haifa	151	95	--			
Tiberias	152	134	69	--		
Beer Sheba	81	105	197	233	--	
Eilat	309	346	438	405	241	--
Nazareth	131	102	35	29	207	488

23. Find a minimum cost spanning tree for the graph you created in problem #3

24. Find a minimum cost spanning tree for the graph you created in problem #22

25. Find a minimum cost spanning tree for the graph from problem #21

---

[7] From http://www.ddtravel-acc.com/Israel-cities-distance.htm

Concepts

26. Can a graph have one vertex with odd degree? If not, are there other values that are not possible? Why?

27. A complete graph is one in which there is an edge connecting every vertex to every other vertex. For what values of n does complete graph with n vertices have an Euler circuit? A Hamiltonian circuit?

28. Create a graph by drawing n vertices in a row, then another n vertices below those. Draw an edge from each vertex in the top row to every vertex in the bottom row. An example when n=3 is shown below. For what values of n will a graph created this way have an Euler circuit? A Hamiltonian circuit?

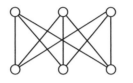

29. Eulerize this graph in the most efficient way possible, considering the weights of the edges.

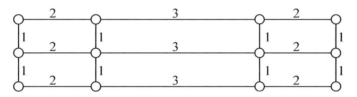

30. Eulerize this graph in the most efficient way possible, considering the weights of the edges.

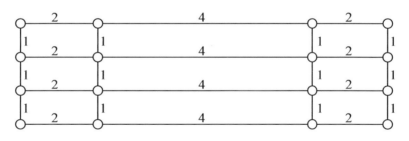

31. Eulerize this graph in the most efficient way possible, considering the weights of the edges.

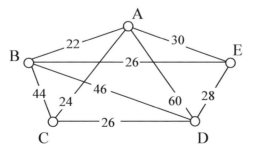

32. Eulerize this graph in the most efficient way possible, considering the weights of the edges.

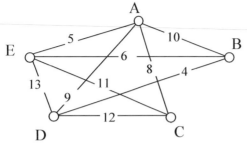

Explorations

33. Social networks such as Facebook and LinkedIn can be represented using graphs in which vertices represent people and edges are drawn between two vertices when those people are "friends." The table below shows a friendship table, where an X shows that two people are friends.

	A	B	C	D	E	F	G	H	I
A		X	X			X	X		
B			X		X				
C					X				
D					X				X
E							X		X
F								X	X
G								X	
H									X

   a. Create a graph of this friendship table
   b. Find the shortest path from A to D. The length of this path is often called the "degrees of separation" of the two people.
   c. Extension: Split into groups. Each group will pick 10 or more movies, and look up their major actors (www.imdb.com is a good source). Create a graph with each actor as a vertex, and edges connecting two actors in the same movie (note the movie name on the edge). Find interesting paths between actors, and quiz the other groups to see if they can guess the connections.

34. A spell checker in a word processing program makes suggestions when it finds a word not in the dictionary. To determine what words to suggest, it tries to find similar words. One measure of word similarity is the Levenshtein distance, which measures the number of substitutions, additions, or deletions that are required to change one word into another. For example, the words spit and spot are a distance of 1 apart; changing spit to spot requires one substitution (i for o). Likewise, spit is distance 1 from pit since the change requires one deletion (the s). The word spite is also distance 1 from spit since it requires one addition (the e). The word soot is distance 2 from spit since two substitutions would be required.
    a. Create a graph using words as vertices, and edges connecting words with a Levenshtein distance of 1. Use the misspelled word "moke" as the center, and try to find at least 10 connected dictionary words. How might a spell checker use this graph?
    b. Improve the method from above by assigning a weight to each edge based on the likelihood of making the substitution, addition, or deletion. You can base the weights on any reasonable approach: proximity of keys on a keyboard, common language errors, etc. Use Dijkstra's algorithm to find the length of the shortest path from each word to "moke". How might a spell checker use these values?

35. The graph below contains two vertices of odd degree. To eulerize this graph, it is necessary to duplicate edges connecting those two vertices.
    a. Use Dijkstra's algorithm to find the shortest path between the two vertices with odd degree. Does this produce the most efficient eulerization and solve the Chinese Postman Problem for this graph?

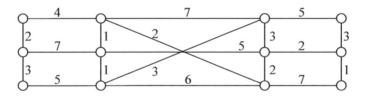

    b. Suppose a graph has n odd vertices. Using the approach from part a, how many shortest paths would need to be considered? Is this approach going to be efficient?

154

Scheduling

An event planner has to juggle many workers completing different tasks, some of which must be completed before others can begin. For example, the banquet tables would need to be arranged in the room before the catering staff could begin setting out silverware. The event planner has to carefully schedule things so that everything gets done in a reasonable amount of time.

The problem of scheduling is fairly universal. Contractors need to schedule workers and subcontractors to build a house as quickly as possible. A magazine needs to schedule writers, editors, photographers, typesetters, and others so that an issue can be completed on time.

Getting Started

To begin thinking about scheduling, let us consider an auto shop that is converting a car from gas to electric. A number of steps are involved. A time estimate for each task is given.

Task 1: Remove engine and gas parts (2 days)
Task 2: Steam clean the inside of the car (0.5 day)
Task 3: Buy an electric motor and speed controller (2 days for travel)
Task 4: Construct the part that connects the motor to the car's transmission (1 day)
Task 5: Construct battery racks (2 days)
Task 6: Install the motor (0.5 day)
Task 7: Install the speed controller (0.5 day)
Task 8: Install the battery racks (0.5 day)
Task 9: Wire the electricity (1 day)

Some tasks have to be completed before others – we certainly can't install the new motor before removing the old engine! There are some tasks, however, that can be worked on simultaneously by two different people, like constructing the battery racks and installing the motor.

To help us visualize the ordering of tasks, we will create a **digraph.**

> **Digraph**
> A diagraph is a graphical representation of a set of tasks in which tasks are represented with dots, called vertices, and arrows between vertices are used to show ordering.

For example, this digraph shows that Task 1, notated T_1 for compactness, needs to be completed before Task 2. The number in parentheses after the task name is the time required for the task.

$$T_1 (2) \qquad T_2 (0.5)$$
$$\circ \longrightarrow \circ$$

Example 1

A complete digraph for our car conversion would look like this:

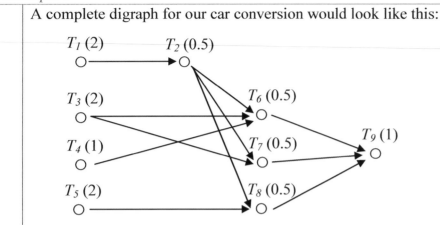

The time it takes to complete this job will partially depend upon how many people are working on the project.

> **Processors**
> In scheduling jargon, the workers completing the tasks are called **processors**. While in this example the processors are humans, in some situations the processors are computers, robots, or other machines.

For simplicity, we are going to make the very big assumptions that every processor can do every task, that they all would take the same time to complete it, and that only one processor can work on a task at a time.

> **Finishing Time**
> The **finishing** time is how long it will take to complete all the tasks. The finishing time will depend upon the number of processors and the specific schedule.

If we had only one processor working on this task, it is easy to determine the finishing time; just add up the individual times. We assume one person can't work on two tasks at the same time, ignore things like drying times during which someone could work on another task. Scheduling with one processor, a possible schedule would look like this, with a finishing time of 10 days.

Time:	0	1	2	3	4	5	6	7	8	9	10
P_1	T_1		T_3		T_4	T_5		T_2 T_6	T_7 T_8	T_9	

In this schedule, all the ordering requirements are met. This is certainly not the only possible schedule for one processor, but no other schedule could complete the job in less time. Because of this, this is an **optimal schedule** with **optimal finishing time** – there is nothing better.

Optimal Schedule
An optimal schedule is the schedule with the shortest possible finishing time.

For two processors, things become more interesting. For small digraphs like this, we probably could fiddle around and guess-and-check a pretty good schedule. Here would be a possibility:

Time:	0	1	2	3	4	5	6	7	8	9	10
P_1	T_1		T_3			T_6	T_9				
P_2	T_5		T_4		T_2	T_8	T_7				

With two processors, the finishing time was reduced to 5.5 days. What was processor 2 doing during the last day? Nothing, because there were no tasks for the processor to do. This is called **idle time**.

Idle Time
Idle time is time in the schedule when there are no tasks available for the processor to work on, so they sit idle.

Is this schedule optimal? Could it have been completed in 5 days? Because every other task had to be completed before task 9 could start, there would be no way that both processors could be busy during task 9, so it is not possible to create a shorter schedule.

So how long will it take if we use three processors? About $10/3 = 3.33$ days? Again we will guess-and-check a schedule:

Time:	0	1	2	3	4	5	6	7	8	9	10
P_1	T_1			T_2	T_6	T_8	T_9				
P_2	T_3			T_7							
P_3	T_4	T_5									

With three processors, the job still took 4.5 days. It is a little harder to tell whether this schedule is optimal. However, it might be helpful to notice that since Task 1, 2, 6, and 9 have to be completed sequentially, there is no way that this job could be completed in less than $2+0.5+0.5+1 = 4$ days, regardless of the number of processors. Four days is, for this digraph, the absolute minimum time to complete the job, called the **critical time**.

Critical Time
The critical time is the absolute minimum time to complete the job, regardless of the number of processors working on the tasks.

Critical time can be determined by looking at the longest sequence of tasks in the digraph, called the critical path

Adding an algorithm

Up until now, we have been creating schedules by guess-and-check, which works well enough for small schedules, but would not work well with dozens or hundreds of tasks. To create a more procedural approach, we might begin by somehow creating a **priority list**. Once we have a priority list, we can begin scheduling using that list and the **list processing algorithm**.

Priority List
A priority list is a list of tasks given in the order in which we desire them to be completed.

The List Processing Algorithm turns a priority list into a schedule

---

List Processing Algorithm
1. On the digraph or priority list, circle all tasks that are ready, meaning that all pre-requisite tasks have been completed.
2. Assign to each available processor, in order, the first ready task. Mark the task as in progress, perhaps by putting a single line through the task.
3. Move forward in time until a task is completed. Mark the task as completed, perhaps by crossing out the task. If any new tasks become ready, mark them as such.
4. Repeat until all tasks have been scheduled.

Example 2

Using our digraph from above, schedule it using the priority list below:
$T_1, T_3, T_4, T_5, T_6, T_7, T_8, T_2, T_9$

Time 0: Mark ready tasks

Priority list: $\widehat{T_1}, \widehat{T_3}, \widehat{T_4}, \widehat{T_5}, T_6, T_7, T_8, T_2, T_9$

We assign the first task, T_1 to the first processor, P_1, and the second ready task, T_3, to the second processor. Making those assignments, we mark those tasks as in progress:

Priority list: $\widehat{T_1}, \widehat{T_3}, \widehat{T_4}, \widehat{T_5}, T_6, T_7, T_8, T_2, T_9$

Schedule up to here:

Time:	0	1	2	3	4	5	6	7	8	9	10
P_1	T_1										
P_2	T_3										

We jump to the time when the next task completes, which is at time 2.

Time 2: Both processors complete their tasks. We mark those tasks as complete. With Task 1 complete, Task 2 becomes ready:

Priority list: $\cancel{(T_1,)}$ $\cancel{(T_3,)}$ $(T_4,)$ $(T_5,)$ $T_6,$ $T_7,$ $T_8,$ $(T_2,)$ T_9
We assign the next ready task on the list, T_4 to P_1, and T_5 to P_2.

Priority list: $\cancel{(T_1,)}$ $\cancel{(T_3,)}$ $\cancel{(T_4,)}$ $\cancel{(T_5,)}$ $T_6,$ $T_7,$ $T_8,$ $(T_2,)$ T_9

Time 3: Processor 1 has completed T_4. Completing T_4 does not make any other tasks ready (note that all the rest require that T_2 be completed first).

Priority list: $\cancel{(T_1,)}$ $\cancel{(T_3,)}$ $\cancel{(T_4,)}$ $\cancel{(T_5,)}$ $T_6,$ $T_7,$ $T_8,$ $(T_2,)$ T_9

Since the next three tasks are not yet ready, we assign the next ready task, T_2 to P_1

Priority list: $\cancel{(T_1,)}$ $\cancel{(T_3,)}$ $\cancel{(T_4,)}$ $\cancel{(T_5,)}$ $T_6,$ $T_7,$ $T_8,$ $\cancel{(T_2,)}$ T_9

Time 3.5: Processor 1 has completed T_2. Completing T_2 causes T_6 and T_7 to become ready. We assign T_6 to P_1

Priority list: $\cancel{(T_1,)}$ $\cancel{(T_3,)}$ $\cancel{(T_4,)}$ $\cancel{(T_5,)}$ $\cancel{(T_6,)}$ $(T_7,)$ $T_8,$ $\cancel{(T_2,)}$ T_9

Time 4: Both processors complete their tasks. The completion of T_5 allows T_8 to become ready. We assign T_7 to P_1 and T_8 to P_2.

Priority list: $\cancel{(T_1,)}$ $\cancel{(T_3,)}$ $\cancel{(T_4,)}$ $\cancel{(T_5,)}$ $\cancel{(T_6,)}$ $\cancel{(T_7,)}$ $\cancel{(T_8,)}$ $\cancel{(T_2,)}$ T_9

Time 4.5: Both processors complete their tasks. T_9 becomes ready, and is assigned to P_1. There is no ready task for P_2 to work on, so P_2 idles.

Priority list: ~~T_1~~ ~~T_3~~ ~~T_4~~ ~~T_5~~ ~~T_6~~ ~~T_7~~ ~~T_8~~ ~~T_2~~ ~~T_9~~

Time:	0	1	2	3	4	5	6	7	8	9	10
P_1	T_1		T_4	T_2 T_6 T_7 T_9							
P_2	T_3		T_5		T_8						

With the last task completed, we have a completed schedule, with finishing time 5.5 days.

Try it Now 1
Using the digraph below, create a schedule using the priority list:
$T_1, T_2, T_3, T_4, T_5, T_6, T_7, T_8, T_9$

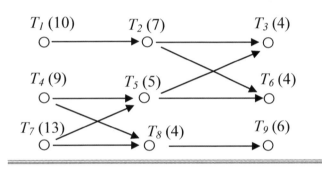

T_1 (10) T_2 (7) T_3 (4)

T_4 (9) T_5 (5) T_6 (4)

T_7 (13) T_8 (4) T_9 (6)

It is important to note that the list processing algorithm itself does not influence the resulting schedule – the schedule is completely determined by the priority list followed. The list processing, while do-able by hand, could just as easily be executed by a computer. The interesting part of scheduling, then, is how to choose a priority list that will create the best possible schedule.

Choosing a priority list

We will explore two algorithms for selecting a priority list.

Decreasing time algorithm

The decreasing time algorithm takes the approach of trying to get the very long tasks out of the way as soon as possible by putting them first on the priority list.

> **Decreasing Time Algorithm**
> Create the priority list by listing the tasks in order from longest completion time to shortest completion time.

Example 3

Consider the scheduling problem represented by the digraph below. Create a priority list using the decreasing time list algorithm, then use it to schedule for two processors using the list processing algorithm.

To use the decreasing time list algorithm, we create our priority list by listing the tasks in order from longest task time to shortest task time. If there is a tie, we will list the task with smaller task number first (not for any good reason, but just for consistency).

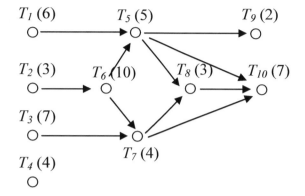

For this digraph, the decreasing time algorithm would create a priority list of:

T_6 (10), T_3 (7), T_{10} (7), T_1 (6), T_5 (5), T_4 (4), T_7 (4), T_2 (3), T_8 (3), T_9 (2)

Once we have the priority list, we can create the schedule using the list processing algorithm. With two processors, we'd get:

Time 0: We identify ready tasks, and assign T_3 to P_1 and T_1 to P_2

Priority list: T_6, $\cancel{T_3}$, T_{10}, $\cancel{T_1}$, T_5, $\boxed{T_4}$, T_7, $\boxed{T_2}$, T_8, T_9

	7	
P_1	T_3	
P_2	T_1	

6

Time 6: P_2 completes T_1. No new tasks become ready, so T_4 is assigned to P_2.

Priority list: T_6, $\cancel{T_3}$, T_{10}, $\cancel{T_1}$, T_5, $\cancel{T_4}$, T_7, $\boxed{T_2}$, T_8, T_9

	7	
P_1	T_3	
P_2	T_1	T_4

6   10

Time 7: P_1 completes T_3. No new tasks become ready, so T_2 is assigned to P_1.

Priority list: T_6, $\cancel{T_3}$, T_{10}, $\cancel{T_1}$, T_5, $\cancel{T_4}$, T_7, $\cancel{T_2}$, T_8, T_9

	7	
P_1	T_3	T_2
P_2	T_1	T_4

6   10

Time 10: Both processors complete their tasks. T_6 becomes ready, and is assigned to P_1. No other tasks are ready, so P_2 idles.

Priority list: ~~T_6~~ ~~T_3~~ T_{10}, ~~T_1~~ T_5, ~~T_4~~ T_7, ~~T_2~~ T_8, T_9

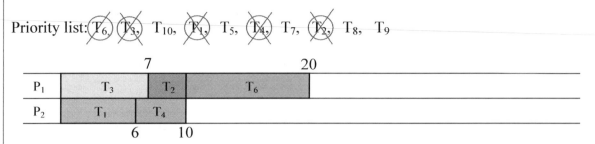

Time 20: With T_6 complete, T_5 and T_7 become ready, and are assigned to P_1 and P_2 respectively.

Priority list: ~~T_6~~ ~~T_3~~ T_{10}, ~~T_1~~ ~~T_5~~ ~~T_4~~ ~~T_7~~ ~~T_2~~ T_8, T_9

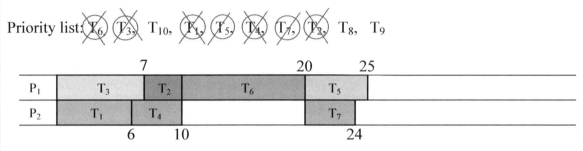

Time 24: P_2 completes T_7. No new items become ready, so P_2 idles.

Time 25: P_1 completes T_5. T_8 and T_9 become ready, and are assigned.

Priority list: ~~T_6~~ ~~T_3~~ T_{10}, ~~T_1~~ ~~T_5~~ ~~T_4~~ ~~T_7~~ ~~T_2~~ ~~T_8~~ ~~T_9~~

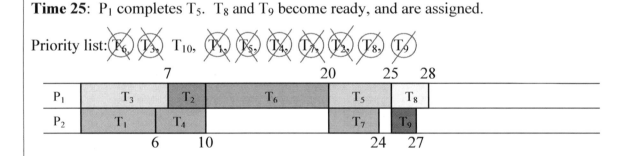

Time 27: T_9 is completed. No items ready, so P_2 idles.

Time 28: T_8 is completed. T_{10} becomes ready, and is assigned to P_1.

Priority list: ~~T_6~~ ~~T_3~~ ~~T_{10}~~ ~~T_1~~ ~~T_5~~ ~~T_4~~ ~~T_7~~ ~~T_2~~ ~~T_8~~ ~~T_9~~

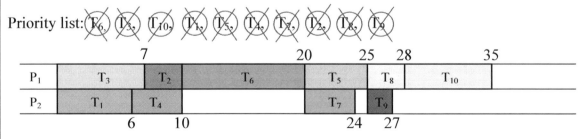

This is our completed schedule, with a finishing time of 35.

Using the decreasing time algorithm, the priority list led to a schedule with a finishing time of 35. Is this good? It certainly looks like there was a lot of idle time in this schedule. To get some idea how good or bad this schedule is, we could compute the critical time, the minimum time to complete the job. To find this, we look for the sequence of tasks with the highest total completion time. For this digraph that sequence would appear to be: T_2, T_6, T_5, T_8, T_{10}, with total sequence time of 28. From this we can conclude that our schedule isn't horrible, but there is a possibility that a better schedule exists.

Try it Now 2
Determine the priority list for the digraph from Try it Now 1 using the decreasing time algorithm.

Critical path algorithm

A sequence of tasks in the digraph is called a **path**. In the previous example, we saw that the critical path dictates the minimum completion time for a schedule. Perhaps, then, it would make sense to consider the critical path when creating our schedule. For example, in the last schedule, the processors began working on tasks 1 and 3 because they were longer tasks, but starting on task 2 earlier would have allowed work to begin on the long task 6 earlier.

The critical path algorithm allows you to create a priority list based on idea of critical paths.

> **Critical Path Algorithm (version 1)**
> 1. Find the critical path.
> 2. The first task in the critical path gets added to the priority list.
> 3. Remove that task from the digraph
> 4. Repeat, finding the new critical path with the revised digraph.

Example 4

The original digraph from Example 3 has critical path T_2, T_6, T_5, T_8, T_{10}, so T_2 gets added first to the priority list. Removing T_2 from the digraph, it now looks like:

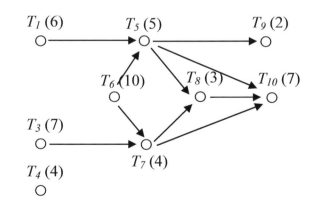

The critical path (longest path) in the remaining digraph is now T_6, T_5, T_8, T_{10}, so T_6 is added to the priority list and removed.

164

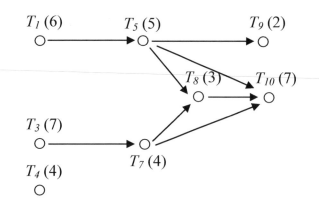

T_1 (6) T_5 (5) T_9 (2)

T_8 (3) T_{10} (7)

T_3 (7)

T_4 (4) T_7 (4)

Now there are two paths with the same longest length: T_1, T_5, T_8, T_{10} and T_3, T_7, T_8, T_{10}. We can add T_1 to the priority list (or T_3 – we usually add the one with smaller item number) and remove it, and continue the process.

I'm sure you can imagine that searching for the critical path every time you remove a task from the digraph would get really tiring, especially for a large digraph. In practice, the critical path algorithm is implementing by first working from the end backwards. This is called the **backflow algorithm**.

Backflow Algorithm
1. Introduce an "end" vertex, and assign it a time of 0, shown in [brackets]
2. Move backwards to every vertex that has an arrow to the end and assign it a critical time
3. From each of those vertices, move backwards and assign those vertices critical times. Notice that the critical time for the earlier vertex will be that task's time plus the critical time for the later vertex.
    Example: Consider this segment of digraph.

T_1 (5) T_2 (4) [10]

In this case, if T2 has already been determined to have a critical time of 10, then T1 will have a critical time of $5 + 10 = 15$

T_1 (5) [15] T_2 (4) [10]

If you have already assigned a critical time to a vertex, replace it only if the new time is larger.
    Example: In the digraph below, T1 should be labeled with a critical time of 16, since it is the longer of $5 + 10$ and $5 + 11$.

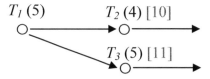

T_1 (5) T_2 (4) [10]

T_3 (5) [11]

4. Repeat until all vertices are labeled with their critical times

One you have completed the backflow algorithm, you can easily create the critical path priority list by using the critical times you just found.

> **Critical Path Algorithm (version 2)**
> 1. Apply the backflow algorithm to the digraph
> 2. Create the priority list by listing the tasks in order from longest critical time to shortest critical time

This version of the Critical Path Algorithm will usually be the easier to implement.

Example 5

Applying this to our digraph from the earlier example, we start applying the backflow algorithm.

We add an end vertex and give it a critical time of 0.

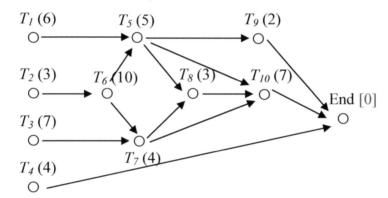

We then move back to T_4, T_9, and T_{10}, labeling them with their critical times

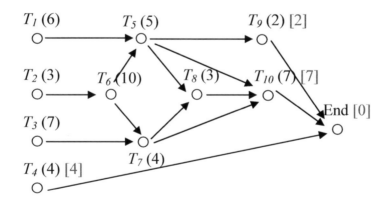

From each vertex marked with a critical time, we go back. T_7, for example, will get labeled with a critical time 11 – the task time of 4 plus the critical time of 7 for T_{10}. For T_5, there are two paths to the end. We use the longer, labeling T_5 with critical time $5+7 = 12$.

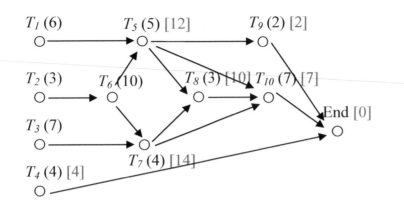

Continue the process until all vertices are labeled. Notice that the critical time for T_5 ended got replaced later with the even longer path through T_8.

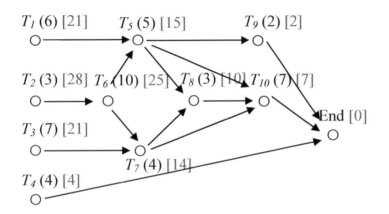

We can now quickly create the critical path priority list by listing the tasks in decreasing order of critical time:

Priority list: T_2, T_6, T_1, T_3, T_5, T_7, T_8, T_{10}, T_4, T_9

Applying this priority list using the list processing algorithm, we get the schedule:

	3		13	18	21	28
P_1	T_2	T_6	T_5	T_8	T_{10}	
P_2	T_1	T_3	T_7	T_4	T_9	
		6		17	23	

In this particular case, we were able to achieve the minimum possible completion time with this schedule, suggesting that this schedule is optimal. This is certainly not always the case.

Try it Now 3
Determine the priority list for the digraph from Try it Now 1 using the critical path algorithm.

Example 6

This example is designed to show that the critical path algorithm doesn't always work wonderfully. Schedule the tasks in the digraph below on three processors using the critical path algorithm.

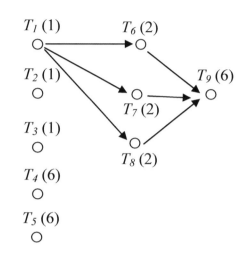

To create a critical path priority list, we could first apply the backflow algorithm:

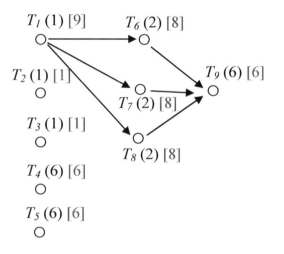

This yields the critical-path priority list: T_1, T_6, T_7, T_8, T_4, T_5, T_9, T_2, T_3.

Applying the list processing algorithm to this priority list leads to the schedule:

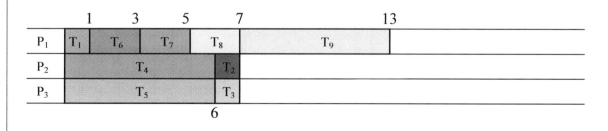

This schedule has finishing time of 13.

168

By observation, we can see that a much better schedule exists for the example above:

	1	3		9
P₁	T₁	T₆	T₉	
P₂	T₂	T₇	T₅	
P₃	T₃	T₈	T₄	

In most cases the critical path algorithm will lead to a very good schedule. There are cases, like this, where it will not. Unfortunately, there is no known algorithm to always produce the optimal schedule.

Try it Now Answers

1. Time 0: Tasks 1, 4, and 7 are ready. Assign Tasks 1 and 4
Time 9: Task 4 completes. Only task 7 is ready; assign task 7
Time 10: Task 1 completes. Task 2 now ready. Assign task 2
Time 17: Task 2 completes. Nothing is ready. Processor 1 idles
Time 22: Task 7 completes. Tasks 5 and 8 are ready. Assign tasks 5 and 8
Time 26: Task 8 completes. Task 9 is ready. Assign task 9
Time 27: Task 5 completes. Tasks 3 and 6 are ready. Assign task 3
Time 31: Task 3 completes. Assign task 6
Time 35: Everything is done. Finishing time is 35 for this schedule.

		10	17		27	31	35
P₁	T₁		T₂		T₅	T₃	
P₂	T₄		T₇		T₈	T₉	
	6	9		22	26	32	

2. T₇, T₁, T₄, T₂, T₉, T₅, T₃, T₆, T₈

3. Applying the backflow algorithm, we get this:

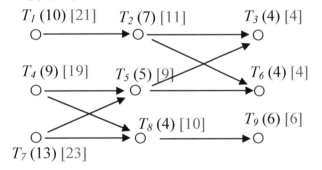

T_1 (10) [21] T_2 (7) [11] T_3 (4) [4]

T_4 (9) [19] T_5 (5) [9] T_6 (4) [4]

T_8 (4) [10] T_9 (6) [6]

T_7 (13) [23]

The critical path priority list is: T₇, T₁, T₄, T₂, T₈, T₅, T₉, T₃, T₆

Exercises

Skills

1. Create a digraph for the following set of tasks:

Task	Time required	Tasks that must be completed first
A	3	
B	4	
C	7	
D	6	A, B
E	5	B
F	5	D, E
G	4	E

2. Create a digraph for the following set of tasks:

Task	Time required	Tasks that must be completed first
A	3	
B	4	
C	7	
D	6	A
E	5	A
F	5	B
G	4	D, E

Use this digraph for the next 2 problems.

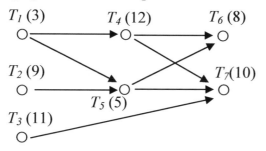

3. Using the priority list T_4, T_1, T_7, T_3, T_6, T_2, T_5, schedule the project with two processors.

4. Using the priority list T_5, T_2, T_3, T_7, T_1, T_4, T_6, schedule the project with two processors.

Use this digraph for the next 4 problems.

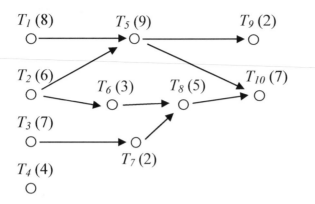

5. Using the priority list T_4, T_3, T_9, T_{10}, T_8, T_5, T_6, T_1, T_7, T_2 schedule the project with two processors.

6. Using the priority list T_2, T_4, T_6, T_8, T_{10}, T_1, T_3, T_5, T_7, T_9 schedule the project with two processors.

7. Using the priority list T_4, T_3, T_9, T_{10}, T_8, T_5, T_6, T_1, T_7, T_2 schedule the project with three processors.

8. Using the priority list T_2, T_4, T_6, T_8, T_{10}, T_1, T_3, T_5, T_7, T_9 schedule the project with three processors.

9. Use the decreasing time algorithm to create a priority list for the digraph from #3, and schedule with two processors.

10. Use the decreasing time algorithm to create a priority list for the digraph from #3, and schedule with three processors.

11. Use the decreasing time algorithm to create a priority list for the digraph from #5, and schedule with two processors.

12. Use the decreasing time algorithm to create a priority list for the digraph from #5, and schedule with three processors.

13. Use the decreasing time algorithm to create a priority list for the problem from #1, and schedule with two processors.

14. Use the decreasing time algorithm to create a priority list for the problem from #2, and schedule with two processors.

15. With the digraph from #3:
    a. Apply the backflow algorithm to find the critical time for each task
    b. Find the critical path for the project and the minimum completion time
    c. Use the critical path algorithm to create a priority list and schedule on two processors.

16. With the digraph from #3, use the critical path algorithm to schedule on three processors.

17. With the digraph from #5:
    a. Apply the backflow algorithm to find the critical time for each task
    b. Find the critical path for the project and the minimum completion time
    c. Use the critical path algorithm to create a priority list and schedule on two processors.

18. With the digraph from #5, use the critical path algorithm to schedule on three processors.

19. Use the critical path algorithm to schedule the problem from #1 on two processors.

20. Use the critical path algorithm to schedule the problem from #2 on two processors.

Concepts

21. If an additional order requirement is added to a digraph, can the optimal finishing time ever become longer? Can the optimal finishing time ever become shorter?

22. Will an optimal schedule always have no idle time?

23. Consider the digraph below.
    a. How many priority lists could be created for these tasks?
    b. How many unique schedules are created by those priority lists?

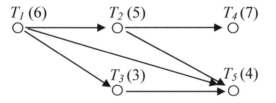

24. Create a digraph and priority list that would lead to the schedule below.

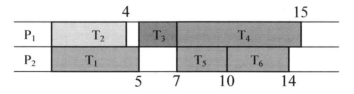

25. Is it possible to create a digraph with three tasks for which every possible priority list creates a different schedule? If so, create it.

26. Is it possible to create a digraph with four tasks for which every possible priority list creates a different schedule? If so, create it.

172

27. Independent tasks are ones that have no order requirements; they can be completed in any order.

    a. Consider three tasks, with completion times 2, 2, and 4 hours respectively. Construct two different schedules on two processors with different completion times to show that the priority list still matters with independent tasks.

    b. Choose a set of independent tasks with different completion times, and implement the decreasing time list algorithm and the critical path algorithm. What do you observe?

    c. Will using the decreasing time list or critical path algorithms with independent tasks always produce an optimal schedule? Why or why not?

    d. Will using the decreasing time list or critical path algorithms with independent tasks always produce the same schedule? Why or why not?

28. In a group, choose ten tasks necessary to throw a birthday party for a friend or child (for example, cleaning the house or buying a cake). Determine order requirements for the tasks, create a digraph, and schedule the tasks for two people.

29-37: At the end of the chapter it was noted that no algorithm exists to determine if an arbitrary schedule is optimal, but there are special cases where we can determine that a schedule is indeed optimal. In each of the following scenarios, determine

    a) If the scenario is even possible

    b) Whether or not the schedule *could* be optimal

    c) Whether or not we can be *sure* that the schedule is optimal

29. A job has a critical time of 30 hours, and the finishing time for the schedule on 2 processors is 30 hours.

30. The sum of all task times for a job is 40 hours, and the finishing time for the schedule on 2 processors was 15 hours

31. The sum of all task times for a job is 100 hours, the critical time of the job was 40 hours, and the finishing time for the schedule on 2 processors was 50 hours.

32. The sum of all task times for a job is 50 hours, and the finishing time for the schedule on 2 processors was 40 hours.

33. A job has a critical time of 30 hours, and the finishing time for the schedule on 3 processors is 20 hours.

34. The sum of all task times for a job is 60 hours, and the finishing time for the schedule on 3 processors was 20 hours.

35. The critical time for a job is 25 hours, and the finishing time for the schedule on 2 processors was 30 hours.

36. The sum of all task times for a job is 20 hours, and the finishing time for the schedule on 2 processors was 25 hours.

37. Based on your observations in the previous scenarios, write guidelines for when you can determine that a schedule is optimal.

Growth Models

Populations of people, animals, and items are growing all around us. By understanding how things grow, we can better understand what to expect in the future. In this chapter, we focus on time-dependant change.

Linear (Algebraic) Growth

Marco is a collector of antique soda bottles. His collection currently contains 437 bottles. Every year, he budgets enough money to buy 32 new bottles. Can we determine how many bottles he will have in 5 years, and how long it will take for his collection to reach 1000 bottles?

While both of these questions you could probably solve without an equation or formal mathematics, we are going to formalize our approach to this problem to provide a means to answer more complicated questions.

Suppose that P_n represents the number, or population, of bottles Marco has after n years. So P_0 would represent the number of bottles now, P_1 would represent the number of bottles after 1 year, P_2 would represent the number of bottles after 2 years, and so on. We could describe how Marco's bottle collection is changing using:

$P_0 = 437$
$P_n = P_{n-1} + 32$

This is called a **recursive relationship**. A recursive relationship is a formula which relates the next value in a sequence to the previous values. Here, the number of bottles in year n can be found by adding 32 to the number of bottles in the previous year, P_{n-1}. Using this relationship, we could calculate:

$P_1 = P_0 + 32 = 437 + 32 = 469$
$P_2 = P_1 + 32 = 469 + 32 = 501$
$P_3 = P_2 + 32 = 501 + 32 = 533$
$P_4 = P_3 + 32 = 533 + 32 = 565$
$P_5 = P_4 + 32 = 565 + 32 = 597$

We have answered the question of how many bottles Marco will have in 5 years. However, solving how long it will take for his collection to reach 1000 bottles would require a lot more calculations.

While recursive relationships are excellent for describing simply and cleanly *how* a quantity is changing, they are not convenient for making predictions or solving problems that stretch far into the future. For that, a closed or explicit form for the relationship is preferred. An **explicit equation** allows us to calculate P_n directly, without needing to know P_{n-1}. While you may already be able to guess the explicit equation, let us derive it from the recursive formula. We can do so by selectively not simplifying as we go:

$P_1 = 437 + 32$ $= 437 + 1(32)$

$$P_2 = P_1 + 32 = 437 + 32 + 32 \qquad = 437 + 2(32)$$
$$P_3 = P_2 + 32 = (437 + 2(32)) + 32 \quad = 437 + 3(32)$$
$$P_4 = P_3 + 32 = (437 + 3(32)) + 32 \quad = 437 + 4(32)$$

You can probably see the pattern now, and generalize that
$$P_n = 437 + n(32) = 437 + 32n$$

Using this equation, we can calculate how many bottles he'll have after 5 years:
$$P_5 = 437 + 32(5) = 437 + 160 = 597$$

We can now also solve for when the collection will reach 1000 bottles by substituting in 1000 for P_n and solving for n

$$1000 = 437 + 32n$$
$$563 = 32n$$
$$n = 563/32 = 17.59$$

So Marco will reach 1000 bottles in 18 years.

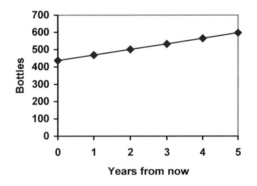

In the previous example, Marco's collection grew by the *same number* of bottles every year. This constant change is the defining characteristic of linear growth. Plotting the values we calculated for Marco's collection, we can see the values form a straight line, the shape of linear growth.

Linear Growth
If a quantity starts at size P_0 and grows by d every time period, then the quantity after n time periods can be determined using either of these relations:

Recursive form:
$$P_n = P_{n-1} + d$$

Explicit form:
$$P_n = P_0 + d\,n$$

In this equation, d represents the **common difference** – the amount that the population changes each time n increases by 1

Connection to prior learning – slope and intercept

You may recognize the common difference, d, in our linear equation as *slope*. In fact, the entire explicit equation should look familiar – it is the same linear equation you learned in algebra, probably stated as $y = mx + b$.

In the standard algebraic equation $y = mx + b$, b was the y-intercept, or the y value when x was zero. In the form of the equation we're using, we are using P_0 to represent that initial amount.

In the $y = mx + b$ equation, recall that m was the slope. You might remember this as "rise over run", or the change in y divided by the change in x. Either way, it represents the same thing as the common difference, d, we are using – the amount the output P_n changes when the input n increases by 1.

The equations $y = mx + b$ and $P_n = P_0 + d\,n$ mean the same thing and can be used the same ways, we're just writing it somewhat differently.

Example 1

The population of elk in a national forest was measured to be 12,000 in 2003, and was measured again to be 15,000 in 2007. If the population continues to grow linearly at this rate, what will the elk population be in 2014?

To begin, we need to define how we're going to measure n. Remember that P_0 is the population when $n = 0$, so we probably don't want to literally use the year 0. Since we already know the population in 2003, let us define $n = 0$ to be the year 2003. Then $P_0 = 12,000$.

Next we need to find d. Remember d is the growth per time period, in this case growth per year. Between the two measurements, the population grew by 15,000-12,000 = 3,000, but it took 2007-2003 = 4 years to grow that much. To find the growth per year, we can divide: 3000 elk / 4 years = 750 elk in 1 year.

Alternatively, you can use the slope formula from algebra to determine the common difference, noting that the population is the output of the formula, and time is the input.

$$d = slope = \frac{\text{change in output}}{\text{change in input}} = \frac{15,000 - 12,000}{2007 - 2003} = \frac{3000}{4} = 750$$

We can now write our equation in whichever form is preferred.

Recursive form:
$P_0 = 12,000$
$P_n = P_{n-1} + 750$

Explicit form:
$P_n = 12,000 + 750n$

To answer the question, we need to first note that the year 2014 will be $n = 11$, since 2014 is 11 years after 2003. The explicit form will be easier to use for this calculation:

$P_{11} = 12,000 + 750(11) = 20,250$ elk

176

Example 2

Gasoline consumption in the US has been increasing steadily. Consumption data from 1992 to 2004 is shown below[1]. Find a model for this data, and use it to predict consumption in 2016. If the trend continues, when will consumption reach 200 billion gallons?

Year	'92	'93	'94	'95	'96	'97	'98	'99	'00	'01	'02	'03	'04
Consumption (billion of gallons)	110	111	113	116	118	119	123	125	126	128	131	133	136

Plotting this data, it appears to have an approximately linear relationship:
While there are more advanced statistical techniques that can be used to find an equation to model the data, to get an idea of what is happening, we can find an equation by using two pieces of the data – perhaps the data from 1993 and 2003.

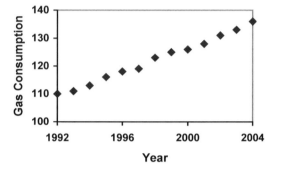

Letting $n = 0$ correspond with 1993 would give $P_0 = 111$ billion gallons.

To find d, we need to know how much the gas consumption increased each year, on average. From 1993 to 2003 the gas consumption increased from 111 billion gallons to 133 billion gallons, a total change of $133 - 111 = 22$ billion gallons, over 10 years. This gives us an average change of 22 billion gallons / 10 year = 2.2 billion gallons per year.

Equivalently,

$$d = slope = \frac{\text{change in output}}{\text{change in input}} = \frac{133 - 111}{10 - 0} = \frac{22}{10} = 2.2 \text{ billion gallons per year}$$

We can now write our equation in whichever form is preferred.

Recursive form:
$P_0 = 111$
$P_n = P_{n-1} + 2.2$

Explicit form:
$P_n = 111 + 2.2n$

Calculating values using the explicit form and plotting them with the original data shows how well our model fits the data.

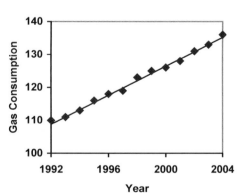

[1] http://www.bts.gov/publications/national_transportation_statistics/2005/html/table_04_10.html

We can now use our model to make predictions about the future, assuming that the previous trend continues unchanged. To predict the gasoline consumption in 2016:

$n = 23$ (2016 – 1993 = 23 years later)
$P_{23} = 111 + 2.2(23) = 161.6$

Our model predicts that the US will consume 161.6 billion gallons of gasoline in 2016 if the current trend continues.

To find when the consumption will reach 200 billion gallons, we would set $P_n = 200$, and solve for n:

$P_n = 200$	Replace P_n with our model
$111 + 2.2n = 200$	Subtract 111 from both sides
$2.2n = 89$	Divide both sides by 2.2
$n = 40.4545$	

This tells us that consumption will reach 200 billion about 40 years after 1993, which would be in the year 2033.

Example 3

The cost, in dollars, of a gym membership for n months can be described by the explicit equation $P_n = 70 + 30n$. What does this equation tell us?

The value for P_0 in this equation is 70, so the initial starting cost is $70. This tells us that there must be an initiation or start-up fee of $70 to join the gym.

The value for d in the equation is 30, so the cost increases by $30 each month. This tells us that the monthly membership fee for the gym is $30 a month.

Try it Now 1

The number of stay-at-home fathers in Canada has been growing steadily[2]. While the trend is not perfectly linear, it is fairly linear. Use the data from 1976 and 2010 to find an explicit formula for the number of stay-at-home fathers, then use it to predict the number if 2020.

Year	1976	1984	1991	2000	2010
Number of stay-at-home fathers	20,610	28,725	43,530	47,665	53,555

When good models go bad

When using mathematical models to predict future behavior, it is important to keep in mind that very few trends will continue indefinitely.

---

[2] http://www.fira.ca/article.php?id=140

Example 4

Suppose a four year old boy is currently 39 inches tall, and you are told to expect him to grow 2.5 inches a year.

We can set up a growth model, with $n = 0$ corresponding to 4 years old.

Recursive form:
$P_0 = 39$
$P_n = P_{n-1} + 2.5$

Explicit form:
$P_n = 39 + 2.5n$

So at 6 years old, we would expect him to be
$P_2 = 39 + 2.5(2) = 44$ inches tall

Any mathematical model will break down eventually. Certainly, we shouldn't expect this boy to continue to grow at the same rate all his life. If he did, at age 50 he would be
$P_{46} = 39 + 2.5(46) = 154$ inches tall $= 12.8$ feet tall!

When using any mathematical model, we have to consider which inputs are reasonable to use. Whenever we **extrapolate**, or make predictions into the future, we are assuming the model will continue to be valid.

Exponential (Geometric) Growth

Suppose that every year, only 10% of the fish in a lake have surviving offspring. If there were 100 fish in the lake last year, there would now be 110 fish. If there were 1000 fish in the lake last year, there would now be 1100 fish. Absent any inhibiting factors, populations of people and animals tend to grow by a percent of the existing population each year.

Suppose our lake began with 1000 fish, and 10% of the fish have surviving offspring each year. Since we start with 1000 fish, $P_0 = 1000$. How do we calculate P_1? The new population will be the old population, plus an additional 10%. Symbolically:

$P_1 = P_0 + 0.10P_0$

Notice this could be condensed to a shorter form by factoring:

$P_1 = P_0 + 0.10P_0 = 1P_0 + 0.10P_0 = (1 + 0.10)P_0 = 1.10P_0$

While 10% is the **growth rate**, 1.10 is the **growth multiplier**. Notice that 1.10 can be thought of as "the original 100% plus an additional 10%"

For our fish population,
$P_1 = 1.10(1000) = 1100$

We could then calculate the population in later years:
$P_2 = 1.10P_1 = 1.10(1100) = 1210$
$P_3 = 1.10P_2 = 1.10(1210) = 1331$

Notice that in the first year, the population grew by 100 fish, in the second year, the population grew by 110 fish, and in the third year the population grew by 121 fish. While there is a constant *percentage* growth, the actual increase in number of fish is increasing each year.

Graphing these values we see that this growth doesn't quite appear linear.

To get a better picture of how this percentage-based growth affects things, we need an explicit form, so we can quickly calculate values further out in the future.

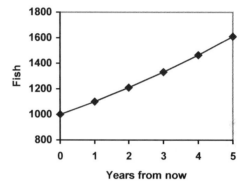

Like we did for the linear model, we will start building from the recursive equation:

$P_1 = 1.10P_0 = 1.10(1000)$
$P_2 = 1.10P_1 = 1.10(1.10(1000)) = 1.10^2(1000)$
$P_3 = 1.10P_2 = 1.10(1.10^2(1000)) = 1.10^3(1000)$
$P_4 = 1.10P_3 = 1.10(1.10^3(1000)) = 1.10^4(1000)$

Observing a pattern, we can generalize the explicit form to be:
$P_n = 1.10^n(1000)$, or equivalently, $P_n = 1000(1.10^n)$

From this, we can quickly calculate the number of fish in 10, 20, or 30 years:

$P_{10} = 1.10^{10}(1000) = 2594$
$P_{20} = 1.10^{20}(1000) = 6727$
$P_{30} = 1.10^{30}(1000) = 17449$

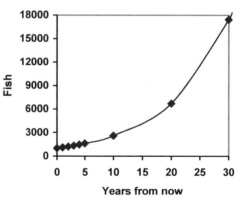

Adding these values to our graph reveals a shape that is definitely not linear. If our fish population had been growing linearly, by 100 fish each year, the population would have only reached 4000 in 30 years compared to almost 18000 with this percent-based growth, called **exponential growth.**

In exponential growth, the population grows proportional to the size of the population, so as the population gets larger, the same percent growth will yield a larger numeric growth.

Exponential Growth

If a quantity starts at size P_0 and grows by $R\%$ (written as a decimal, r) every time period, then the quantity after n time periods can be determined using either of these relations:

Recursive form:
$P_n = (1+r)\, P_{n-1}$

Explicit form:
$P_n = (1+r)^n\, P_0$ or equivalently, $P_n = P_0\,(1+r)^n$

We call r the **growth rate**.
The term $(1+r)$ is called the **growth multiplier**, or common ratio.

Example 5

Between 2007 and 2008, Olympia, WA grew almost 3% to a population of 245 thousand people. If this growth rate was to continue, what would the population of Olympia be in 2014?

As we did before, we first need to define what year will correspond to $n = 0$. Since we know the population in 2008, it would make sense to have 2008 correspond to $n = 0$, so $P_0 = 245,000$. The year 2014 would then be $n = 6$.

We know the growth rate is 3%, giving $r = 0.03$.

Using the explicit form:
$P_6 = (1+0.03)^6 (245,000) = 1.19405(245,000) = 292,542.25$

The model predicts that in 2014, Olympia would have a population of about 293 thousand people.

Evaluating exponents on the calculator

To evaluate expressions like $(1.03)^6$, it will be easier to use a calculator than multiply 1.03 by itself six times. Most scientific calculators have a button for exponents. It is typically either labeled like:

$\boxed{\wedge}$, $\boxed{y^x}$,   or $\boxed{x^y}$.

To evaluate 1.03^6 we'd type 1.03 $\boxed{\wedge}$ 6, or 1.03 $\boxed{y^x}$ 6. Try it out - you should get an answer around 1.1940523.

Try it Now 2

India is the second most populous country in the world, with a population in 2008 of about 1.14 billion people. The population is growing by about 1.34% each year. If this trend continues, what will India's population grow to by 2020?

Example 6

A friend is using the equation $P_n = 4600(1.072)^n$ to predict the annual tuition at a local college. She says the formula is based on years after 2010. What does this equation tell us?

In the equation, $P_0 = 4600$, which is the starting value of the tuition when $n = 0$. This tells us that the tuition in 2010 was $4,600.

The growth multiplier is 1.072, so the growth rate is 0.072, or 7.2%. This tells us that the tuition is expected to grow by 7.2% each year.

Putting this together, we could say that the tuition in 2010 was $4,600, and is expected to grow by 7.2% each year.

Example 7

In 1990, the residential energy use in the US was responsible for 962 million metric tons of carbon dioxide emissions. By the year 2000, that number had risen to 1182 million metric tons[3]. If the emissions grow exponentially and continue at the same rate, what will the emissions grow to by 2050?

Similar to before, we will correspond $n = 0$ with 1990, as that is the year for the first piece of data we have. That will make $P_0 = 962$ (million metric tons of CO_2). In this problem, we are not given the growth rate, but instead are given that $P_{10} = 1182$.

When $n = 10$, the explicit equation looks like:
$P_{10} = (1+r)^{10} P_0$

We know the value for P_0, so we can put that into the equation:
$P_{10} = (1+r)^{10} 962$

We also know that $P_{10} = 1182$, so substituting that in, we get
$1182 = (1+r)^{10} 962$

We can now solve this equation for the growth rate, r. Start by dividing by 962.

$$\frac{1182}{962} = (1+r)^{10}$$ Take the 10th root of both sides

$$\sqrt[10]{\frac{1182}{962}} = 1+r$$ Subtract 1 from both sides

---

[3] http://www.eia.doe.gov/oiaf/1605/ggrpt/carbon.html

182

$$r = \sqrt[10]{\frac{1182}{962}} - 1 = 0.0208 = 2.08\%$$

So if the emissions are growing exponentially, they are growing by about 2.08% per year. We can now predict the emissions in 2050 by finding P_{60}

$P_{60} = (1+0.0208)^{60} \, 962 = 3308.4$ million metric tons of CO_2 in 2050

Rounding

As a note on rounding, notice that if we had rounded the growth rate to 2.1%, our calculation for the emissions in 2050 would have been 3347. Rounding to 2% would have changed our result to 3156. A very small difference in the growth rates gets magnified greatly in exponential growth. For this reason, it is recommended to round the growth rate as little as possible.

If you need to round, **keep at least three significant digits** - numbers after any leading zeros. So 0.4162 could be reasonably rounded to 0.416. A growth rate of 0.001027 could be reasonably rounded to 0.00103.

Evaluating roots on the calculator

In the previous example, we had to calculate the 10th root of a number. This is different than taking the basic square root, $\sqrt{}$. Many scientific calculators have a button for general roots. It is typically labeled like:

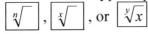

To evaluate the 3rd root of 8, for example, we'd either type 3 $\boxed{\sqrt[x]{}}$ 8, or 8 $\boxed{\sqrt[x]{}}$ 3, depending on the calculator. Try it on yours to see which to use – you should get an answer of 2.

If your calculator does not have a general root button, all is not lost. You can instead use the property of exponents which states that $\sqrt[n]{a} = a^{1/n}$. So, to compute the 3rd root of 8, you could use your calculator's exponent key to evaluate $8^{1/3}$. To do this, type:

8 $\boxed{y^x}$ (1 $\boxed{\div}$ 3)

The parentheses tell the calculator to divide 1/3 before doing the exponent.

Try it Now 3

The number of users on a social networking site was 45 thousand in February when they officially went public, and grew to 60 thousand by October. If the site is growing exponentially, and growth continues at the same rate, how many users should then expect two years after they went public?

Example 8

Looking back at the last example, for the sake of comparison, what would the carbon emissions be in 2050 if emissions grow linearly at the same rate?

Again we will get $n = 0$ correspond with 1990, giving $P_0 = 962$. To find d, we could take the same approach as earlier, noting that the emissions increased by 220 million metric tons in 10 years, giving a common difference of 22 million metric tons each year.

Alternatively, we could use an approach similar to that which we used to find the exponential equation. When $n = 10$, the explicit linear equation looks like:
$P_{10} = P_0 + 10d$

We know the value for P_0, so we can put that into the equation:
$P_{10} = 962 + 10d$

Since we know that $P_{10} = 1182$, substituting that in we get
$1182 = 962 + 10d$

We can now solve this equation for the common difference, d.
$1182 - 962 = 10d$
$220 = 10d$
$d = 22$

This tells us that if the emissions are changing linearly, they are growing by 22 million metric tons each year. Predicting the emissions in 2050,

$P_{60} = 962 + 22(60) = 2282$ million metric tons.

You will notice that this number is substantially smaller than the prediction from the exponential growth model. Calculating and plotting more values helps illustrate the differences.

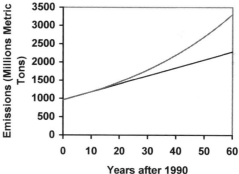

So how do we know which growth model to use when working with data? There are two approaches which should be used together whenever possible:

1) Find more than two pieces of data. Plot the values, and look for a trend. Does the data appear to be changing like a line, or do the values appear to be curving upwards?
2) Consider the factors contributing to the data. Are they things you would expect to change linearly or exponentially? For example, in the case of carbon emissions, we could expect that, absent other factors, they would be tied closely to population values, which tend to change exponentially.

Solving Exponentials for Time: Logarithms

Earlier, we found that since Olympia, WA had a population of 245 thousand in 2008 and had been growing at 3% per year, the population could be modeled by the equation

$$P_n = (1+0.03)^n (245,000), \text{ or equivalently, } P_n = 245,000(1.03)^n.$$

Using this equation, we were able to predict the population in the future.

Suppose we wanted to know when the population of Olympia would reach 400 thousand. Since we are looking for the year n when the population will be 400 thousand, we would need to solve the equation

$400,000 = 245,000(1.03)^n$ dividing both sides by 245,000 gives
$1.6327 = 1.03^n$

One approach to this problem would be to create a table of values, or to use technology to draw a graph to estimate the solution.

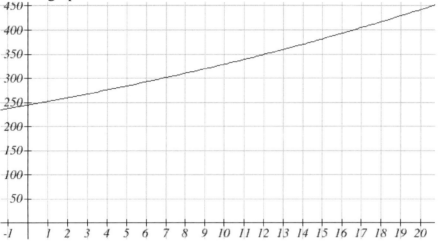

From the graph, we can estimate that the solution will be around 16 to 17 years after 2008 (2024 to 2025). This is pretty good, but we'd really like to have an algebraic tool to answer this question. To do that, we need to introduce a new function that will undo exponentials, similar to how a square root undoes a square. For exponentials, the function we need is called a logarithm. It is the inverse of the exponential, meaning it undoes the exponential. While there is a whole family of logarithms with different bases, we will focus on the common log, which is based on the exponential 10^x.

Common Logarithm
The common logarithm, written $\log(x)$, undoes the exponential 10^x
This means that $\log(10^x) = x$, and likewise $10^{\log(x)} = x$
This also means the statement $10^a = b$ is equivalent to the statement $\log(b) = a$

$\log(x)$ is read as "log of x", and means "the logarithm of the value x". It is important to note that this is *not* multiplication – the log doesn't mean anything by itself, just like $\sqrt{}$ doesn't mean anything by itself; it has to be applied to a number.

Example 9

Evaluate each of the following

a) log(100)   b) log(1000)   c) log(10000)   d) log(1/100)   e) log(1)

a) log(100) can be written as $\log(10^2)$. Since the log undoes the exponential, $\log(10^2) = 2$
b) $\log(1000) = \log(10^3) = 3$
c) $\log(10000) = \log(10^4) = 4$

d) Recall that $x^{-n} = \dfrac{1}{x^n}$. $\log\left(\dfrac{1}{100}\right) = \log(10^{-2}) = -2$

e) Recall that $x^0 = 1$. $\log(1) = \log(10^0) = 0$

It is helpful to note that from the first three parts of the previous example that the number we're taking the log of has to get *10 times bigger* for the log to increase in value by 1. Of course, most numbers cannot be written as a nice simple power of 10. For those numbers, we can evaluate the log using a scientific calculator with a log button.

Example 10

Evaluate log(300)

Using a calculator, log(300) is approximately 2.477121

With an equation, just like we can add a number to both sides, multiply both sides by a number, or square both sides, we can also take the logarithm of both sides of the equation and end up with an equivalent equation. This will allow us to solve some simple equations.

Example 11

a) Solve $10^x = 1000$ b) Solve $10^x = 3$ c) Solve $2(10^x) = 8$

a) Taking the log of both sides gives $\log(10^x) = \log(1000)$

Since the log undoes the exponential, $\log(10^x) = x$. Similarly $\log(1000) = \log(10^3) = 3$. The equation simplifies then to $x = 3$.

b) Taking the log of both sides gives $\log(10^x) = \log(3)$.

On the left side, $\log(10^x) = x$, so $x = \log(3)$. We can approximate this value with a calculator. $x \approx 0.477$

c) Here we would first want to isolate the exponential by dividing both sides of the equation by 2, giving
$10^x = 4$.

Now we can take the log of both sides, giving $\log(10^x) = \log(4)$, which simplifies to $x = \log(4) \approx 0.602$

This approach allows us to solve exponential equations with powers of 10, but what about problems like $2 = 1.03^n$ from earlier, which have a base of 1.03? For that, we need the exponent property for logs.

Properties of Logs: Exponent Property

$$\log(A^r) = r\log(A)$$

To show why this is true, we offer a proof.

Since the logarithm and exponential undo each other, $10^{\log A} = A$.

So $A^r = \left(10^{\log A}\right)^r$

Utilizing the exponential rule that states $\left(x^a\right)^b = x^{ab}$,

$$A^r = \left(10^{\log A}\right)^r = 10^{r\log A}$$

So then $\log(A^r) = \log\left(10^{r\log A}\right)$

Again utilizing the property that the log undoes the exponential on the right side yields the result

$$\log(A^r) = r\log A$$

Example 12

Rewrite $\log(25)$ using the exponent property for logs

$\log(25) = \log(5^2) = 2\log(5)$

This property will finally allow us to answer our original question.

Solving exponential equations with logarithms
1. Isolate the exponential. In other words, get it by itself on one side of the equation. This usually involves dividing by a number multiplying it.
2. Take the log of both sides of the equation.
3. Use the exponent property of logs to rewrite the exponential with the variable exponent multiplying the logarithm.
4. Divide as needed to solve for the variable.

Example 13

If Olympia is growing according to the equation, $P_n = 245(1.03)^n$, where n is years after 2008, and the population is measured in thousands. Find when the population will be 400 thousand.

We need to solve the equation

$400 = 245(1.03)^n$ Begin by dividing both sides by 245 to isolate the exponential

$1.633 = 1.03^n$ — Now take the log of both sides

$\log(1.633) = \log(1.03^n)$ — Use the exponent property of logs on the right side

$\log(1.633) = n \log(1.03)$ — Now we can divide by $\log(1.03)$

$\dfrac{\log(1.633)}{\log(1.03)} = n$ — We can approximate this value on a calculator

$n \approx 16.591$

Alternatively, after applying the exponent property of logs on the right side, we could have evaluated the logarithms to decimal approximations and completed our calculations using those approximations, as you'll see in the next example. While the final answer may come out slightly differently, as long as we keep enough significant values during calculation, our answer will be close enough for most purposes.

Example 14

Polluted water is passed through a series of filters. Each filter removes 90% of the remaining impurities from the water. If you have 10 million particles of pollutant per gallon originally, how many filters would the water need to be passed through to reduce the pollutant to 500 particles per gallon?

In this problem, our "population" is the number of particles of pollutant per gallon. The initial pollutant is 10 million particles per gallon, so $P_0 = 10,000,000$. Instead of changing with time, the pollutant changes with the number of filters, so n will represent the number of filters the water passes through.

Also, since the amount of pollutant is *decreasing* with each filter instead of increasing, our "growth" rate will be negative, indicating that the population is decreasing instead of increasing, so $r = -0.90$.

We can then write the explicit equation for the pollutant:
$P_n = 10,000,000(1 - 0.90)^n = 10,000,000(0.10)^n$

To solve the question of how many filters are needed to lower the pollutant to 500 particles per gallon, we can set P_n equal to 500, and solve for n.

$500 = 10,000,000(0.10)^n$ — Divide both sides by 10,000,000

$0.00005 = 0.10^n$ — Take the log of both sides

$\log(0.00005) = \log(0.10^n)$ — Use the exponent property of logs on the right side

$\log(0.00005) = n \log(0.10)$ — Evaluate the logarithms to a decimal approximation

$-4.301 = n(-1)$ — Divide by -1, the value multiplying n

$4.301 = n$

It would take about 4.301 filters. Of course, since we probably can't install 0.3 filters, we would need to use 5 filters to bring the pollutant below the desired level.

188

Logistic Growth

In our basic exponential growth scenario, we had a recursive equation of the form
$$P_n = P_{n-1} + r\, P_{n-1}$$

In a confined environment, however, the growth rate may not remain constant. In a lake, for
example, there is some *maximum sustainable population* of fish, also called a **carrying
capacity**.

> **Carrying Capacity**
> The **carrying capacity**, or **maximum sustainable population**, is the largest
> population that an environment can support.

For our fish, the carrying capacity is the largest population that the resources in the lake can
sustain. If the population in the lake is far below the carrying capacity, then we would expect
the population to grow essentially exponentially. However, as the population approaches the
carrying capacity, there will be a scarcity of food and space available, and the growth rate
will decrease. If the population exceeds the carrying capacity, there won't be enough
resources to sustain all the fish and there will be a negative growth rate, causing the
population to decrease back to the carrying capacity.

If the carrying capacity was 5000, the growth rate
might vary something like that in the graph
shown. Note that this is a linear equation with
intercept at 0.1 and slope $-\dfrac{0.1}{5000}$, so we could
write an equation for this adjusted growth rate as:

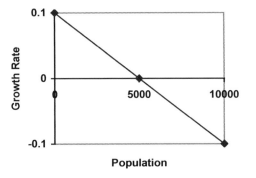

$$r_{adjusted} = 0.1 - \frac{0.1}{5000}P = 0.1\left(1 - \frac{P}{5000}\right)$$

Substituting this in to our original exponential growth model for r gives

$$P_n = P_{n-1} + 0.1\left(1 - \frac{P_{n-1}}{5000}\right)P_{n-1}$$

Logistic Growth

If a population is growing in a constrained environment with carrying capacity K, and absent constraint would grow exponentially with growth rate r, then the population behavior can be described by the logistic growth model:

$$P_n = P_{n-1} + r\left(1 - \frac{P_{n-1}}{K}\right)P_{n-1}$$

Unlike linear and exponential growth, logistic growth behaves differently if the populations grow steadily throughout the year or if they have one breeding time per year. The recursive formula provided above models generational growth, where there is one breeding time per year (or, at least a finite number); there is no explicit formula for this type of logistic growth.

Example 15

A forest is currently home to a population of 200 rabbits. The forest is estimated to be able to sustain a population of 2000 rabbits. Absent any restrictions, the rabbits would grow by 50% per year. Predict the future population using the logistic growth model.

Modeling this with a logistic growth model, $r = 0.50$, $K = 2000$, and $P_0 = 200$. Calculating the next year:

$$P_1 = P_0 + 0.50\left(1 - \frac{P_0}{2000}\right)P_0 = 200 + 0.50\left(1 - \frac{200}{2000}\right)200 = 290$$

We can use this to calculate the following year:

$$P_2 = P_1 + 0.50\left(1 - \frac{P_1}{2000}\right)P_1 = 290 + 0.50\left(1 - \frac{290}{2000}\right)290 \approx 414$$

A calculator was used to compute several more values:

n	0	1	2	3	4	5	6	7	8	9	10
P_n	200	290	414	578	784	1022	1272	1503	1690	1821	1902

Plotting these values, we can see that the population starts to increase faster and the graph curves upwards during the first few years, like exponential growth, but then the growth slows down as the population approaches the carrying capacity.

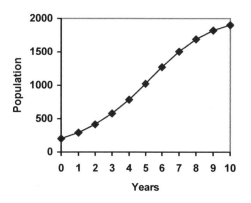

Example 16

On an island that can support a population of 1000 lizards, there is currently a population of 600. These lizards have a lot of offspring and not a lot of natural predators, so have very high growth rate, around 150%. Calculating out the next couple generations:

$$P_1 = P_0 + 1.50\left(1 - \frac{P_0}{1000}\right)P_0 = 600 + 1.50\left(1 - \frac{600}{1000}\right)600 = 960$$

$$P_2 = P_1 + 1.50\left(1 - \frac{P_1}{1000}\right)P_1 = 960 + 1.50\left(1 - \frac{960}{1000}\right)960 = 1018$$

Interestingly, even though the factor that limits the growth rate slowed the growth a lot, the population still overshot the carrying capacity. We would expect the population to decline the next year.

$$P_3 = P_2 + 1.50\left(1 - \frac{P_3}{1000}\right)P_3 = 1018 + 1.50\left(1 - \frac{1018}{1000}\right)1018 = 991$$

Calculating out a few more years and plotting the results, we see the population wavers above and below the carrying capacity, but eventually settles down, leaving a steady population near the carrying capacity.

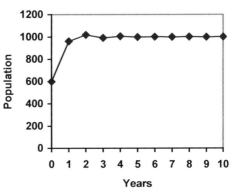

Try it Now 5

A field currently contains 20 mint plants. Absent constraints, the number of plants would increase by 70% each year, but the field can only support a maximum population of 300 plants. Use the logistic model to predict the population in the next three years.

Example 17

On a neighboring island to the one from the previous example, there is another population of lizards, but the growth rate is even higher – about 205%.

Calculating out several generations and plotting the results, we get a surprise: the population seems to be oscillating between two values, a pattern called a 2-cycle.

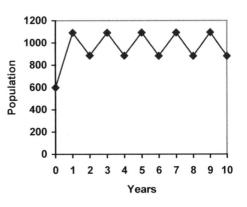

While it would be tempting to treat this only as a strange side effect of mathematics, this has actually been observed in nature. Researchers from the University of California observed a stable 2-cycle in a lizard population in California[4].

Taking this even further, we get more and more extreme behaviors as the growth rate increases higher.   It is possible to get stable 4-cycles, 8-cycles, and higher. Quickly, though, the behavior approaches chaos (remember the movie Jurassic Park?).

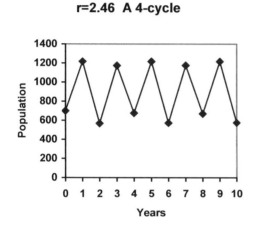
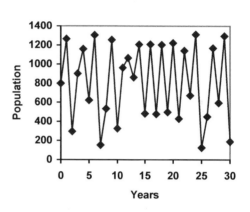

---

[4] http://users.rcn.com/jkimball.ma.ultranet/BiologyPages/P/Populations2.html

Try it Now Answers

1. Letting $n = 0$ correspond with 1976, then $P_0 = 20,610$.
From 1976 to 2010 the number of stay-at-home fathers increased by
$53,555 - 20,610 = 32,945$
This happened over 34 years, giving a common different d of $32,945 / 34 = 969$.
$P_n = 20,610 + 969n$

Predicting for 2020, we use $n = 44$
$P_{44} = 20,610 + 969(44) = 63,246$ stay-at-home fathers in 2020.

2. Using $n = 0$ corresponding with 2008,
$P_{12} = (1+0.0134)^{12}(1.14) =$ about 1.337 billion people in 2020

3. Here we will measure n in months rather than years, with $n = 0$ corresponding to the February when they went public. This gives $P_0 = 45$ thousand. October is 8 months later, so
$P_8 = 60$.

$P_8 = (1+r)^8 P_0$
$60 = (1+r)^8 45$

$\dfrac{60}{45} = (1+r)^8$

$\sqrt[8]{\dfrac{60}{45}} = 1+r$

$r = \sqrt[8]{\dfrac{60}{45}} - 1 = 0.0366$, or 3.66%

The general explicit equation is $P_n = (1.0366)^n 45$
Predicting 24 months (2 years) after they went public:
$P_{24} = (1.0366)^{24} 45 = 106.63$ thousand users.

4. $1.14(1.0134)^n = 1.2$. $n = 3.853$, which is during 2011

5. $P_1 = P_0 + 0.70\left(1 - \dfrac{P_0}{300}\right)P_0 = 20 + 0.70\left(1 - \dfrac{20}{300}\right)20 = 33$

$P_2 = 54$
$P_3 = 85$

Exercises

Skills

1. Marko currently has 20 tulips in his yard. Each year he plants 5 more.
   a. Write a recursive formula for the number of tulips Marko has
   b. Write an explicit formula for the number of tulips Marko has

2. Pam is a Disc Jockey. Every week she buys 3 new albums to keep her collection current. She currently owns 450 albums.
   a. Write a recursive formula for the number of albums Pam has
   b. Write an explicit formula for the number of albums Pam has

3. A store's sales (in thousands of dollars) grow according to the recursive rule $P_n=P_{n-1} + 15$, with initial population $P_0=40$.
   a. Calculate P_1 and P_2
   b. Find an explicit formula for P_n
   c. Use your formula to predict the store's sales in 10 years
   d. When will the store's sales exceed $100,000?

4. The number of houses in a town has been growing according to the recursive rule $P_n=P_{n-1} + 30$, with initial population $P_0=200$.
   a. Calculate P_1 and P_2
   b. Find an explicit formula for P_n
   c. Use your formula to predict the number of houses in 10 years
   d. When will the number of houses reach 400 houses?

5. A population of beetles is growing according to a linear growth model. The initial population (week 0) was $P_0=3$, and the population after 8 weeks is $P_8=67$.
   a. Find an explicit formula for the beetle population in week n
   b. After how many weeks will the beetle population reach 187?

6. The number of streetlights in a town is growing linearly. Four months ago ($n = 0$) there were 130 lights. Now ($n = 4$) there are 146 lights. If this trend continues,
   a. Find an explicit formula for the number of lights in month n
   b. How many months will it take to reach 200 lights?

7. Tacoma's population in 2000 was about 200 thousand, and had been growing by about 9% each year.
   a. Write a recursive formula for the population of Tacoma
   b. Write an explicit formula for the population of Tacoma
   c. If this trend continues, what will Tacoma's population be in 2016?
   d. When does this model predict Tacoma's population to exceed 400 thousand?

8. Portland's population in 2007 was about 568 thousand, and had been growing by about 1.1% each year.
    a. Write a recursive formula for the population of Portland
    b. Write an explicit formula for the population of Portland
    c. If this trend continues, what will Portland's population be in 2016?
    d. If this trend continues, when will Portland's population reach 700 thousand?

9. Diseases tend to spread according to the exponential growth model. In the early days of AIDS, the growth rate was around 190%. In 1983, about 1700 people in the U.S. died of AIDS. If the trend had continued unchecked, how many people would have died from AIDS in 2005?

10. The population of the world in 1987 was 5 billion and the annual growth rate was estimated at 2 percent per year. Assuming that the world population follows an exponential growth model, find the projected world population in 2015.

11. A bacteria culture is started with 300 bacteria. After 4 hours, the population has grown to 500 bacteria. If the population grows exponentially,
    a. Write a recursive formula for the number of bacteria
    b. Write an explicit formula for the number of bacteria
    c. If this trend continues, how many bacteria will there be in 1 day?
    d. How long does it take for the culture to triple in size?

12. A native wolf species has been reintroduced into a national forest. Originally 200 wolves were transplanted. After 3 years, the population had grown to 270 wolves. If the population grows exponentially,
    a. Write a recursive formula for the number of wolves
    b. Write an explicit formula for the number of wolves
    c. If this trend continues, how many wolves will there be in 10 years?
    d. If this trend continues, how long will it take the population to grow to 1000 wolves?

13. One hundred trout are seeded into a lake. Absent constraint, their population will grow by 70% a year. The lake can sustain a maximum of 2000 trout. Using the logistic growth model,
    a. Write a recursive formula for the number of trout
    b. Calculate the number of trout after 1 year and after 2 years.

14. Ten blackberry plants started growing in my yard. Absent constraint, blackberries will spread by 200% a month. My yard can only sustain about 50 plants. Using the logistic growth model,
    a. Write a recursive formula for the number of blackberry plants in my yard
    b. Calculate the number of plants after 1, 2, and 3 months

15. In 1968, the U.S. minimum wage was $1.60 per hour. In 1976, the minimum wage was $2.30 per hour. Assume the minimum wage grows according to an exponential model where n represents the time in years after 1960.
    a. Find an explicit formula for the minimum wage.
    b. What does the model predict for the minimum wage in 1960?
    c. If the minimum wage was $5.15 in 1996, is this above, below or equal to what the model predicts?

Concepts

16. The population of a small town can be described by the equation $P_n = 4000 + 70n$, where n is the number of years after 2005. Explain in words what this equation tells us about how the population is changing.

17. The population of a small town can be described by the equation $P_n = 4000(1.04)^n$, where n is the number of years after 2005. Explain in words what this equation tells us about how the population is changing.

Exploration

Most of the examples in the text examined growing quantities, but linear and exponential equations can also describe decreasing quantities, as the next few problems will explore.

18. A new truck costs $32,000. The car's value will depreciate over time, which means it will lose value. For tax purposes, depreciation is usually calculated linearly. If the truck is worth $24,500 after three years, write an explicit formula for the value of the car after n years.

19. Inflation causes things to cost more, and for our money to buy less (hence your grandparents saying, "In my day, you could buy a cup of coffee for a nickel"). Suppose inflation decreases the value of money by 5% each year. In other words, if you have $1 this year, next year it will only buy you $0.95 worth of stuff. How much will $100 buy you in 20 years?

20. Suppose that you have a bowl of 500 M&M candies, and each day you eat ¼ of the candies you have. Is the number of candies left changing linearly or exponentially? Write an equation to model the number of candies left after n days.

21. A warm object in a cooler room will decrease in temperature exponentially, approaching the room temperature according to the formula $T_n = a(1-r)^n + T_r$ where T_n is the temperature after n minutes, r is the rate at which temperature is changing, a is a constant, and T_r is the temperature of the room. Forensic investigators can use this to predict the time of death of a homicide victim. Suppose that when a body was discovered ($n = 0$) it was 85 degrees. After 20 minutes, the temperature was measured again to be 80 degrees. The body was in a 70 degree room.
    a. Use the given information with the formula provided to find a formula for the temperature of the body.
    b. When did the victim die, if the body started at 98.6 degrees?

22. Recursive equations can be very handy for modeling complicated situations for which explicit equations would be hard to interpret. As an example, consider a lake in which 2000 fish currently reside. The fish population grows by 10% each year, but every year 100 fish are harvested from the lake by people fishing.
    a. Write a recursive equation for the number of fish in the lake after n years.
    b. Calculate the population after 1 and 2 years. Does the population appear to be increasing or decreasing?
    c. What is the maximum number of fish that could be harvested each year without causing the fish population to decrease in the long run?

23. The number of Starbucks stores grew after first opened. The number of stores from 1990-2007, as reported on their corporate website[5], is shown below.
    a. Carefully plot the data. Does is appear to be changing linearly or exponentially?
    b. Try finding an equation to model the data by picking two points to work from. How well does the equation model the data?
    c. Try using an equation of the form $P_n = P_0 n^k$, where k is a constant, to model the data. This type of model is called a Power model. Compare your results to the results from part b. *Note: to use this model, you will need to have 1990 correspond with $n = 1$ rather than $n = 0$.*

Year	Number of Starbucks stores	Year	Number of Starbucks stores
1990	84	1999	2498
1991	116	2000	3501
1992	165	2001	4709
1993	272	2002	5886
1994	425	2003	7225
1995	677	2004	8569
1996	1015	2005	10241
1997	1412	2006	12440
1998	1886	2007	15756

24. Thomas Malthus was an economist who put forth the principle that population grows based on an exponential growth model, while food and resources grow based on a linear growth model. Based on this, Malthus predicted that eventually demand for food and resources would out outgrow supply, with doom-and-gloom consequences. Do some research about Malthus to answer these questions.
    a. What societal changes did Malthus propose to avoid the doom-and-gloom outcome he was predicting?
    b. Why do you think his predictions did not occur?
    c. What are the similarities and differences between Malthus's theory and the logistic growth model?

---

[5] http://www.starbucks.com/aboutus/Company_Timeline.pdf retrieved May 2009

Finance

We have to work with money every day. While balancing your checkbook or calculating your monthly expenditures on espresso requires only arithmetic, when we start saving, planning for retirement, or need a loan, we need more mathematics.

Simple Interest

Discussing interest starts with the **principal**, or amount your account starts with. This could be a starting investment, or the starting amount of a loan. Interest, in its most simple form, is calculated as a percent of the principal. For example, if you borrowed $100 from a friend and agree to repay it with 5% interest, then the amount of interest you would pay would just be 5% of 100: $100(0.05) = $5. The total amount you would repay would be $105, the original principal plus the interest.

Simple One-time Interest

$$I = P_0 r$$

$$A = P_0 + I = P_0 + P_0 r = P_0(1+r)$$

I is the interest
A is the end amount: principal plus interest
P_0 is the principal (starting amount)
r is the interest rate (in decimal form. Example: 5% = 0.05)

Example 1

A friend asks to borrow $300 and agrees to repay it in 30 days with 3% interest. How much interest will you earn?

$P_0 = 300 the principal
$r = 0.03$ 3% rate
$I = $300(0.03) = $9.$ You will earn $9 interest.

One-time simple interest is only common for extremely short-term loans. For longer term loans, it is common for interest to be paid on a daily, monthly, quarterly, or annual basis. In that case, interest would be earned regularly. For example, bonds are essentially a loan made to the bond issuer (a company or government) by you, the bond holder. In return for the loan, the issuer agrees to pay interest, often annually. Bonds have a maturity date, at which time the issuer pays back the original bond value.

Example 2

Suppose your city is building a new park, and issues bonds to raise the money to build it. You obtain a $1,000 bond that pays 5% interest annually that matures in 5 years. How much interest will you earn?

Each year, you would earn 5% interest: $1000(0.05) = $50 in interest. So over the course of five years, you would earn a total of $250 in interest. When the bond matures, you would receive back the $1,000 you originally paid, leaving you with a total of $1,250.

We can generalize this idea of simple interest over time.

Simple Interest over Time

$I = P_0 rt$

$A = P_0 + I = P_0 + P_0 rt = P_0(1 + rt)$

I is the interest
A is the end amount: principal plus interest
P_0 is the principal (starting amount)
r is the interest rate in decimal form
t is time

The units of measurement (years, months, etc.) for the time should match the time period for the interest rate.

APR – Annual Percentage Rate

Interest rates are usually given as an **annual percentage rate (APR)** – the total interest that will be paid in the year. If the interest is paid in smaller time increments, the APR will be divided up.

For example, a 6% APR paid monthly would be divided into twelve 0.5% payments. A 4% annual rate paid quarterly would be divided into four 1% payments.

Example 3

Treasury Notes (T-notes) are bonds issued by the federal government to cover its expenses. Suppose you obtain a $1,000 T-note with a 4% annual rate, paid semi-annually, with a maturity in 4 years. How much interest will you earn?

Since interest is being paid semi-annually (twice a year), the 4% interest will be divided into two 2% payments.

$P_0 = 1000 the principal
$r = 0.02$ 2% rate per half-year
$t = 8$ 4 years = 8 half-years
$I = $1000(0.02)(8) = 160. You will earn $160 interest total over the four years.

Try it Now 1

A loan company charges $30 interest for a one month loan of $500. Find the annual interest rate they are charging.

Compound Interest

With simple interest, we were assuming that we pocketed the interest when we received it. In a standard bank account, any interest we earn is automatically added to our balance, and we earn interest on that interest in future years. This reinvestment of interest is called **compounding**.

Suppose that we deposit $1000 in a bank account offering 3% interest, compounded monthly. How will our money grow?

The 3% interest is an annual percentage rate (APR) – the total interest to be paid during the year. Since interest is being paid monthly, each month, we will earn $\frac{3\%}{12} = 0.25\%$ per month.

In the first month,
$P_0 = \$1000$
$r = 0.0025\ (0.25\%)$
$I = \$1000\ (0.0025) = \$2.50$
$A = \$1000 + \$2.50 = \$1002.50$

In the first month, we will earn $2.50 in interest, raising our account balance to $1002.50. In the second month,

$P_0 = \$1002.50$
$I = \$1002.50\ (0.0025) = \$2.51$ (rounded)
$A = \$1002.50 + \$2.51 = \$1005.01$

Notice that in the second month we earned more interest than we did in the first month. This is because we earned interest not only on the original $1000 we deposited, but we also earned interest on the $2.50 of interest we earned the first month. This is the key advantage that **compounding** of interest gives us.

Calculating out a few more months:

Month	Starting balance	Interest earned	Ending Balance
1	1000.00	2.50	1002.50
2	1002.50	2.51	1005.01
3	1005.01	2.51	1007.52
4	1007.52	2.52	1010.04
5	1010.04	2.53	1012.57
6	1012.57	2.53	1015.10
7	1015.10	2.54	1017.64
8	1017.64	2.54	1020.18
9	1020.18	2.55	1022.73
10	1022.73	2.56	1025.29
11	1025.29	2.56	1027.85
12	1027.85	2.57	1030.42

To find an equation to represent this, if P_m represents the amount of money after m months, then we could write the recursive equation:

$P_0 = \$1000$
$P_m = (1+0.0025)P_{m-1}$

You probably recognize this as the recursive form of exponential growth. If not, we could go through the steps to build an explicit equation for the growth:
$P_0 = \$1000$
$P_1 = 1.0025P_0 = 1.0025\,(1000)$
$P_2 = 1.0025P_1 = 1.0025\,(1.0025\,(1000)) = 1.0025^2(1000)$
$P_3 = 1.0025P_2 = 1.0025\,(1.0025^2(1000)) = 1.0025^3(1000)$
$P_4 = 1.0025P_3 = 1.0025\,(1.0025^3(1000)) = 1.0025^4(1000)$

Observing a pattern, we could conclude

$P_m = (1.0025)^m(\$1000)$

Notice that the $1000 in the equation was P_0, the starting amount. We found 1.0025 by adding one to the growth rate divided by 12, since we were compounding 12 times per year. Generalizing our result, we could write

$$P_m = P_0\left(1+\frac{r}{k}\right)^m$$

In this formula:
m is the number of compounding periods (months in our example)
r is the annual interest rate
k is the number of compounds per year.

While this formula works fine, it is more common to use a formula that involves the number of years, rather than the number of compounding periods. If N is the number of years, then $m = N\,k$. Making this change gives us the standard formula for compound interest.

Compound Interest

$$P_N = P_0\left(1+\frac{r}{k}\right)^{Nk}$$

P_N is the balance in the account after N years.
P_0 is the starting balance of the account (also called initial deposit, or principal)
r is the annual interest rate in decimal form
k is the number of compounding periods in one year.

If the compounding is done annually (once a year), $k = 1$.
If the compounding is done quarterly, $k = 4$.
If the compounding is done monthly, $k = 12$.
If the compounding is done daily, $k = 365$.

The most important thing to remember about using this formula is that it assumes that we put money in the account <u>once</u> and let it sit there earning interest.

Example 4

A certificate of deposit (CD) is a savings instrument that many banks offer. It usually gives a higher interest rate, but you cannot access your investment for a specified length of time. Suppose you deposit $3000 in a CD paying 6% interest, compounded monthly. How much will you have in the account after 20 years?

In this example,

$P_0 = \$3000$ the initial deposit

$r = 0.06$ 6% annual rate

$k = 12$ 12 months in 1 year

$N = 20$ since we're looking for how much we'll have after 20 years

So $P_{20} = 3000\left(1 + \dfrac{0.06}{12}\right)^{20 \times 12} = \$9930.61$ (round your answer to the nearest penny)

Let us compare the amount of money earned from compounding against the amount you would earn from simple interest

Years	Simple Interest ($15 per month)	6% compounded monthly = 0.5% each month.
5	$3900	$4046.55
10	$4800	$5458.19
15	$5700	$7362.28
20	$6600	$9930.61
25	$7500	$13394.91
30	$8400	$18067.73
35	$9300	$24370.65

As you can see, over a long period of time, compounding makes a large difference in the account balance. You may recognize this as the difference between linear growth and exponential growth.

202

> **Evaluating exponents on the calculator**
> When we need to calculate something like 5^3 it is easy enough to just multiply
> $5 \cdot 5 \cdot 5 = 125$. But when we need to calculate something like 1.005^{240}, it would be very
> tedious to calculate this by multiplying 1.005 by itself 240 times! So to make things
> easier, we can harness the power of our scientific calculators.
>
> Most scientific calculators have a button for exponents. It is typically either labeled
> like:
>
> $\boxed{\wedge}$, $\boxed{y^x}$, or $\boxed{x^y}$.
>
> To evaluate 1.005^{240} we'd type 1.005 $\boxed{\wedge}$ 240, or 1.005 $\boxed{y^x}$ 240. Try it out - you should
> get something around 3.3102044758.

Example 5

> You know that you will need \$40,000 for your child's education in 18 years. If your account
> earns 4% compounded quarterly, how much would you need to deposit now to reach your
> goal?
>
> In this example,
> We're looking for P_0.
> $r = 0.04$ 4%
> $k = 4$ 4 quarters in 1 year
> $N = 18$ Since we know the balance in 18 years
> $P_{18} = \$40,000$ The amount we have in 18 years
>
> In this case, we're going to have to set up the equation, and solve for P_0.
> $$40000 = P_0 \left(1 + \frac{0.04}{4}\right)^{18 \times 4}$$
> $$40000 = P_0 (2.0471)$$
> $$P_0 = \frac{40000}{2.0471} = \$19539.84$$
>
> So you would need to deposit \$19,539.84 now to have \$40,000 in 18 years.

> **Rounding**
> It is important to be very careful about rounding when calculating things with
> exponents. In general, you want to keep as many decimals during calculations as you
> can. Be sure to **keep at least 3 significant digits** (numbers after any leading zeros).
> Rounding 0.00012345 to 0.000123 will usually give you a "close enough" answer, but
> keeping more digits is always better.

Example 6

To see why not over-rounding is so important, suppose you were investing $1000 at 5% interest compounded monthly for 30 years.

$P_0 = \$1000$ the initial deposit
$r = 0.05$ 5%
$k = 12$ 12 months in 1 year
$N = 30$ since we're looking for the amount after 30 years
If we first compute r/k, we find $0.05/12 = 0.00416666666667$

Here is the effect of rounding this to different values:

r/k rounded to:	Gives P_{30} to be:	Error
0.004	$4208.59	$259.15
0.0042	$4521.45	$53.71
0.00417	$4473.09	$5.35
0.004167	$4468.28	$0.54
0.0041667	$4467.80	$0.06
no rounding	$4467.74	

If you're working in a bank, of course you wouldn't round at all. For our purposes, the answer we got by rounding to 0.00417, three significant digits, is close enough - $5 off of $4500 isn't too bad. Certainly keeping that fourth decimal place wouldn't have hurt.

Using your calculator

In many cases, you can avoid rounding completely by how you enter things in your calculator. For example, in the example above, we needed to calculate

$$P_{30} = 1000\left(1 + \frac{0.05}{12}\right)^{12 \times 30}$$

We can quickly calculate $12 \times 30 = 360$, giving $P_{30} = 1000\left(1 + \frac{0.05}{12}\right)^{360}$.

Now we can use the calculator.

Type this	Calculator shows
0.05 $\div$ 12 $=$	0.00416666666667
$+$ 1 $=$	1.00416666666667
y^x 360 $=$	4.46774431400613
$\times$ 1000 $=$	4467.74431400613

Using your calculator continued

The previous steps were assuming you have a "one operation at a time" calculator; a more advanced calculator will often allow you to type in the entire expression to be evaluated. If you have a calculator like this, you will probably just need to enter:

1000 $\boxed{\times}$ (1 $\boxed{+}$ 0.05 $\boxed{\div}$ 12) $\boxed{y^x}$ 360 $\boxed{=}$

Annuities

For most of us, we aren't able to put a large sum of money in the bank today. Instead, we save for the future by depositing a smaller amount of money from each paycheck into the bank. This idea is called a **savings annuity**. Most retirement plans like 401k plans or IRA plans are examples of savings annuities.

An annuity can be described recursively in a fairly simple way. Recall that basic compound interest follows from the relationship

$$P_m = \left(1+\frac{r}{k}\right)P_{m-1}$$

For a savings annuity, we simply need to add a deposit, d, to the account with each compounding period:

$$P_m = \left(1+\frac{r}{k}\right)P_{m-1} + d$$

Taking this equation from recursive form to explicit form is a bit trickier than with compound interest. It will be easiest to see by working with an example rather than working in general.

Suppose we will deposit $100 each month into an account paying 6% interest. We assume that the account is compounded with the same frequency as we make deposits unless stated otherwise. In this example:

$r = 0.06$ (6%)
$k = 12$ (12 compounds/deposits per year)
$d = \$100$ (our deposit per month)

Writing out the recursive equation gives

$$P_m = \left(1+\frac{0.06}{12}\right)P_{m-1} + 100 = (1.005)P_{m-1} + 100$$

Assuming we start with an empty account, we can begin using this relationship:

$$P_0 = 0$$

$$P_1 = (1.005) P_0 + 100 = 100$$

$$P_2 = (1.005) P_1 + 100 = (1.005)(100) + 100 = 100(1.005) + 100$$

$$P_3 = (1.005) P_2 + 100 = (1.005)(100(1.005) + 100) + 100 = 100(1.005)^2 + 100(1.005) + 100$$

Continuing this pattern, after m deposits, we'd have saved:

$$P_m = 100(1.005)^{m-1} + 100(1.005)^{m-2} + \cdots + 100(1.005) + 100$$

In other words, after m months, the first deposit will have earned compound interest for m-1 months. The second deposit will have earned interest for m-2 months. Last months deposit would have earned only one month worth of interest. The most recent deposit will have earned no interest yet.

This equation leaves a lot to be desired, though – it doesn't make calculating the ending balance any easier! To simplify things, multiply both sides of the equation by 1.005:

$$1.005 P_m = 1.005 \left(100(1.005)^{m-1} + 100(1.005)^{m-2} + \cdots + 100(1.005) + 100 \right)$$

Distributing on the right side of the equation gives

$$1.005 P_m = 100(1.005)^m + 100(1.005)^{m-1} + \cdots + 100(1.005)^2 + 100(1.005)$$

Now we'll line this up with like terms from our original equation, and subtract each side

$$
\begin{aligned}
1.005 P_m &= 100(1.005)^m + 100(1.005)^{m-1} + \cdots + 100(1.005) \\
P_m &= \qquad\qquad\quad 100(1.005)^{m-1} + \cdots + 100(1.005) + 100
\end{aligned}
$$

Almost all the terms cancel on the right hand side when we subtract, leaving

$$1.005 P_m - P_m = 100(1.005)^m - 100$$

Solving for P_m

$$0.005 P_m = 100 \left((1.005)^m - 1 \right)$$

$$P_m = \frac{100 \left((1.005)^m - 1 \right)}{0.005}$$

Replacing m months with $12N$, where N is measured in years, gives

$$P_N = \frac{100\left((1.005)^{12N} - 1\right)}{0.005}$$

Recall 0.005 was r/k and 100 was the deposit d. 12 was k, the number of deposit each year. Generalizing this result, we get the saving annuity formula.

Annuity Formula

$$P_N = \frac{d\left(\left(1 + \frac{r}{k}\right)^{Nk} - 1\right)}{\left(\frac{r}{k}\right)}$$

P_N is the balance in the account after N years.
d is the regular deposit (the amount you deposit each year, each month, etc.)
r is the annual interest rate in decimal form.
k is the number of compounding periods in one year.

If the compounding frequency is not explicitly stated, assume there are the same number of compounds in a year as there are deposits made in a year.

For example, if the compounding frequency isn't stated:
If you make your deposits every month, use monthly compounding, $k = 12$.
If you make your deposits every year, use yearly compounding, $k = 1$.
If you make your deposits every quarter, use quarterly compounding, $k = 4$.
Etc.

When do you use this
Annuities assume that you put money in the account <u>on a regular schedule (every month, year, quarter, etc.)</u> and let it sit there earning interest.

Compound interest assumes that you put money in the account <u>once</u> and let it sit there earning interest.

Compound interest: <u>One</u> deposit
Annuity: <u>Many</u> deposits.

Example 7

A traditional individual retirement account (IRA) is a special type of retirement account in which the money you invest is exempt from income taxes until you withdraw it. If you deposit $100 each month into an IRA earning 6% interest, how much will you have in the account after 20 years?

In this example,
$d = \$100$ the monthly deposit
$r = 0.06$ 6% annual rate
$k = 12$ since we're doing monthly deposits, we'll compound monthly
$N = 20$ we want the amount after 20 years

Putting this into the equation:

$$P_{20} = \frac{100\left(\left(1+\dfrac{0.06}{12}\right)^{20(12)}-1\right)}{\left(\dfrac{0.06}{12}\right)}$$

$$P_{20} = \frac{100\left((1.005)^{240}-1\right)}{(0.005)}$$

$$P_{20} = \frac{100(3.310-1)}{(0.005)}$$

$$P_{20} = \frac{100(2.310)}{(0.005)} = \$46200$$

The account will grow to $46,200 after 20 years.

Notice that you deposited into the account a total of $24,000 ($100 a month for 240 months). The difference between what you end up with and how much you put in is the <u>interest earned</u>. In this case it is $46,200 - $24,000 = $22,200.

Example 8

You want to have $200,000 in your account when you retire in 30 years. Your retirement account earns 8% interest. How much do you need to deposit each month to meet your retirement goal?

In this example,
We're looking for d
$r = 0.08$ 8% annual rate
$k = 12$ since we're depositing monthly
$N = 30$ 30 years
$P_{30} = \$200,000$ The amount we want to have in 30 years

In this case, we're going to have to set up the equation, and solve for d.

$$200,000 = \frac{d\left(\left(1+\dfrac{0.08}{12}\right)^{30(12)} - 1\right)}{\left(\dfrac{0.08}{12}\right)}$$

$$200,000 = \frac{d\left((1.00667)^{360} - 1\right)}{(0.00667)}$$

$$200,000 = d(1491.57)$$

$$d = \frac{200,000}{1491.57} = \$134.09$$

So you would need to deposit $134.09 each month to have $200,000 in 30 years if your account earns 8% interest

---

Try it Now 2
A more conservative investment account pays 3% interest. If you deposit $5 a day into this account, how much will you have after 10 years? How much is from interest?

Payout Annuities

In the last section you learned about annuities. In an annuity, you start with nothing, put money into an account on a regular basis, and end up with money in your account.

In this section, we will learn about a variation called a **Payout Annuity**. With a payout annuity, you start with money in the account, and pull money out of the account on a regular basis. Any remaining money in the account earns interest. After a fixed amount of time, the account will end up empty.

Payout annuities are typically used after retirement. Perhaps you have saved $500,000 for retirement, and want to take money out of the account each month to live on. You want the money to last you 20 years. This is a payout annuity. The formula is derived in a similar way as we did for savings annuities. The details are omitted here.

Payout Annuity Formula

$$P_0 = \frac{d\left(1-\left(1+\dfrac{r}{k}\right)^{-Nk}\right)}{\left(\dfrac{r}{k}\right)}$$

P_0 is the balance in the account at the beginning (starting amount, or principal).
d is the regular withdrawal (the amount you take out each year, each month, etc.)
r is the annual interest rate (in decimal form. Example: 5% = 0.05)
k is the number of compounding periods in one year.
N is the number of years we plan to take withdrawals

Like with annuities, the compounding frequency is not always explicitly given, but is determined by how often you take the withdrawals.

When do you use this
Payout annuities assume that you <u>take</u> money from the account <u>on a regular schedule</u> <u>(every month, year, quarter, etc.)</u> and let the rest sit there earning interest.

Compound interest: <u>One</u> deposit
Annuity: <u>Many</u> deposits.
Payout Annuity: <u>Many withdrawals</u>

Example 9

After retiring, you want to be able to take $1000 every month for a total of 20 years from your retirement account. The account earns 6% interest. How much will you need in your account when you retire?

In this example,
$d = \$1000$	the monthly withdrawal
$r = 0.06$	6% annual rate
$k = 12$	since we're doing monthly withdrawals, we'll compound monthly
$N = 20$	since were taking withdrawals for 20 years

We're looking for P_0; how much money needs to be in the account at the beginning.
Putting this into the equation:

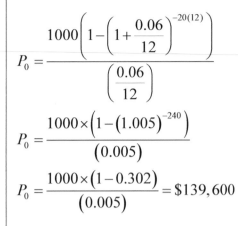

$$P_0 = \frac{1000 \times \left(1-(1.005)^{-240}\right)}{(0.005)}$$

$$P_0 = \frac{1000 \times (1-0.302)}{(0.005)} = \$139,600$$

You will need to have $139,600 in your account when you retire.

Notice that you withdrew a total of $240,000 ($1000 a month for 240 months). The difference between what you pulled out and what you started with is the interest earned. In this case it is $240,000 - $139,600 = $100,400 in interest.

Evaluating negative exponents on your calculator

With these problems, you need to raise numbers to negative powers. Most calculators have a separate button for negating a number that is different than the subtraction button. Some calculators label this (-), some with +/- . The button is often near the = key or the decimal point.

If your calculator displays operations on it (typically a calculator with multiline display), to calculate 1.005^{-240} you'd type something like: 1.005 [^] [(-)] 240

If your calculator only shows one value at a time, then usually you hit the (-) key after a number to negate it, so you'd hit: 1.005 [y^x] 240 [(-)] =

Give it a try - you should get $1.005^{-240} = 0.302096$

Example 10

You know you will have $500,000 in your account when you retire. You want to be able to take monthly withdrawals from the account for a total of 30 years. Your retirement account earns 8% interest. How much will you be able to withdraw each month?

In this example,
We're looking for d.
$r = 0.08$ 8% annual rate
$k = 12$ since we're withdrawing monthly
$N = 30$ 30 years
$P_0 = \$500,000$ we are beginning with $500,000

In this case, we're going to have to set up the equation, and solve for d.

$$500,000 = \frac{d\left(1-\left(1+\dfrac{0.08}{12}\right)^{-30(12)}\right)}{\left(\dfrac{0.08}{12}\right)}$$

$$500,000 = \frac{d\left(1-(1.00667)^{-360}\right)}{(0.00667)}$$

$$500,000 = d(136.232)$$

$$d = \frac{500,000}{136.232} = \$3670.21$$

You would be able to withdraw $3,670.21 each month for 30 years.

Try it Now 3
A donor gives $100,000 to a university, and specifies that it is to be used to give annual scholarships for the next 20 years. If the university can earn 4% interest, how much can they give in scholarships each year?

Loans

In the last section, you learned about payout annuities.

In this section, you will learn about conventional loans (also called amortized loans or installment loans). Examples include auto loans and home mortgages. These techniques do not apply to payday loans, add-on loans, or other loan types where the interest is calculated up front.

One great thing about loans is that they use exactly the same formula as a payout annuity. To see why, imagine that you had $10,000 invested at a bank, and started taking out payments while earning interest as part of a payout annuity, and after 5 years your balance was zero. Flip that around, and imagine that you are acting as the bank, and a car lender is acting as you. The car lender invests $10,000 in you. Since you're acting as the bank, you pay interest. The car lender takes payments until the balance is zero.

Loans Formula

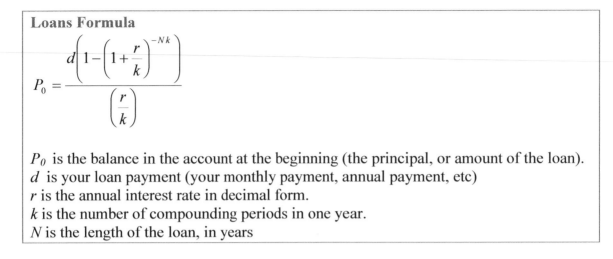

$$P_0 = \dfrac{d\left(1-\left(1+\dfrac{r}{k}\right)^{-Nk}\right)}{\left(\dfrac{r}{k}\right)}$$

P_0 is the balance in the account at the beginning (the principal, or amount of the loan).
d is your loan payment (your monthly payment, annual payment, etc)
r is the annual interest rate in decimal form.
k is the number of compounding periods in one year.
N is the length of the loan, in years

Like before, the compounding frequency is not always explicitly given, but is determined by how often you make payments.

When do you use this
The loan formula assumes that you make loan payments <u>on a regular schedule (every month, year, quarter, etc.)</u> and are paying interest on the loan.

Compound interest: <u>One</u> deposit
Annuity: <u>Many</u> deposits.
Payout Annuity: <u>Many withdrawals</u>
Loans: <u>Many payments</u>

Example 11

You can afford $200 per month as a car payment. If you can get an auto loan at 3% interest for 60 months (5 years), how expensive of a car can you afford? In other words, what amount loan can you pay off with $200 per month?

In this example,
$d = \$200$ the monthly loan payment
$r = 0.03$ 3% annual rate
$k = 12$ since we're doing monthly payments, we'll compound monthly
$N = 5$ since we're making monthly payments for 5 years

We're looking for P_0, the starting amount of the loan.

$$P_0 = \frac{200\left(1-\left(1+\dfrac{0.03}{12}\right)^{-5(12)}\right)}{\left(\dfrac{0.03}{12}\right)}$$

$$P_0 = \frac{200\left(1-(1.0025)^{-60}\right)}{(0.0025)}$$

$$P_0 = \frac{200(1-0.861)}{(0.0025)} = \$11,120$$

You can afford a $11,120 loan.

You will pay a total of $12,000 ($200 per month for 60 months) to the loan company. The difference between the amount you pay and the amount of the loan is the <u>interest paid</u>. In this case, you're paying $12,000-$11,120 = $880 interest total.

Example 12

You want to take out a $140,000 mortgage (home loan). The interest rate on the loan is 6%, and the loan is for 30 years. How much will your monthly payments be?

In this example,
We're looking for d
$r = 0.06$ 6% annual rate
$k = 12$ since we're paying monthly
$N = 30$ 30 years
$P_0 = \$140,000$ the starting loan amount

In this case, we're going to have to set up the equation, and solve for d.

$$140,000 = \frac{d\left(1-\left(1+\dfrac{0.06}{12}\right)^{-30(12)}\right)}{\left(\dfrac{0.06}{12}\right)}$$

$$140,000 = \frac{d\left(1-(1.005)^{-360}\right)}{(0.005)}$$

$$140,000 = d(166.792)$$

$$d = \frac{140,000}{166.792} = \$839.37$$

You will make payments of $839.37 per month for 30 years.

You're paying a total of $302,173.20 to the loan company: $839.37 per month for 360 months. You are paying a total of $302,173.20 - $140,000 = $162,173.20 in interest over the life of the loan.

Janine bought $3,000 of new furniture on credit. Because her credit score isn't very good, the store is charging her a fairly high interest rate on the loan: 16%. If she agreed to pay off the furniture over 2 years, how much will she have to pay each month?

Remaining Loan Balance

With loans, it is often desirable to determine what the remaining loan balance will be after some number of years. For example, if you purchase a home and plan to sell it in five years, you might want to know how much of the loan balance you will have paid off and how much you have to pay from the sale.

To determine the remaining loan balance after some number of years, we first need to know the loan payments, if we don't already know them. Remember that only a portion of your loan payments go towards the loan balance; a portion is going to go towards interest. For example, if your payments were $1,000 a month, after a year you will *not* have paid off $12,000 of the loan balance.

To determine the remaining loan balance, we can think "how much loan will these loan payments be able to pay off in the remaining time on the loan?"

Example 13

If a mortgage at a 6% interest rate has payments of $1,000 a month, how much will the loan balance be 10 years from the end the loan?

To determine this, we are looking for the amount of the loan that can be paid off by $1,000 a month payments in 10 years. In other words, we're looking for P0 when

$d = \$1,000$	the monthly loan payment
$r = 0.06$	6% annual rate
$k = 12$	since we're doing monthly payments, we'll compound monthly
$N = 10$	since we're making monthly payments for 10 more years

$$P_0 = \frac{1000\left(1 - \left(1 + \frac{0.06}{12}\right)^{-10(12)}\right)}{\left(\frac{0.06}{12}\right)}$$

$$P_0 = \frac{1000\left(1 - (1.005)^{-120}\right)}{(0.005)}$$

$$P_0 = \frac{1000(1 - 0.5496)}{(0.005)} = \$90,073.45$$

The loan balance with 10 years remaining on the loan will be $90,073.45

Often times answering remaining balance questions requires two steps:
1) Calculating the monthly payments on the loan
2) Calculating the remaining loan balance based on the *remaining time* on the loan

Example 14

A couple purchases a home with a $180,000 mortgage at 4% for 30 years with monthly payments. What will the remaining balance on their mortgage be after 5 years?

First we will calculate their monthly payments.
We're looking for d.
$r = 0.04$ 4% annual rate
$k = 12$ since they're paying monthly
$N = 30$ 30 years
$P_0 = \$180,000$ the starting loan amount

We set up the equation and solve for d.

$$180,000 = \frac{d\left(1 - \left(1 + \frac{0.04}{12}\right)^{-30(12)}\right)}{\left(\frac{0.04}{12}\right)}$$

$$180,000 = \frac{d\left(1 - (1.00333)^{-360}\right)}{(0.00333)}$$

$$180,000 = d(209.562)$$

$$d = \frac{180,000}{209.562} = \$858.93$$

Now that we know the monthly payments, we can determine the remaining balance. We want the remaining balance after 5 years, when 25 years will be remaining on the loan, so we calculate the loan balance that will be paid off with the monthly payments over those 25 years.

$d = \$858.93$ — the monthly loan payment we calculated above
$r = 0.04$ — 4% annual rate
$k = 12$ — since they're doing monthly payments
$N = 25$ — since they'd be making monthly payments for 25 more years

$$P_0 = \frac{858.93\left(1 - \left(1 + \frac{0.04}{12}\right)^{-25(12)}\right)}{\left(\frac{0.04}{12}\right)}$$

$$P_0 = \frac{858.93\left(1 - (1.00333)^{-300}\right)}{(0.00333)}$$

$$P_0 = \frac{858.93(1 - 0.369)}{(0.00333)} = \$155,793.91$$

The loan balance after 5 years, with 25 years remaining on the loan, will be $155,793.91

Over that 5 years, the couple has paid off $180,000 - $155,793.91 = $24,206.09 of the loan balance. They have paid a total of $858.93 a month for 5 years (60 months), for a total of $51,535.80, so $51,535.80 - $24,206.09 = $27,329.71 of what they have paid so far has been interest.

Which equation to use?

When presented with a finance problem (on an exam or in real life), you're usually not told what type of problem it is or which equation to use. Here are some hints on deciding which equation to use based on the wording of the problem.

The easiest types of problem to identify are loans. Loan problems almost always include words like: "loan", "amortize" (the fancy word for loans), "finance (a car)", or "mortgage" (a home loan). Look for these words. If they're there, you're probably looking at a loan problem. To make sure, see if you're given what your monthly (or annual) payment is, or if you're trying to find a monthly payment.

If the problem is not a loan, the next question you want to ask is: "Am I putting money in an account and letting it sit, or am I making regular (monthly/annually/quarterly) payments or withdrawals?" If you're letting the money sit in the account with nothing but interest changing the balance, then you're looking at a compound interest problem. The exception would be bonds and other investments where the interest is not reinvested; in those cases you're looking at simple interest.

If you're making regular payments or withdrawals, the next questions is: "Am I putting money into the account, or am I pulling money out?" If you're putting money into the account on a regular basis (monthly/annually/quarterly) then you're looking at a basic Annuity problem. Basic annuities are when you are saving money. Usually in an annuity problem, your account starts empty, and has money in the future.

If you're pulling money out of the account on a regular basis, then you're looking at a Payout Annuity problem. Payout annuities are used for things like retirement income, where you start with money in your account, pull money out on a regular basis, and your account ends up empty in the future.

Remember, the most important part of answering any kind of question, money or otherwise, is first to correctly identify what the question is really asking, and to determine what approach will best allow you to solve the problem.

Try it Now 5

For each of the following scenarios, determine if it is a compound interest problem, a savings annuity problem, a payout annuity problem, or a loans problem. Then solve each problem.

    a. Marcy received an inheritance of $20,000, and invested it at 6% interest. She is going to use it for college, withdrawing money for tuition and expenses each quarter. How much can she take out each quarter if she has 3 years of school left?

    b. Paul wants to buy a new car. Rather than take out a loan, he decides to save $200 a month in an account earning 3% interest compounded monthly. How much will he have saved up after 3 years?

    c. Keisha is managing investments for a non-profit company. They want to invest some money in an account earning 5% interest compounded annually with the goal to have $30,000 in the account in 6 years. How much should Keisha deposit into the account?

    d. Miao is going to finance new office equipment at a 2% rate over a 4 year term. If she can afford monthly payments of $100, how much new equipment can she buy?

    e. How much would you need to save every month in an account earning 4% interest to have $5,000 saved up in two years?

Solving for time

Often we are interested in how long it will take to accumulate money or how long we'd need to extend a loan to bring payments down to a reasonable level.

Note: This section assumes you've covered solving exponential equations using logarithms, either in prior classes or in the growth models chapter.

218

Example 15

If you invest $2000 at 6% compounded monthly, how long will it take the account to double in value?

This is a compound interest problem, since we are depositing money once and allowing it to grow. In this problem,

$P_0 = \$2000$ the initial deposit
$r = 0.06$ 6% annual rate
$k = 12$ 12 months in 1 year

So our general equation is $P_N = 2000\left(1 + \dfrac{0.06}{12}\right)^{N \times 12}$. We also know that we want our ending

amount to be double of $2000, which is $4000, so we're looking for N so that $P_N = 4000$. To solve this, we set our equation for P_N equal to 4000.

$4000 = 2000\left(1 + \dfrac{0.06}{12}\right)^{N \times 12}$ Divide both sides by 2000

$2 = (1.005)^{12N}$ To solve for the exponent, take the log of both sides

$\log(2) = \log\left((1.005)^{12N}\right)$ Use the exponent property of logs on the right side

$\log(2) = 12N \log(1.005)$ Now we can divide both sides by 12log(1.005)

$\dfrac{\log(2)}{12 \log(1.005)} = N$ Approximating this to a decimal

$N = 11.581$

It will take about 11.581 years for the account to double in value. Note that your answer may come out slightly differently if you had evaluated the logs to decimals and rounded during your calculations, but your answer should be close. For example if you rounded log(2) to 0.301 and log(1.005) to 0.00217, then your final answer would have been about 11.577 years.

Example 16

If you invest $100 each month into an account earning 3% compounded monthly, how long will it take the account to grow to $10,000?

This is a savings annuity problem since we are making regular deposits into the account.
$d = \$100$ the monthly deposit
$r = 0.03$ 3% annual rate
$k = 12$ since we're doing monthly deposits, we'll compound monthly
We don't know N, but we want P_N to be $10,000.

Putting this into the equation:

$$10,000 = \frac{100\left(\left(1+\dfrac{0.03}{12}\right)^{N(12)}-1\right)}{\left(\dfrac{0.03}{12}\right)}$$ Simplifying the fractions a bit

$$10,000 = \frac{100\left((1.0025)^{12N}-1\right)}{0.0025}$$

We want to isolate the exponential term, 1.0025^{12N}, so multiply both sides by 0.0025

$25 = 100\left((1.0025)^{12N}-1\right)$ Divide both sides by 100

$0.25 = (1.0025)^{12N}-1$ Add 1 to both sides

$1.25 = (1.0025)^{12N}$ Now take the log of both sides

$\log(1.25) = \log\left((1.0025)^{12N}\right)$ Use the exponent property of logs

$\log(1.25) = 12N\log(1.0025)$ Divide by 12log(1.0025)

$\dfrac{\log(1.25)}{12\log(1.0025)} = N$ Approximating to a decimal

$N = 7.447$ years

It will take about 7.447 years to grow the account to $10,000.

Try it Now 6
Joel is considering putting a $1,000 laptop purchase on his credit card, which has an interest rate of 12% compounded monthly. How long will it take him to pay off the purchase if he makes payments of $30 a month?

Try it Now Answers

1.

I = \$30 of interest

P_0 = \$500 principal

r = unknown

t = 1 month

Using $I = P_0rt$, we get $30 = 500 \cdot r \cdot 1$. Solving, we get $r = 0.06$, or 6%. Since the time was monthly, this is the monthly interest. The annual rate would be 12 times this: 72% interest.

2.

d = \$5 — the daily deposit

r = 0.03 — 3% annual rate

k = 365 — since we're doing daily deposits, we'll compound daily

N = 10 — we want the amount after 10 years

$$P_{10} = \frac{5\left(\left(1 + \dfrac{0.03}{365}\right)^{365\times10} - 1\right)}{\dfrac{0.03}{365}} = \$21{,}282.07$$

We would have deposited a total of $\$5 \cdot 365 \cdot 10 = \$18{,}250$, so $3,032.07$ is from interest

3.

d = unknown

r = 0.04 — 4% annual rate

k = 1 — since we're doing annual scholarships

N = 20 — 20 years

P_0 = 100,000 — we're starting with \$100,000

$$100{,}000 = \frac{d\left(1 - \left(1 + \dfrac{0.04}{1}\right)^{-20\times1}\right)}{\dfrac{0.04}{1}}$$

Solving for d gives \$7,358.18 each year that they can give in scholarships.

It is worth noting that usually donors instead specify that only interest is to be used for scholarship, which makes the original donation last indefinitely. If this donor had specified that, $\$100{,}000(0.04) = \$4{,}000$ a year would have been available.

4.

d = unknown

$r = 0.16$ 16% annual rate

$k = 12$ since we're making monthly payments

$N = 2$ 2 years to repay

$P_0 = 3{,}000$ we're starting with a $3,000 loan

$$3{,}000 = \dfrac{d\left(1 - \left(1 + \dfrac{0.16}{12}\right)^{-2 \times 12}\right)}{\dfrac{0.16}{12}}$$

Solving for d gives $146.89 as monthly payments.

In total, she will pay $3,525.36 to the store, meaning she will pay $525.36 in interest over the two years.

5.

     a. This is a payout annuity problem. She can pull out $1833.60 a quarter.

     b. This is a savings annuity problem. He will have saved up $7,524.11/

     c. This is compound interest problem. She would need to deposit $22,386.46.

     d. This is a loans problem. She can buy $4,609.33 of new equipment.

     e. This is a savings annuity problem. You would need to save $200.46 each month

6.

$d = \$30$ The monthly payments

$r = 0.12$ 12% annual rate

$k = 12$ since we're making monthly payments

$P_0 = 1{,}000$ we're starting with a $1,000 loan

We are solving for N, the time to pay off the loan

$$1{,}000 = \dfrac{30\left(1 - \left(1 + \dfrac{0.12}{12}\right)^{-N(12)}\right)}{\dfrac{0.12}{12}}$$

Solving for N gives 3.396. It will take about 3.4 years to pay off the purchase.

Exercises

Skills

1. A friend lends you $200 for a week, which you agree to repay with 5% one-time interest. How much will you have to repay?

2. Suppose you obtain a $3,000 T-note with a 3% annual rate, paid quarterly, with maturity in 5 years. How much interest will you earn?

3. A T-bill is a type of bond that is sold at a discount over the face value. For example, suppose you buy a 13-week T-bill with a face value of $10,000 for $9,800. This means that in 13 weeks, the government will give you the face value, earning you $200. What annual interest rate have you earned?

4. Suppose you are looking to buy a $5000 face value 26-week T-bill. If you want to earn at least 1% annual interest, what is the most you should pay for the T-bill?

5. You deposit $300 in an account earning 5% interest compounded annually. How much will you have in the account in 10 years?

6. How much will $1000 deposited in an account earning 7% interest compounded annually be worth in 20 years?

7. You deposit $2000 in an account earning 3% interest compounded monthly.
   a. How much will you have in the account in 20 years?
   b. How much interest will you earn?

8. You deposit $10,000 in an account earning 4% interest compounded monthly.
   a. How much will you have in the account in 25 years?
   b. How much interest will you earn?

9. How much would you need to deposit in an account now in order to have $6,000 in the account in 8 years? Assume the account earns 6% interest compounded monthly.

10. How much would you need to deposit in an account now in order to have $20,000 in the account in 4 years? Assume the account earns 5% interest.

11. You deposit $200 each month into an account earning 3% interest compounded monthly.
   a. How much will you have in the account in 30 years?
   b. How much total money will you put into the account?
   c. How much total interest will you earn?

12. You deposit $1000 each year into an account earning 8% compounded annually.
   a. How much will you have in the account in 10 years?
   b. How much total money will you put into the account?
   c. How much total interest will you earn?

13. Jose has determined he needs to have $800,000 for retirement in 30 years.   His account earns 6% interest.
    a.   How much would you need to deposit in the account each month?
    b.   How much total money will you put into the account?
    c.   How much total interest will you earn?

14. You wish to have $3000 in 2 years to buy a fancy new stereo system. How much should you deposit each quarter into an account paying 8% compounded quarterly?

15. You want to be able to withdraw $30,000 each year for 25 years. Your account earns 8% interest.
    a.   How much do you need in your account at the beginning
    b.   How much total money will you pull out of the account?
    c.   How much of that money is interest?

16. How much money will I need to have at retirement so I can withdraw $60,000 a year for 20 years from an account earning 8% compounded annually?
    a.   How much do you need in your account at the beginning
    b.   How much total money will you pull out of the account?
    c.   How much of that money is interest?

17. You have $500,000 saved for retirement. Your account earns 6% interest. How much will you be able to pull out each month, if you want to be able to take withdrawals for 20 years?

18. Loren already knows that he will have $500,000 when he retires. If he sets up a payout annuity for 30 years in an account paying 10% interest, how much could the annuity provide each month?

19. You can afford a $700 per month mortgage payment. You've found a 30 year loan at 5% interest.
    a.   How big of a loan can you afford?
    b.   How much total money will you pay the loan company?
    c.   How much of that money is interest?

20. Marie can afford a $250 per month car payment. She's found a 5 year loan at 7% interest.
    a.   How expensive of a car can she afford?
    b.   How much total money will she pay the loan company?
    c.   How much of that money is interest?

21. You want to buy a $25,000 car. The company is offering a 2% interest rate for 48 months (4 years). What will your monthly payments be?

22. You decide finance a $12,000 car at 3% compounded monthly for 4 years. What will your monthly payments be? How much interest will you pay over the life of the loan?

23. You want to buy a $200,000 home. You plan to pay 10% as a down payment, and take out a 30 year loan for the rest.
    a. How much is the loan amount going to be?
    b. What will your monthly payments be if the interest rate is 5%?
    c. What will your monthly payments be if the interest rate is 6%?

24. Lynn bought a $300,000 house, paying 10% down, and financing the rest at 6% interest for 30 years.
    a. Find her monthly payments.
    b. How much interest will she pay over the life of the loan?

25. Emile bought a car for $24,000 three years ago. The loan had a 5 year term at 3% interest rate. How much does he still owe on the car?

26. A friend bought a house 15 years ago, taking out a $120,000 mortgage at 6% for 30 years. How much does she still owe on the mortgage?

27. Pat deposits $6,000 into an account earning 4% compounded monthly. How long will it take the account to grow to $10,000?

28. Kay is saving $200 a month into an account earning 5% interest. How long will it take her to save $20,000?

29. James has $3,000 in credit card debt, which charges 14% interest. How long will it take to pay off the card if he makes the minimum payment of $60 a month?

30. Chris has saved $200,000 for retirement, and it is in an account earning 6% interest. If she withdraws $3,000 a month, how long will the money last?

Concepts

31. Suppose you invest $50 a month for 5 years into an account earning 8% compounded monthly. After 5 years, you leave the money, without making additional deposits, in the account for another 25 years. How much will you have in the end?

32. Suppose you put off making investments for the first 5 years, and instead made deposits of $50 a month for 25 years into an account earning 8% compounded monthly. How much will you have in the end?

33. Mike plans to make contributions to his retirement account for 15 years. After the last contribution, he will start withdrawing $10,000 a quarter for 10 years. Assuming Mike's account earns 8% compounded quarterly, how large must his quarterly contributions be during the first 15 years, in order to accomplish his goal?

34. Kendra wants to be able to make withdrawals of $60,000 a year for 30 years after retiring in 35 years. How much will she have to save each year up until retirement if her account earns 7% interest?

35. You have $2,000 to invest, and want it to grow to $3,000 in two years. What interest rate would you need to find to make this possible?

36. You have $5,000 to invest, and want it to grow to $20,000 in ten years. What interest rate would you need to find to make this possible?

37. You plan to save $600 a month for the next 30 years for retirement. What interest rate would you need to have $1,000,000 at retirement?

38. You really want to buy a used car for $11,000, but can only afford $200 a month. What interest rate would you need to find to be able to afford the car, assuming the loan is for 60 months?

Exploration

39. Pay day loans are short term loans that you take out against future paychecks: The company advances you money against a future paycheck. Either visit a pay day loan company, or look one up online. Be forewarned that many companies do not make their fees obvious, so you might need to do some digging or look at several companies.
    a. Explain the general method by which the loan works.
    b. We will assume that we need to borrow $500 and that we will pay back the loan in 14 days. Determine the total amount that you would need to pay back and the effective loan rate. The effective loan rate is the percentage of the original loan amount that you pay back. It is not the same as the APR (annual rate) that is probably published.
    c. If you cannot pay back the loan after 14 days, you will need to get an extension for another 14 days. Determine the fees for an extension, determine the total amount you will be paying for the now 28 day loan, and compute the effective loan rate.

40. Suppose that 10 years ago you bought a home for $110,000, paying 10% as a down payment, and financing the rest at 9% interest for 30 years.
   a. Let's consider your existing mortgage:
      i. How much money did you pay as your down payment?
      ii. How much money was your mortgage (loan) for?
      iii. What is your current monthly payment?
      iv. How much total interest will you pay over the life of the loan?
   b. This year, you check your loan balance. Only part of your payments have been going to pay down the loan; the rest has been going towards interest. You see that you still have $88,536 left to pay on your loan. Your house is now valued at $150,000.
      i. How much of the loan have you paid off? (i.e., how much have you reduced the loan balance by? Keep in mind that interest is charged each month - it's not part of the loan balance.)
      ii. How much money have you paid to the loan company so far?
      iii. How much interest have you paid so far?
      iv. How much equity do you have in your home (equity is value minus remaining debt)
   c. Since interest rates have dropped, you consider refinancing your mortgage at a lower 6% rate.
      i. If you took out a new 30 year mortgage at 6% for your remaining loan balance, what would your new monthly payments be?
      ii. How much interest will you pay over the life of the new loan?
   d. Notice that if you refinance, you are going to be making payments on your home for another 30 years. In addition to the 10 years you've already been paying, that's 40 years total.
      i. How much will you save each month because of the lower monthly payment?
      ii. How much total interest will you be paying (you need to consider the amount from 2c and 3b)
      iii. Does it make sense to refinance? (there isn't a correct answer to this question. Just give your opinion and your reason)

Statistics

Like most people, you probably feel that it is important to "take control of your life." But what does this mean? Partly it means being able to properly evaluate the data and claims that bombard you every day. If you cannot distinguish good from faulty reasoning, then you are vulnerable to manipulation and to decisions that are not in your best interest. Statistics provides tools that you need in order to react intelligently to information you hear or read. In this sense, Statistics is one of the most important things that you can study.

To be more specific, here are some claims that we have heard on several occasions. (We are *not* saying that each one of these claims is true!)

- 4 out of 5 dentists recommend Dentyne.
- Almost 85% of lung cancers in men and 45% in women are tobacco-related.
- Condoms are effective 94% of the time.
- Native Americans are significantly more likely to be hit crossing the streets than are people of other ethnicities.
- People tend to be more persuasive when they look others directly in the eye and speak loudly and quickly.
- Women make 75 cents to every dollar a man makes when they work the same job.
- A surprising new study shows that eating egg whites can increase one's life span.
- People predict that it is very unlikely there will ever be another baseball player with a batting average over 400.
- There is an 80% chance that in a room full of 30 people that at least two people will share the same birthday.
- 79.48% of all statistics are made up on the spot.

All of these claims are statistical in character. We suspect that some of them sound familiar; if not, we bet that you have heard other claims like them. Notice how diverse the examples are; they come from psychology, health, law, sports, business, etc. Indeed, data and data-interpretation show up in discourse from virtually every facet of contemporary life.

Statistics are often presented in an effort to add credibility to an argument or advice. You can see this by paying attention to television advertisements. Many of the numbers thrown about in this way do not represent careful statistical analysis. They can be misleading, and push you into decisions that you might find cause to regret. For these reasons, learning about statistics is a long step towards taking control of your life. (It is not, of course, the only step needed for this purpose.) These chapters will help you learn statistical essentials. It will make you into an intelligent consumer of statistical claims.

You can take the first step right away. To be an intelligent consumer of statistics, your first reflex must be to question the statistics that you encounter. The British Prime Minister Benjamin Disraeli famously said, "There are three kinds of lies -- lies, damned lies, and statistics." This quote reminds us why it is so important to understand statistics. So let us invite you to reform your statistical habits from now on. No longer will you blindly accept numbers or findings. Instead, you will begin to think about the numbers, their sources, and most importantly, the procedures used to generate them.

We have put the emphasis on defending ourselves against fraudulent claims wrapped up as statistics. Just as important as detecting the deceptive use of statistics is the appreciation of the proper use of statistics. You must also learn to recognize statistical evidence that supports a stated conclusion. When a research team is testing a new treatment for a disease, statistics allows them to conclude based on a relatively small trial that there is good evidence their drug is effective. Statistics allowed prosecutors in the 1950's and 60's to demonstrate racial bias existed in jury panels. Statistics are all around you, sometimes used well, sometimes not. We must learn how to distinguish the two cases.

Populations and samples

Before we begin gathering and analyzing data we need to characterize the **population** we are studying. If we want to study the amount of money spent on textbooks by a typical first-year college student, our population might be all first-year students at your college. Or it might be:

- All first-year community college students in the state of Washington.
- All first-year students at public colleges and universities in the state of Washington.
- All first-year students at all colleges and universities in the state of Washington.
- All first-year students at all colleges and universities in the entire United States.
- And so on.

> **Population**
> The **population** of a study is the group the collected data is intended to describe.

Sometimes the intended population is called the **target population**, since if we design our study badly, the collected data might not actually be representative of the intended population.

Why is it important to specify the population? We might get different answers to our question as we vary the population we are studying. First-year students at the University of Washington might take slightly more diverse courses than those at your college, and some of these courses may require less popular textbooks that cost more; or, on the other hand, the University Bookstore might have a larger pool of used textbooks, reducing the cost of these books to the students. Whichever the case (and it is likely that some combination of these and other factors are in play), the data we gather from your college will probably not be the same as that from the University of Washington. Particularly when conveying our results to others, we want to be clear about the population we are describing with our data.

Example 1

A newspaper website contains a poll asking people their opinion on a recent news article. What is the population?

While the target (intended) population may have been all people, the real population of the survey is readers of the website.

If we were able to gather data on every member of our population, say the average (we will define "average" more carefully in a subsequent section) amount of money spent on textbooks by each first-year student at your college during the 2009-2010 academic year, the resulting number would be called a **parameter**.

> **Parameter**
> A **parameter** is a value (average, percentage, etc.) calculated using all the data from a population

We seldom see parameters, however, since surveying an entire population is usually very time-consuming and expensive, unless the population is very small or we already have the data collected.

> **Census**
> A survey of an entire population is called a **census**.

You are probably familiar with two common censuses: the official government Census that attempts to count the population of the U.S. every ten years, and voting, which asks the opinion of all eligible voters in a district. The first of these demonstrates one additional problem with a census: the difficulty in finding and getting participation from everyone in a large population, which can bias, or skew, the results.

There are occasionally times when a census is appropriate, usually when the population is fairly small. For example, if the manager of Starbucks wanted to know the average number of hours her employees worked last week, she should be able to pull up payroll records or ask each employee directly.

Since surveying an entire population is often impractical, we usually select a **sample** to study;

> **Sample**
> A **sample** is a smaller subset of the entire population, ideally one that is fairly representative of the whole population.

We will discuss sampling methods in greater detail in a later section. For now, let us assume that samples are chosen in an appropriate manner. If we survey a sample, say 100 first-year students at your college, and find the average amount of money spent by these students on textbooks, the resulting number is called a **statistic**.

> **Statistic**
> A **statistic** is a value (average, percentage, etc.) calculated using the data from a sample.

Example 2

A researcher wanted to know how citizens of Tacoma felt about a voter initiative. To study this, she goes to the Tacoma Mall and randomly selects 500 shoppers and asks them their opinion. 60% indicate they are supportive of the initiative. What is the sample and population? Is the 60% value a parameter or a statistic?

The sample is the 500 shoppers questioned. The population is less clear. While the intended population of this survey was Tacoma citizens, the effective population was mall shoppers. There is no reason to assume that the 500 shoppers questioned would be representative of all Tacoma citizens.

The 60% value was based on the sample, so it is a statistic.

Try it Now 1

To determine the average length of trout in a lake, researchers catch 20 fish and measure them. What is the sample and population in this study?

Try it Now 2

A college reports that the average age of their students is 28 years old. Is this a statistic or a parameter?

Categorizing data

Once we have gathered data, we might wish to classify it. Roughly speaking, data can be classified as categorical data or quantitative data.

> **Quantitative and categorical data**
> **Categorical (qualitative) data** are pieces of information that allow us to classify the objects under investigation into various categories.
>
> **Quantitative data** are responses that are numerical in nature and with which we can perform meaningful arithmetic calculations.

Example 3

We might conduct a survey to determine the name of the favorite movie that each person in a math class saw in a movie theater.

When we conduct such a survey, the responses would look like: *Finding Nemo*, *The Hulk*, or *Terminator 3: Rise of the Machines*. We might count the number of people who give each answer, but the answers themselves do not have any numerical values: we cannot perform computations with an answer like "*Finding Nemo*." This would be categorical data.

Example 4

A survey could ask the number of movies you have seen in a movie theater in the past 12 months (0, 1, 2, 3, 4, ...)

This would be quantitative data.

Other examples of quantitative data would be the running time of the movie you saw most recently (104 minutes, 137 minutes, 104 minutes, ...) or the amount of money you paid for a movie ticket the last time you went to a movie theater ($5.50, $7.75, $9, ...).

Sometimes, determining whether or not data is categorical or quantitative can be a bit trickier.

Example 5

Suppose we gather respondents' ZIP codes in a survey to track their geographical location.

ZIP codes are numbers, but we can't do any meaningful mathematical calculations with them (it doesn't make sense to say that 98036 is "twice" 49018 — that's like saying that Lynnwood, WA is "twice" Battle Creek, MI, which doesn't make sense at all), so ZIP codes are really categorical data.

Example 6

A survey about the movie you most recently attended includes the question "How would you rate the movie you just saw?" with these possible answers:

1 - it was awful
2 - it was just OK
3 - I liked it
4 - it was great
5 - best movie ever!

Again, there are numbers associated with the responses, but we can't really do any calculations with them: a movie that rates a 4 is not necessarily twice as good as a movie that rates a 2, whatever that means; if two people see the movie and one of them thinks it stinks and the other thinks it's the best ever it doesn't necessarily make sense to say that "on average they liked it."

As we study movie-going habits and preferences, we shouldn't forget to specify the population under consideration. If we survey 3-7 year-olds the runaway favorite might be *Finding Nemo*. 13-17 year-olds might prefer *Terminator 3*. And 33-37 year-olds might prefer...well, *Finding Nemo*.

Try it Now 3
Classify each measurement as categorical or quantitative
a. Eye color of a group of people
b. Daily high temperature of a city over several weeks
c. Annual income

Sampling methods

As we mentioned in a previous section, the first thing we should do before conducting a survey is to identify the population that we want to study. Suppose we are hired by a politician to determine the amount of support he has among the electorate should he decide to run for another term. What population should we study? Every person in the district? Not every person is eligible to vote, and regardless of how strongly someone likes or dislikes the candidate, they don't have much to do with him being re-elected if they are not able to vote.

What about eligible voters in the district? That might be better, but if someone is eligible to vote but does not register by the deadline, they won't have any say in the election either. What about registered voters? Many people are registered but choose not to vote. What about "likely voters?"

This is the criteria used in much political polling, but it is sometimes difficult to define a "likely voter." Is it someone who voted in the last election? In the last general election? In the last presidential election? Should we consider someone who just turned 18 a "likely voter?" They weren't eligible to vote in the past, so how do we judge the likelihood that they will vote in the next election?

In November 1998, former professional wrestler Jesse "The Body" Ventura was elected governor of Minnesota. Up until right before the election, most polls showed he had little chance of winning. There were several contributing factors to the polls not reflecting the actual intent of the electorate:

- Ventura was running on a third-party ticket and most polling methods are better suited to a two-candidate race.
- Many respondents to polls may have been embarrassed to tell pollsters that they were planning to vote for a professional wrestler.
- The mere fact that the polls showed Ventura had little chance of winning might have prompted some people to vote for him in protest to send a message to the major-party candidates.

But one of the major contributing factors was that Ventura recruited a substantial amount of support from young people, particularly college students, who had never voted before and who registered specifically to vote in the gubernatorial election. The polls did not deem these young people likely voters (since in most cases young people have a lower rate of voter registration and a turnout rate for elections) and so the polling samples were subject to **sampling bias**: they omitted a portion of the electorate that was weighted in favor of the winning candidate.

> **Sampling bias**
> A sampling method is biased if every member of the population doesn't have equal likelihood of being in the sample.

So even identifying the population can be a difficult job, but once we have identified the population, how do we choose an appropriate sample? Remember, although we would prefer to survey all members of the population, this is usually impractical unless the population is very small, so we choose a sample. There are many ways to sample a population, but there is one goal we need to keep in mind: we would like the sample to be *representative of the population*.

Returning to our hypothetical job as a political pollster, we would not anticipate very accurate results if we drew all of our samples from among the customers at a Starbucks, nor would we expect that a sample drawn entirely from the membership list of the local Elks club would provide a useful picture of district-wide support for our candidate.

One way to ensure that the sample has a reasonable chance of mirroring the population is to employ *randomness*. The most basic random method is simple random sampling.

> **Simple random sample**
> A **random sample** is one in which each member of the population has an equal probability of being chosen. A **simple random sample** is one in which every member of the population and any group of members has an equal probability of being chosen.

Example 7

> If we could somehow identify all likely voters in the state, put each of their names on a piece of paper, toss the slips into a (very large) hat and draw 1000 slips out of the hat, we would have a simple random sample.

In practice, computers are better suited for this sort of endeavor than millions of slips of paper and extremely large headgear.

It is always possible, however, that even a random sample might end up not being totally representative of the population. If we repeatedly take samples of 1000 people from among the population of likely voters in the state of Washington, some of these samples might tend to have a slightly higher percentage of Democrats (or Republicans) than does the general population; some samples might include more older people and some samples might include more younger people; etc. In most cases, this **sampling variability** is not significant.

> **Sampling variability**
> The natural variation of samples is called **sampling variability**.
> This is unavoidable and expected in random sampling, and in most cases is not an issue.

234

To help account for variability, pollsters might instead use a **stratified sample**.

> **Stratified sampling**
> In **stratified sampling**, a population is divided into a number of subgroups (or strata). Random samples are then taken from each subgroup with sample sizes proportional to the size of the subgroup in the population.

Example 8

Suppose in a particular state that previous data indicated that the electorate was comprised of 39% Democrats, 37% Republicans and 24% independents. In a sample of 1000 people, they would then expect to get about 390 Democrats, 370 Republicans and 240 independents. To accomplish this, they could randomly select 390 people from among those voters known to be Democrats, 370 from those known to be Republicans, and 240 from those with no party affiliation.

Stratified sampling can also be used to select a sample with people in desired age groups, a specified mix ratio of males and females, etc. A variation on this technique is called **quota sampling**.

> **Quota sampling**
> **Quota sampling** is a variation on stratified sampling, wherein samples are collected in each subgroup until the desired quota is met.

Example 9

Suppose the pollsters call people at random, but once they have met their quota of 390 Democrats, they only gather people who do not identify themselves as a Democrat.

You may have had the experience of being called by a telephone pollster who started by asking you your age, income, etc. and then thanked you for your time and hung up before asking any "real" questions. Most likely, they already had contacted enough people in your demographic group and were looking for people who were older or younger, richer or poorer, etc. Quota sampling is usually a bit easier than stratified sampling, but also does not ensure the same level of randomness.

Another sampling method is **cluster sampling**, in which the population is divided into groups, and one or more groups are randomly selected to be in the sample.

> **Cluster sampling**
> In **cluster sampling**, the population is divided into subgroups (clusters), and a set of subgroups are selected to be in the sample

Example 10

> If the college wanted to survey students, since students are already divided into classes, they could randomly select 10 classes and give the survey to all the students in those classes. This would be cluster sampling.

Other sampling methods include **systematic sampling**.

> **Systematic sampling**
> In **systematic sampling**, every n^{th} member of the population is selected to be in the sample.

Example 11

> To select a sample using systematic sampling, a pollster calls every 100th name in the phone book.
>
> Systematic sampling is not as random as a simple random sample (if your name is Albert Aardvark and your sister Alexis Aardvark is right after you in the phone book, there is no way you could both end up in the sample) but it can yield acceptable samples.

Perhaps the worst types of sampling methods are **convenience samples** and **voluntary response samples**.

> **Convenience sampling and voluntary response sampling**
> **Convenience sampling** is samples chosen by selecting whoever is convenient.
> **Voluntary response sampling** is allowing the sample to volunteer.

Example 12

> A pollster stands on a street corner and interviews the first 100 people who agree to speak to him. This is a convenience sample.

Example 13

> A website has a survey asking readers to give their opinion on a tax proposal. This is a self-selected sample, or voluntary response sample, in which respondents volunteer to participate.

Usually voluntary response samples are skewed towards people who have a particularly strong opinion about the subject of the survey or who just have way too much time on their hands and enjoy taking surveys.

Try it Now 4

In each case, indicate what sampling method was used
a. Every 4th person in the class was selected
b. A sample was selected to contain 25 men and 35 women
c. Viewers of a new show are asked to vote on the show's website
d. A website randomly selects 50 of their customers to send a satisfaction survey to
e. To survey voters in a town, a polling company randomly selects 10 city blocks, and interviews everyone who lives on those blocks.

How to mess things up before you start

There are number of ways that a study can be ruined before you even start collecting data. The first we have already explored – **sampling** or **selection bias**, which is when the sample is not representative of the population. One example of this is **voluntary response bias**, which is bias introduced by only collecting data from those who volunteer to participate. This is not the only potential source of bias.

Sources of bias

Sampling bias – when the sample is not representative of the population

Voluntary response bias – the sampling bias that often occurs when the sample is volunteers

Self-interest study – bias that can occur when the researchers have an interest in the outcome

Response bias – when the responder gives inaccurate responses for any reason

Perceived lack of anonymity – when the responder fears giving an honest answer might negatively affect them

Loaded questions – when the question wording influences the responses

Non-response bias – when people refusing to participate in the study can influence the validity of the outcome

---

Example 14

Consider a recent study which found that chewing gum may raise math grades in teenagers[1]. This study was conducted by the Wrigley Science Institute, a branch of the Wrigley chewing gum company. This is an example of a **self-interest study**; one in which the researches have a vested interest in the outcome of the study. While this does not necessarily ensure that the study was biased, it certainly suggests that we should subject the study to extra scrutiny.

Example 15

A survey asks people "when was the last time you visited your doctor?" This might suffer from **response bias**, since many people might not remember exactly when they last saw a doctor and give inaccurate responses.

---

[1] Reuters. http://news.yahoo.com/s/nm/20090423/od_uk_nm/oukoe_uk_gum_learning. Retrieved 4/27/09

Sources of response bias may be innocent, such as bad memory, or as intentional as pressuring by the pollster. Consider, for example, how many voting initiative petitions people sign without even reading them.

Example 16

A survey asks participants a question about their interactions with members of other races. Here, a **perceived lack of anonymity** could influence the outcome. The respondent might not want to be perceived as racist even if they are, and give an untruthful answer.

Example 17

An employer puts out a survey asking their employees if they have a drug abuse problem and need treatment help. Here, answering truthfully might have consequences; responses might not be accurate if the employees do not feel their responses are anonymous or fear retribution from their employer.

Example 18

A survey asks "do you support funding research of alternative energy sources to reduce our reliance on high-polluting fossil fuels?" This is an example of a **loaded** or **leading question** – questions whose wording leads the respondent towards an answer.

Loaded questions can occur intentionally by pollsters with an agenda, or accidentally through poor question wording. Also a concern is **question order**, where the order of questions changes the results. A psychology researcher provides an example[2]:

> "My favorite finding is this: we did a study where we asked students, 'How satisfied are you with your life? How often do you have a date?' The two answers were not statistically related - you would conclude that there is no relationship between dating frequency and life satisfaction. But when we reversed the order and asked, 'How often do you have a date? How satisfied are you with your life?' the statistical relationship was a strong one. You would now conclude that there is nothing as important in a student's life as dating frequency."

Example 19

A telephone poll asks the question "Do you often have time to relax and read a book?", and 50% of the people called refused to answer the survey. It is unlikely that the results will be representative of the entire population. This is an example of **non-response bias**, introduced by people refusing to participate in a study or dropping out of an experiment. When people refuse to participate, we can no longer be so certain that our sample is representative of the population.

---

[2] Swartz, Norbert. http://www.umich.edu/~newsinfo/MT/01/Fal01/mt6f01.html. Retrieved 3/31/2009

238

Try it Now 5

In each situation, identify a potential source of bias

a. A survey asks how many sexual partners a person has had in the last year

b. A radio station asks readers to phone in their choice in a daily poll.

c. A substitute teacher wants to know how students in the class did on their last test. The teacher asks the 10 students sitting in the front row to state their latest test score.

d. High school students are asked if they have consumed alcohol in the last two weeks.

e. The Beef Council releases a study stating that consuming red meat poses little cardiovascular risk.

f. A poll asks "Do you support a new transportation tax, or would you prefer to see our public transportation system fall apart?"

Experiments

So far, we have primarily discussed **observational studies** – studies in which conclusions would be drawn from observations of a sample or the population. In some cases these observations might be unsolicited, such as studying the percentage of cars that turn right at a red light even when there is a "no turn on red" sign. In other cases the observations are solicited, like in a survey or a poll.

In contrast, it is common to use **experiments** when exploring how subjects react to an outside influence. In an experiment, some kind of **treatment** is applied to the subjects and the results are measured and recorded.

> **Observational studies and experiments**
> An **observational study** is a study based on observations or measurements
> An **experiment** is a study in which the effects of a **treatment** are measured

Here are some examples of experiments:

Example 20

a. A pharmaceutical company tests a new medicine for treating Alzheimer's disease by administering the drug to 50 elderly patients with recent diagnoses. The treatment here is the new drug.

b. A gym tests out a new weight loss program by enlisting 30 volunteers to try out the program. The treatment here is the new program.

c. You test a new kitchen cleaner by buying a bottle and cleaning your kitchen. The new cleaner is the treatment.

d. A psychology researcher explores the effect of music on temperament by measuring people's temperament while listening to different types of music. The music is the treatment.

Is each scenario describing an observational study or an experiment?
a. The weights of 30 randomly selected people are measured
b. Subjects are asked to do 20 jumping jacks, and then their heart rates are measured
c. Twenty coffee drinkers and twenty tea drinkers are given a concentration test

When conducting experiments, it is essential to isolate the treatment being tested.

Example 21

Suppose a middle school (junior high) finds that their students are not scoring well on the state's standardized math test. They decide to run an experiment to see if an alternate curriculum would improve scores. To run the test, they hire a math specialist to come in and teach a class using the new curriculum. To their delight, they see an improvement in test scores.

The difficulty with this scenario is that it is not clear whether the curriculum is responsible for the improvement, or whether the improvement is due to a math specialist teaching the class. This is called **confounding** – when it is not clear which factor or factors caused the observed effect. Confounding is the downfall of many experiments, though sometimes it is hidden.

> **Confounding**
> **Confounding** occurs when there are two potential variables that could have caused the outcome and it is not possible to determine which actually caused the result.

Example 22

A drug company study about a weight loss pill might report that people lost an average of 8 pounds while using their new drug. However, in the fine print you find a statement saying that participants were encouraged to also diet and exercise. It is not clear in this case whether the weight loss is due to the pill, to diet and exercise, or a combination of both. In this case confounding has occurred.

Example 23

Researchers conduct an experiment to determine whether students will perform better on an arithmetic test if they listen to music during the test. They first give the student a test without music, then give a similar test while the student listens to music. In this case, the student might perform better on the second test, regardless of the music, simply because it was the second test and they were warmed up.

There are a number of measures that can be introduced to help reduce the likelihood of confounding. The primary measure is to use a **control group**.

> **Control group**
> When using a control group, the participants are divided into two or more groups, typically a **control group** and a treatment group. The treatment group receives the treatment being tested; the control group does not receive the treatment.

Ideally, the groups are otherwise as similar as possible, isolating the treatment as the only potential source of difference between the groups. For this reason, the method of dividing groups is important. Some researchers attempt to ensure that the groups have similar characteristics (same number of females, same number of people over 50, etc.), but it is nearly impossible to control for every characteristic. Because of this, random assignment is very commonly used.

Example 24

To determine if a two day prep course would help high school students improve their scores on the SAT test, a group of students was randomly divided into two subgroups. The first group, the treatment group, was given a two day prep course. The second group, the control group, was not given the prep course. Afterwards, both groups were given the SAT.

Example 25

A company testing a new plant food grows two crops of plants in adjacent fields, the treatment group receiving the new plant food and the control group not. The crop yield would then be compared. By growing them at the same time in adjacent fields, they are controlling for weather and other confounding factors.

Sometimes not giving the control group anything does not completely control for confounding variables. For example, suppose a medicine study is testing a new headache pill by giving the treatment group the pill and the control group nothing. If the treatment group showed improvement, we would not know whether it was due to the medicine in the pill, or a response to have taken any pill. This is called a **placebo effect**.

> **Placebo effect**
> The **placebo effect** is when the effectiveness of a treatment is influenced by the patient's perception of how effective they think the treatment will be, so a result might be seen even if the treatment is ineffectual.

Example 26

A study found that when doing painful dental tooth extractions, patients told they were receiving a strong painkiller while actually receiving a saltwater injection found as much pain relief as patients receiving a dose of morphine.[3]

---

[3] Levine JD, Gordon NC, Smith R, Fields HL. (1981) Analgesic responses to morphine and placebo in individuals with postoperative pain. Pain. 10:379-89.

To control for the placebo effect, a **placebo**, or dummy treatment, is often given to the control group. This way, both groups are truly identical except for the specific treatment given.

Placebo and Placebo controlled experiments
A **placebo** is a dummy treatment given to control for the placebo effect.
An experiment that gives the control group a placebo is called a **placebo controlled experiment**.

Example 27

a. In a study for a new medicine that is dispensed in a pill form, a sugar pill could be used as a placebo.
b. In a study on the effect of alcohol on memory, a non-alcoholic beer might be given to the control group as a placebo.
c. In a study of a frozen meal diet plan, the treatment group would receive the diet food, and the control could be given standard frozen meals stripped of their original packaging.

In some cases, it is more appropriate to compare to a conventional treatment than a placebo. For example, in a cancer research study, it would not be ethical to deny any treatment to the control group or to give a placebo treatment. In this case, the currently acceptable medicine would be given to the second group, called a **comparison group** in this case. In our SAT test example, the non-treatment group would most likely be encouraged to study on their own, rather than be asked to not study at all, to provide a meaningful comparison.

When using a placebo, it would defeat the purpose if the participant knew they were receiving the placebo.

Blind studies
A **blind study** is one in which the participant does not know whether or not they are receiving the treatment or a placebo.

A **double-blind study** is one in which those interacting with the participants don't know who is in the treatment group and who is in the control group.

Example 28

In a study about anti-depression medicine, you would not want the psychological evaluator to know whether the patient is in the treatment or control group either, as it might influence their evaluation, so the experiment should be conducted as a double-blind study.

It should be noted that not every experiment needs a control group.

242

Example 29

If a researcher is testing whether a new fabric can withstand fire, she simply needs to torch multiple samples of the fabric – there is no need for a control group.

Try it Now 7

To test a new lie detector, two groups of subjects are given the new test. One group is asked to answer all the questions truthfully, and the second group is asked to lie on one set of questions. The person administering the lie detector test does not know what group each subject is in.

Does this experiment have a control group? Is it blind, double-blind, or neither?

Try it Now Answers

1. The sample is the 20 fish caught. The population is all fish in the lake. The sample may be somewhat unrepresentative of the population since not all fish may be large enough to catch the bait.

2. This is a parameter, since the college would have access to data on all students (the population)

3. a. Categorical.   b. Quantitative   c. Quantitative

4. a. Systematic
   b. Stratified or Quota
   c. Voluntary response
   d. Simple random
   e. Cluster

5. a. Response bias – historically, men are likely to over-report, and women are likely to under-report to this question.
   b. Voluntary response bias – the sample is self-selected
   c. Sampling bias – the sample may not be representative of the whole class
   d. Lack of anonymity
   e. Self-interest study
   f. Loaded question

6. a. Observational study
   b. Experiment; the treatment is the jumping jacks
   c. Experiment; the treatments are coffee and tea

7. The truth-telling group could be considered the control group, but really both groups are treatment groups here, since it is important for the lie detector to be able to correctly identify lies, and also not identify truth telling as lying. This study is blind, since the person running the test does not know what group each subject is in.

Exercises

Skills

1. A political scientist surveys 28 of the current 106 representatives in a state's congress. Of them, 14 said they were supporting a new education bill, 12 said there were not supporting the bill, and 2 were undecided.
   a. What is the population of this survey?
   b. What is the size of the population?
   c. What is the size of the sample?
   d. Give the sample statistic for the proportion of voters surveyed who said they were supporting the education bill.
   e. Based on this sample, we might expect how many of the representatives to support the education bill?

2. The city of Raleigh has 9500 registered voters. There are two candidates for city council in an upcoming election: Brown and Feliz. The day before the election, a telephone poll of 350 randomly selected registered voters was conducted. 112 said they'd vote for Brown, 207 said they'd vote for Feliz, and 31 were undecided.
   a. What is the population of this survey?
   b. What is the size of the population?
   c. What is the size of the sample?
   d. Give the sample statistic for the proportion of voters surveyed who said they'd vote for Brown.
   e. Based on this sample, we might expect how many of the 9500 voters to vote for Brown?

3. Identify the most relevant source of bias in this situation: A survey asks the following: Should the mall prohibit loud and annoying rock music in clothing stores catering to teenagers?

4. Identify the most relevant source of bias in this situation: To determine opinions on voter support for a downtown renovation project, a surveyor randomly questions people working in downtown businesses.

5. Identify the most relevant source of bias in this situation: A survey asks people to report their actual income and the income they reported on their IRS tax form.

6. Identify the most relevant source of bias in this situation: A survey randomly calls people from the phone book and asks them to answer a long series of questions.

7. Identify the most relevant source of bias in this situation: A survey asks the following: Should the death penalty be permitted if innocent people might die?

8. Identify the most relevant source of bias in this situation: A study seeks to investigate whether a new pain medication is safe to market to the public. They test by randomly selecting 300 men from a set of volunteers.

9. In a study, you ask the subjects their age in years. Is this data qualitative or quantitative?

10. In a study, you ask the subjects their gender. Is this data qualitative or quantitative?

11. Does this describe an observational study or an experiment: The temperature on randomly selected days throughout the year was measured.

12. Does this describe an observational study or an experiment? A group of students are told to listen to music while taking a test and their results are compared to a group not listening to music.

13. In a study, the sample is chosen by separating all cars by size, and selecting 10 of each size grouping. What is the sampling method?

14. In a study, the sample is chosen by writing everyone's name on a playing card, shuffling the deck, then choosing the top 20 cards. What is the sampling method?

15. A team of researchers is testing the effectiveness of a new HPV vaccine. They randomly divide the subjects into two groups. Group 1 receives new HPV vaccine, and Group 2 receives the existing HPV vaccine. The patients in the study do not know which group they are in.
    a. Which is the treatment group?
    b. Which is the control group (if there is one)?
    c. Is this study blind, double-blind, or neither?
    d. Is this best described as an experiment, a controlled experiment, or a placebo controlled experiment?

16. For the clinical trials of a weight loss drug containing *Garcinia cambogia* the subjects were randomly divided into two groups. The first received an inert pill along with an exercise and diet plan, while the second received the test medicine along with the same exercise and diet plan. The patients do not know which group they are in, nor do the fitness and nutrition advisors.
    a. Which is the treatment group?
    b. Which is the control group (if there is one)?
    c. Is this study blind, double-blind, or neither?
    d. Is this best described as an experiment, a controlled experiment, or a placebo controlled experiment?

Concepts

17. A teacher wishes to know whether the males in his/her class have more conservative attitudes than the females. A questionnaire is distributed assessing attitudes.
    a. Is this a sampling or a census?
    b. Is this an observational study or an experiment?
    c. Are there any possible sources of bias in this study?

18. A study is conducted to determine whether people learn better with spaced or massed practice. Subjects volunteer from an introductory psychology class. At the beginning of the semester 12 subjects volunteer and are assigned to the massed-practice group. At the end of the semester 12 subjects volunteer and are assigned to the spaced-practice condition.

    a. Is this a sampling or a census?

    b. Is this an observational study or an experiment?

    c. This study involves two kinds of non-random sampling: (1) Subjects are not randomly sampled from some specified population and (2) Subjects are not randomly assigned to groups. Which problem is more serious? What affect on the results does each have?

19. A farmer believes that playing Barry Manilow songs to his peas will increase their yield. Describe a controlled experiment the farmer could use to test his theory.

20. A sports psychologist believes that people are more likely to be extroverted as adults if they played team sports as children. Describe two possible studies to test this theory. Design one as an observational study and the other as an experiment. Which is more practical?

Exploration

21. Studies are often done by pharmaceutical companies to determine the effectiveness of a treatment program. Suppose that a new AIDS antibody drug is currently under study. It is given to patients once the AIDS symptoms have revealed themselves. Of interest is the average length of time in months patients live once starting the treatment. Two researchers each follow a different set of 50 AIDS patients from the start of treatment until their deaths.

    a. What is the population of this study?

    b. List two reasons why the data may differ.

    c. Can you tell if one researcher is correct and the other one is incorrect? Why?

    d. Would you expect the data to be identical? Why or why not?

    e. If the first researcher collected her data by randomly selecting 40 states, then selecting 1 person from each of those states. What sampling method is that?

    f. If the second researcher collected his data by choosing 40 patients he knew. What sampling method would that researcher have used? What concerns would you have about this data set, based upon the data collection method?

22. Find a newspaper or magazine article, or the online equivalent, describing the results of a recent study (the results of a poll are not sufficient). Give a summary of the study's findings, then analyze whether the article provided enough information to determine the validity of the conclusions. If not, produce a list of things that are missing from the article that would help you determine the validity of the study. Look for the things discussed in the text: population, sample, randomness, blind, control, placebos, etc.

Describing Data

Once we have collected data from surveys or experiments, we need to summarize and present the data in a way that will be meaningful to the reader. We will begin with graphical presentations of data then explore numerical summaries of data.

Presenting Categorical Data Graphically

Categorical, or qualitative, data are pieces of information that allow us to classify the objects under investigation into various categories. We usually begin working with categorical data by summarizing the data into a **frequency table.**

> **Frequency Table**
> A frequency table is a table with two columns. One column lists the categories, and another for the frequencies with which the items in the categories occur (how many items fit into each category).

Example 1

An insurance company determines vehicle insurance premiums based on known risk factors. If a person is considered a higher risk, their premiums will be higher. One potential factor is the color of your car. The insurance company believes that people with some color cars are more likely to get in accidents. To research this, they examine police reports for recent total-loss collisions. The data is summarized in the frequency table below.

Color	Frequency
Blue	25
Green	52
Red	41
White	36
Black	39
Grey	23

Sometimes we need an even more intuitive way of displaying data. This is where charts and graphs come in. There are many, many ways of displaying data graphically, but we will concentrate on one very useful type of graph called a bar graph. In this section we will work with bar graphs that display categorical data; the next section will be devoted to bar graphs that display quantitative data.

> **Bar graph**
> A **bar graph** is a graph that displays a bar for each category with the length of each bar indicating the frequency of that category.

248

To construct a bar graph, we need to draw a vertical axis and a horizontal axis. The vertical direction will have a scale and measure the frequency of each category; the horizontal axis has no scale in this instance. The construction of a bar chart is most easily described by use of an example.

Example 2

Using our car data from above, note the highest frequency is 52, so our vertical axis needs to go from 0 to 52, but we might as well use 0 to 55, so that we can put a hash mark every 5 units:

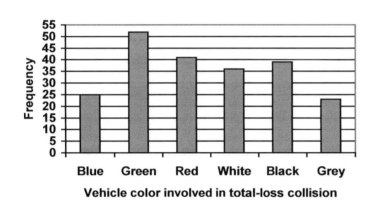

Notice that the height of each bar is determined by the frequency of the corresponding color. The horizontal gridlines are a nice touch, but not necessary. In practice, you will find it useful to draw bar graphs using graph paper, so the gridlines will already be in place, or using technology. Instead of gridlines, we might also list the frequencies at the top of each bar, like this:

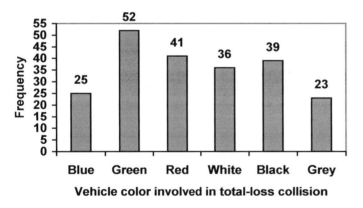

In this case, our chart might benefit from being reordered from largest to smallest frequency values. This arrangement can make it easier to compare similar values in the chart, even without gridlines. When we arrange the categories in decreasing frequency order like this, it is called a **Pareto chart**.

Pareto chart
A **Pareto chart** is a bar graph ordered from highest to lowest frequency

Example 3

Transforming our bar graph from earlier into a Pareto chart, we get:

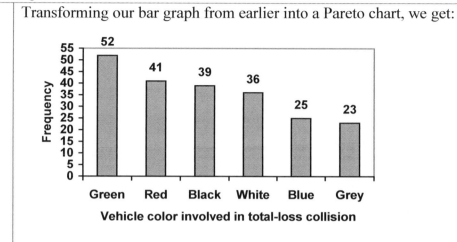

Example 4

In a survey[1], adults were asked whether they personally worried about a variety of environmental concerns. The numbers (out of 1012 surveyed) who indicated that they worried "a great deal" about some selected concerns are summarized below.

Environmental Issue	Frequency
Pollution of drinking water	597
Contamination of soil and water by toxic waste	526
Air pollution	455
Global warming	354

This data could be shown graphically in a bar graph:

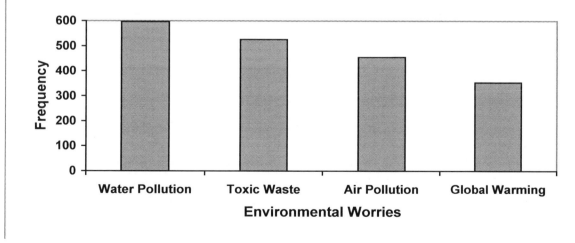

[1] Gallup Poll. March 5-8, 2009. http://www.pollingreport.com/enviro.htm

To show relative sizes, it is common to use a pie chart.

> **Pie Chart**
> A **pie chart** is a circle with wedges cut of varying sizes marked out like slices of pie or pizza. The relative sizes of the wedges correspond to the relative frequencies of the categories.

Example 5

For our vehicle color data, a pie chart might look like this:

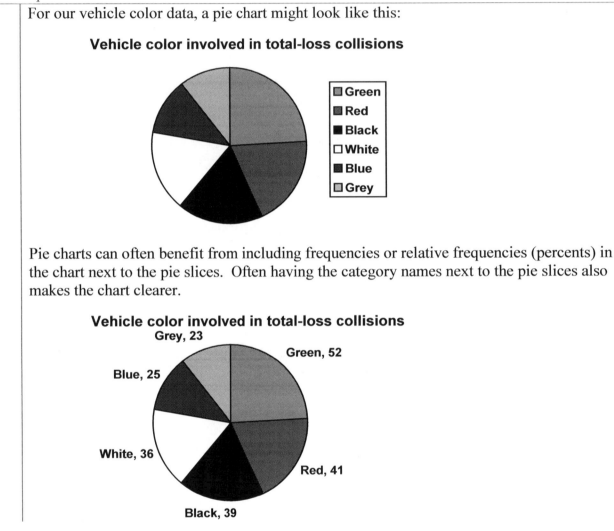

Vehicle color involved in total-loss collisions

Pie charts can often benefit from including frequencies or relative frequencies (percents) in the chart next to the pie slices. Often having the category names next to the pie slices also makes the chart clearer.

Vehicle color involved in total-loss collisions

Example 6

The pie chart to the right shows the percentage of voters supporting each candidate running for a local senate seat.

If there are 20,000 voters in the district, the pie chart shows that about 11% of those, about 2,200 voters, support Reeves.

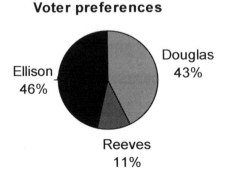

Voter preferences

Pie charts look nice, but are harder to draw by hand than bar charts since to draw them accurately we would need to compute the angle each wedge cuts out of the circle, then measure the angle with a protractor. Computers are much better suited to drawing pie charts. Common software programs like Microsoft Word or Excel, OpenOffice.org Write or Calc, or Google Docs are able to create bar graphs, pie charts, and other graph types. There are also numerous online tools that can create graphs[2].

Try it Now 1

Create a bar graph and a pie chart to illustrate the grades on a history exam below.
A: 12 students, B: 19 students, C: 14 students, D: 4 students, F: 5 students

---

Don't get fancy with graphs! People sometimes add features to graphs that don't help to convey their information. For example, 3-dimensional bar charts like the one shown below are usually not as effective as their two-dimensional counterparts.

Here is another way that fanciness can lead to trouble. Instead of plain bars, it is tempting to substitute meaningful images. This type of graph is called a **pictogram**.

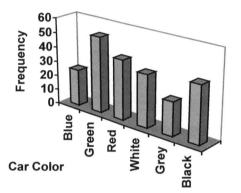

Pictogram
A **pictogram** is a statistical graphic in which the size of the picture is intended to represent the frequencies or size of the values being represented.

Example 7

A labor union might produce the graph to the right to show the difference between the average manager salary and the average worker salary.

Looking at the picture, it would be reasonable to guess that the manager salaries is 4 times as large as the worker salaries – the area of the bag looks about 4 times as large. However, the manager salaries are in fact only twice as large as worker salaries, which were reflected in the picture by making the manager bag twice as tall.

Manager Salaries   **Worker Salaries**

---

[2] For example: http://nces.ed.gov/nceskids/createAgraph/ or http://docs.google.com

252

Another distortion in bar charts results from setting the baseline to a value other than zero. The baseline is the bottom of the vertical axis, representing the least number of cases that could have occurred in a category. Normally, this number should be zero.

Example 8

Compare the two graphs below showing support for same-sex marriage rights from a poll taken in December 2008[3]. The difference in the vertical scale on the first graph suggests a different story than the true differences in percentages; the second graph makes it look like twice as many people oppose marriage rights as support it.

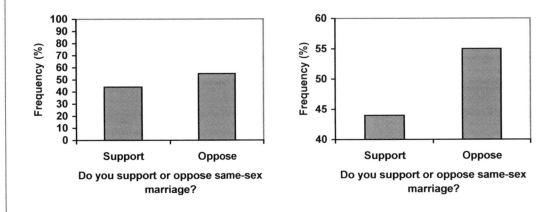

Try it Now 2

A poll was taken asking people if they agreed with the positions of the 4 candidates for a county office. Does the pie chart present a good representation of this data? Explain.

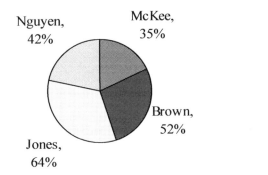

[3]CNN/Opinion Research Corporation Poll. Dec 19-21, 2008, from http://www.pollingreport.com/civil.htm

Presenting Quantitative Data Graphically

Quantitative, or numerical, data can also be summarized into frequency tables.

Example 9

A teacher records scores on a 20-point quiz for the 30 students in his class. The scores are:

19 20 18 18 17 18 19 17 20 18 20 16 20 15 17 12 18 19 18 19 17 20 18 16 15 18 20 5 0 0

These scores could be summarized into a frequency table by grouping like values:

Score	Frequency
0	2
5	1
12	1
15	2
16	2
17	4
18	8
19	4
20	6

Using this table, it would be possible to create a standard bar chart from this summary, like we did for categorical data:

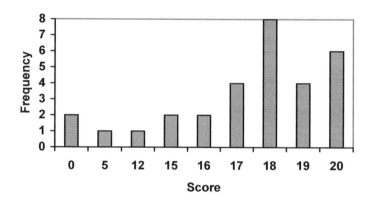

However, since the scores are numerical values, this chart doesn't really make sense; the first and second bars are five values apart, while the later bars are only one value apart. It would be more correct to treat the horizontal axis as a number line. This type of graph is called a **histogram**.

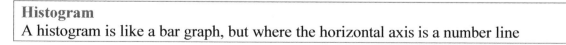

Histogram
A histogram is like a bar graph, but where the horizontal axis is a number line

Example 10

For the values above, a histogram would look like:

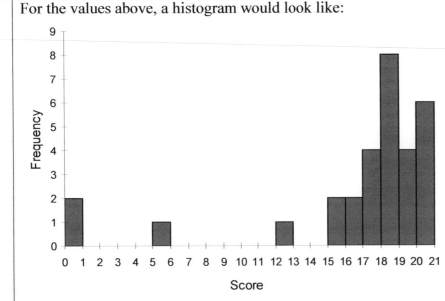

Notice that in the histogram, a bar represents values on the horizontal axis from that on the left hand-side of the bar up to, but not including, the value on the right hand side of the bar. Some people choose to have bars start at ½ values to avoid this ambiguity.

Unfortunately, not a lot of common software packages can correctly graph a histogram. About the best you can do in Excel or Word is a bar graph with no gap between the bars and spacing added to simulate a numerical horizontal axis.

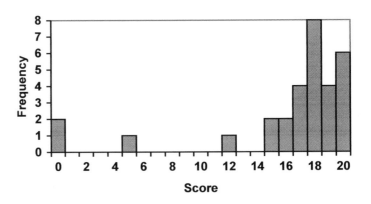

If we have a large number of widely varying data values, creating a frequency table that lists every possible value as a category would lead to an exceptionally long frequency table, and probably would not reveal any patterns. For this reason, it is common with quantitative data to group data into **class intervals**.

Class Intervals
Class intervals are groupings of the data. In general, we define class intervals so that:
- Each interval is equal in size. For example, if the first class contains values from 120-129, the second class should include values from 130-139.
- We have somewhere between 5 and 20 classes, typically, depending upon the number of data we're working with.

Example 11

Suppose that we have collected weights from 100 male subjects as part of a nutrition study. For our weight data, we have values ranging from a low of 121 pounds to a high of 263 pounds, giving a total span of 263-121 = 142. We could create 7 intervals with a width of around 20, 14 intervals with a width of around 10, or somewhere in between. Often time we have to experiment with a few possibilities to find something that represents the data well. Let us try using an interval width of 15. We could start at 121, or at 120 since it is a nice round number.

Interval	Frequency
120 - 134	4
135 – 149	14
150 – 164	16
165 – 179	28
180 – 194	12
195 – 209	8
210 – 224	7
225 – 239	6
240 – 254	2
255 - 269	3

A histogram of this data would look like:

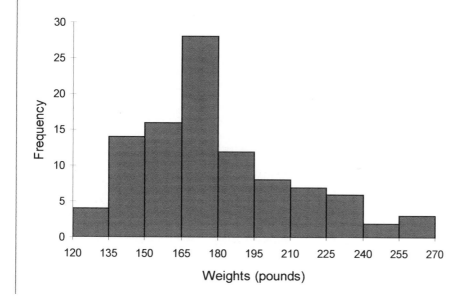

In many software packages, you can create a graph similar to a histogram by putting the class intervals as the labels on a bar chart.

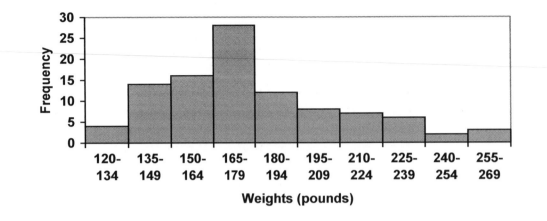

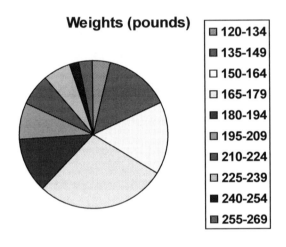

Other graph types such as pie charts are possible for quantitative data. The usefulness of different graph types will vary depending upon the number of intervals and the type of data being represented. For example, a pie chart of our weight data is difficult to read because of the quantity of intervals we used.

Weights (pounds)

▨	120-134
▦	135-149
☐	150-164
☐	165-179
■	180-194
▨	195-209
▦	210-224
▨	225-239
■	240-254
▨	255-269

Try it Now 3
The total cost of textbooks for the term was collected from 36 students. Create a histogram for this data.

$140	$160	$160	$165	$180	$220	$235	$240	$250	$260	$280	$285
$285	$285	$290	$300	$300	$305	$310	$310	$315	$315	$320	$320
$330	$340	$345	$350	$355	$360	$360	$380	$395	$420	$460	$460

When collecting data to compare two groups, it is desirable to create a graph that compares quantities.

Example 12

The data below came from a task in which the goal is to move a computer mouse to a target on the screen as fast as possible. On 20 of the trials, the target was a small rectangle; on the other 20, the target was a large rectangle. Time to reach the target was recorded on each trial.

Interval (milliseconds)	Frequency small target	Frequency large target
300-399	0	0
400-499	1	5
500-599	3	10
600-699	6	5
700-799	5	0
800-899	4	0
900-999	0	0
1000-1099	1	0
1100-1199	0	0

One option to represent this data would be a comparative histogram or bar chart, in which bars for the small target group and large target group are placed next to each other.

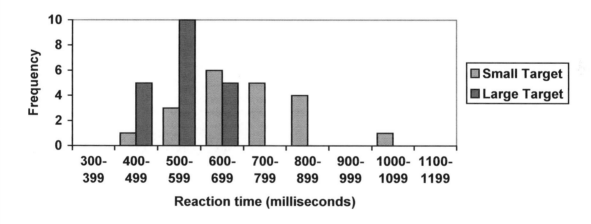

Frequency polygon
An alternative representation is a **frequency polygon**. A frequency polygon starts out like a histogram, but instead of drawing a bar, a point is placed in the midpoint of each interval at height equal to the frequency. Typically the points are connected with straight lines to emphasize the distribution of the data.

Example 13

This graph makes it easier to see that reaction times were generally shorter for the larger target, and that the reaction times for the smaller target were more spread out.

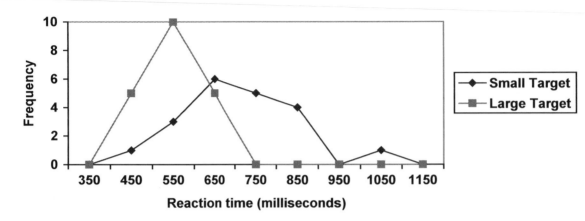

Numerical Summaries of Data

It is often desirable to use a few numbers to summarize a distribution. One important aspect of a distribution is where its center is located. Measures of central tendency are discussed first. A second aspect of a distribution is how spread out it is. In other words, how much the data in the distribution vary from one another. The second section describes measures of variability.

Measures of Central Tendency

Let's begin by trying to find the most "typical" value of a data set.

Note that we just used the word "typical" although in many cases you might think of using the word "average." We need to be careful with the word "average" as it means different things to different people in different contexts. One of the most common uses of the word "average" is what mathematicians and statisticians call the **arithmetic mean**, or just plain old **mean** for short. "Arithmetic mean" sounds rather fancy, but you have likely calculated a mean many times without realizing it; the mean is what most people think of when they use the word "average".

> **Mean**
> The **mean** of a set of data is the sum of the data values divided by the number of values.

Example 14

Marci's exam scores for her last math class were: 79, 86, 82, 94. The mean of these values would be:

$$\frac{79 + 86 + 82 + 94}{4} = 85.25.$$ Typically we round means to one more decimal place than the original data had. In this case, we would round 85.25 to 85.3.

Example 15

The number of touchdown (TD) passes thrown by each of the 31 teams in the National Football League in the 2000 season are shown below.

37 33 33 32 29 28 28 23 22 22 22 21 21 21 20
20 19 19 18 18 18 18 16 15 14 14 14 12 12 9 6

Adding these values, we get 634 total TDs. Dividing by 31, the number of data values, we get $634/31 = 20.4516$.   It would be appropriate to round this to 20.5.

It would be most correct for us to report that "The mean number of touchdown passes thrown in the NFL in the 2000 season was 20.5 passes," but it is not uncommon to see the more casual word "average" used in place of "mean."

Try it Now 4

The price of a jar of peanut butter at 5 stores was: $3.29, $3.59, $3.79, $3.75, and $3.99. Find the mean price.

Example 16

The one hundred families in a particular neighborhood are asked their annual household income, to the nearest $5 thousand dollars. The results are summarized in a frequency table below.

Income (thousands of dollars)	Frequency
15	6
20	8
25	11
30	17
35	19
40	20
45	12
50	7

Calculating the mean by hand could get tricky if we try to type in all 100 values:

$$\frac{\overbrace{15+\cdots+15}^{6 \text{ terms}}+\overbrace{20+\cdots+20}^{8 \text{ terms}}+\overbrace{25+\cdots+25}^{11 \text{ terms}}+\cdots}{100}$$

We could calculate this more easily by noticing that adding 15 to itself six times is the same as $15 \cdot 6 = 90$. Using this simplification, we get

$$\frac{15 \cdot 6 + 20 \cdot 8 + 25 \cdot 11 + 30 \cdot 17 + 35 \cdot 19 + 40 \cdot 20 + 45 \cdot 12 + 50 \cdot 7}{100} = \frac{3390}{100} = 33.9$$

The mean household income of our sample is 33.9 thousand dollars ($33,900).

Example 17

Extending off the last example, suppose a new family moves into the neighborhood example that has a household income of $5 million ($5000 thousand). Adding this to our sample, our mean is now:

$$\frac{15 \cdot 6 + 20 \cdot 8 + 25 \cdot 11 + 30 \cdot 17 + 35 \cdot 19 + 40 \cdot 20 + 45 \cdot 12 + 50 \cdot 7 + 5000 \cdot 1}{101} = \frac{8390}{101} = 83.069$$

While 83.1 thousand dollars ($83,069) is the correct mean household income, it no longer represents a "typical" value.

Imagine the data values on a see-saw or balance scale. The mean is the value that keeps the data in balance, like in the picture below.

If we graph our household data, the $5 million data value is so far out to the right that the mean has to adjust up to keep things in balance

For this reason, when working with data that have **outliers** – values far outside the primary grouping – it is common to use a different measure of center, the **median**.

Median

The **median** of a set of data is the value in the middle when the data is in order

To find the median, begin by listing the data in order from smallest to largest, or largest to smallest.

If the number of data values, N, is odd, then the median is the middle data value. This value can be found by rounding $N/2$ up to the next whole number.

If the number of data values is even, there is no one middle value, so we find the mean of the two middle values (values $N/2$ and $N/2 + 1$)

Example 18

Returning to the football touchdown data, we would start by listing the data in order. Luckily, it was already in decreasing order, so we can work with it without needing to reorder it first.

37 33 33 32 29 28 28 23 22 22 22 21 21 21 20
20 19 19 18 18 18 18 16 15 14 14 14 12 12 9 6

Since there are 31 data values, an odd number, the median will be the middle number, the 16th data value ($31/2 = 15.5$, round up to 16, leaving 15 values below and 15 above). The 16th data value is 20, so the median number of touchdown passes in the 2000 season was 20 passes. Notice that for this data, the median is fairly close to the mean we calculated earlier, 20.5.

Example 19

Find the median of these quiz scores: 5 10 8 6 4 8 2 5 7 7

We start by listing the data in order: 2 4 5 5 6 7 7 8 8 10

Since there are 10 data values, an even number, there is no one middle number. So we find the mean of the two middle numbers, 6 and 7, and get $(6+7)/2 = 6.5$.

The median quiz score was 6.5.

Try it Now 5
The price of a jar of peanut butter at 5 stores were: $3.29, $3.59, $3.79, $3.75, and $3.99. Find the median price.

Example 20

Let us return now to our original household income data

Income (thousands of dollars)	Frequency
15	6
20	8
25	11
30	17
35	19
40	20
45	12
50	7

Here we have 100 data values. If we didn't already know that, we could find it by adding the frequencies. Since 100 is an even number, we need to find the mean of the middle two data values - the 50th and 51st data values. To find these, we start counting up from the bottom:

There are 6 data values of $15, so	Values 1 to 6 are $15 thousand
The next 8 data values are $20, so	Values 7 to (6+8)=14 are $20 thousand
The next 11 data values are $25, so	Values 15 to (14+11)=25 are $25 thousand
The next 17 data values are $30, so	Values 26 to (25+17)=42 are $30 thousand
The next 19 data values are $35, so	Values 43 to (42+19)=61 are $35 thousand

From this we can tell that values 50 and 51 will be $35 thousand, and the mean of these two values is $35 thousand. The median income in this neighborhood is $35 thousand.

Example 21

If we add in the new neighbor with a $5 million household income, then there will be 101 data values, and the 51st value will be the median. As we discovered in the last example, the 51st value is $35 thousand. Notice that the new neighbor did not affect the median in this case. The median is not swayed as much by outliers as the mean is.

In addition to the mean and the median, there is one other common measurement of the "typical" value of a data set: the **mode**.

> **Mode**
> The **mode** is the element of the data set that occurs most frequently.

The mode is fairly useless with data like weights or heights where there are a large number of possible values. The mode is most commonly used for categorical data, for which median and mean cannot be computed.

Example 22

In our vehicle color survey, we collected the data

Color	Frequency
Blue	3
Green	5
Red	4
White	3
Black	2
Grey	3

For this data, Green is the mode, since it is the data value that occurred the most frequently.

It is possible for a data set to have more than one mode if several categories have the same frequency, or no modes if each every category occurs only once.

Try it Now 6
Reviewers were asked to rate a product on a scale of 1 to 5. Find
a. The mean rating
b. The median rating
c. The mode rating

Rating	Frequency
1	4
2	8
3	7
4	3
5	1

Measures of Variation

Consider these three sets of quiz scores:

Section A: 5 5 5 5 5 5 5 5 5 5

Section B: 0 0 0 0 0 10 10 10 10 10

Section C: 4 4 4 5 5 5 5 6 6 6

All three of these sets of data have a mean of 5 and median of 5, yet the sets of scores are clearly quite different. In section A, everyone had the same score; in section B half the class got no points and the other half got a perfect score, assuming this was a 10-point quiz. Section C was not as consistent as section A, but not as widely varied as section B.

In addition to the mean and median, which are measures of the "typical" or "middle" value, we also need a measure of how "spread out" or varied each data set is.

There are several ways to measure this "spread" of the data. The first is the simplest and is called the **range**.

> **Range**
> The range is the difference between the maximum value and the minimum value of the data set.

Example 23

Using the quiz scores from above,

For section A, the range is 0 since both maximum and minimum are 5 and $5 - 5 = 0$
For section B, the range is 10 since $10 - 0 = 10$
For section C, the range is 2 since $6 - 4 = 2$

In the last example, the range seems to be revealing how spread out the data is. However, suppose we add a fourth section, Section D, with scores 0 5 5 5 5 5 5 5 10.

This section also has a mean and median of 5. The range is 10, yet this data set is quite different than Section B. To better illuminate the differences, we'll have to turn to more sophisticated measures of variation.

> **Standard deviation**
> The standard deviation is a measure of variation based on measuring how far each data value deviates, or is different, from the mean. A few important characteristics:
> - Standard deviation is always positive. Standard deviation will be zero if all the data values are equal, and will get larger as the data spreads out.
> - Standard deviation has the same units as the original data.
> - Standard deviation, like the mean, can be highly influenced by outliers.

Using the data from section D, we could compute for each data value the difference between the data value and the mean:

data value	deviation: data value - mean
0	$0-5 = -5$
5	$5-5 = 0$
5	$5-5 = 0$
5	$5-5 = 0$
5	$5-5 = 0$
5	$5-5 = 0$
5	$5-5 = 0$
5	$5-5 = 0$
5	$5-5 = 0$
10	$10-5 = 5$

We would like to get an idea of the "average" deviation from the mean, but if we find the average of the values in the second column the negative and positive values cancel each other out (this will always happen), so to prevent this we square every value in the second column:

data value	deviation: data value - mean	deviation squared
0	$0-5 = -5$	$(-5)^2 = 25$
5	$5-5 = 0$	$0^2 = 0$
5	$5-5 = 0$	$0^2 = 0$
5	$5-5 = 0$	$0^2 = 0$
5	$5-5 = 0$	$0^2 = 0$
5	$5-5 = 0$	$0^2 = 0$
5	$5-5 = 0$	$0^2 = 0$
5	$5-5 = 0$	$0^2 = 0$
5	$5-5 = 0$	$0^2 = 0$
10	$10-5 = 5$	$(5)^2 = 25$

We then add the squared deviations up to get $25 + 0 + 0 + 0 + 0 + 0 + 0 + 0 + 0 + 25 =$ 50. Ordinarily we would then divide by the number of scores, n, (in this case, 10) to find the mean of the deviations. But we only do this if the data set represents a population; if the data set represents a sample (as it almost always does), we instead divide by n - 1 (in this case, $10 - 1 = 9$).[4]

So in our example, we would have $50/10 = 5$ if section D represents a population and $50/9 =$ about 5.56 if section D represents a sample. These values (5 and 5.56) are called, respectively, the **population variance** and the **sample variance** for section D.

Variance can be a useful statistical concept, but note that the units of variance in this instance would be points-squared since we squared all of the deviations. What are points-squared? Good question. We would rather deal with the units we started with (points in this case), so to convert back we take the square root and get:

$$\text{population standard deviation} = \sqrt{\frac{50}{10}} = \sqrt{5} \approx 2.2$$

or

$$\text{sample standard deviation} = \sqrt{\frac{50}{9}} \approx 2.4$$

If we are unsure whether the data set is a sample or a population, we will usually assume it is a sample, and we will round answers to one more decimal place than the original data, as we have done above.

> **To compute standard deviation:**
> 1. Find the deviation of each data from the mean. In other words, subtract the mean from the data value.
> 2. Square each deviation.
> 3. Add the squared deviations.
> 4. Divide by n, the number of data values, if the data represents a whole population; divide by $n - 1$ if the data is from a sample.
> 5. Compute the square root of the result.

---

[4] The reason we do this is highly technical, but we can see how it might be useful by considering the case of a small sample from a population that contains an outlier, which would increase the average deviation: the outlier very likely won't be included in the sample, so the mean deviation of the sample would underestimate the mean deviation of the population; thus we divide by a slightly smaller number to get a slightly bigger average deviation.

266

Example 24

Computing the standard deviation for Section B above, we first calculate that the mean is 5. Using a table can help keep track of your computations for the standard deviation:

data value	deviation: data value - mean	deviation squared
0	0-5 = -5	$(-5)^2 = 25$
0	0-5 = -5	$(-5)^2 = 25$
0	0-5 = -5	$(-5)^2 = 25$
0	0-5 = -5	$(-5)^2 = 25$
0	0-5 = -5	$(-5)^2 = 25$
10	10-5 = 5	$(5)^2 = 25$
10	10-5 = 5	$(5)^2 = 25$
10	10-5 = 5	$(5)^2 = 25$
10	10-5 = 5	$(5)^2 = 25$
10	10-5 = 5	$(5)^2 = 25$

Assuming this data represents a population, we will add the squared deviations, divide by 10, the number of data values, and compute the square root:

$$\sqrt{\frac{25+25+25+25+25+25+25+25+25+25}{10}} = \sqrt{\frac{250}{10}} = 5$$

Notice that the standard deviation of this data set is much larger than that of section D since the data in this set is more spread out.

For comparison, the standard deviations of all four sections are:

Section A: 5 5 5 5 5 5 5 5 5 5	Standard deviation: 0
Section B: 0 0 0 0 0 10 10 10 10 10	Standard deviation: 5
Section C: 4 4 4 5 5 5 5 6 6 6	Standard deviation: 0.8
Section D: 0 5 5 5 5 5 5 5 5 10	Standard deviation: 2.2

Try it Now 7
The price of a jar of peanut butter at 5 stores were: $3.29, $3.59, $3.79, $3.75, and $3.99. Find the standard deviation of the prices.

Where standard deviation is a measure of variation based on the mean, **quartiles** are based on the median.

Quartiles
Quartiles are values that divide the data in quarters.

The first quartile (Q_1) is the value so that 25% of the data values are below it; the third quartile (Q_3) is the value so that 75% of the data values are below it. You may have guessed that the second quartile is the same as the median, since the median is the value so that 50% of the data values are below it.

This divides the data into quarters; 25% of the data is between the minimum and Q_1, 25% is between Q_1 and the median, 25% is between the median and Q_3, and 25% is between Q_3 and the maximum value

---

While quartiles are not a 1-number summary of variation like standard deviation, the quartiles are used with the median, minimum, and maximum values to form a **5 number summary** of the data.

Five number summary
The five number summary takes this form:
Minimum, Q_1, Median, Q_3, Maximum

To find the first quartile, we need to find the data value so that 25% of the data is below it. If *n* is the number of data values, we compute a locator by finding 25% of *n*. If this locator is a decimal value, we round up, and find the data value in that position. If the locator is a whole number, we find the mean of the data value in that position and the next data value. This is identical to the process we used to find the median, except we use 25% of the data values rather than half the data values as the locator.

To find the first quartile, Q_1
Begin by ordering the data from smallest to largest
Compute the locator: $L = 0.25n$
If L is a decimal value:
    Round up to $L+$
    Use the data value in the $L+^{th}$ position
If L is a whole number:
    Find the mean of the data values in the Lth and $L+1^{th}$ positions.

To find the third quartile, Q_3
Use the same procedure as for Q_1, but with locator: $L = 0.75n$

Examples should help make this clearer.

Example 25

Suppose we have measured 9 females and their heights (in inches), sorted from smallest to largest are:

59 60 62 64 66 67 69 70 72

To find the first quartile we first compute the locator: 25% of 9 is $L = 0.25(9) = 2.25$. Since this value is not a whole number, we round up to 3. The first quartile will be the third data value: 62 inches.

To find the third quartile, we again compute the locator: 75% of 9 is $0.75(9) = 6.75$. Since this value is not a whole number, we round up to 7. The third quartile will be the seventh data value: 69 inches.

Example 26

Suppose we had measured 8 females and their heights (in inches), sorted from smallest to largest are:

59 60 62 64 66 67 69 70

To find the first quartile we first compute the locator: 25% of 8 is $L = 0.25(8) = 2$. Since this value *is* a whole number, we will find the mean of the 2nd and 3rd data values: $(60+62)/2 = 61$, so the first quartile is 61 inches.

The third quartile is computed similarly, using 75% instead of 25%. $L = 0.75(8) = 6$. This is a whole number, so we will find the mean of the 6th and 7th data values: $(67+69)/2 = 68$, so Q_3 is 68.

Note that the median could be computed the same way, using 50%.

The 5-number summary combines the first and third quartile with the minimum, median, and maximum values.

Example 27

For the 9 female sample, the median is 66, the minimum is 59, and the maximum is 72. The 5 number summary is: 59, 62, 66, 69, 72.

For the 8 female sample, the median is 65, the minimum is 59, and the maximum is 70, so the 5 number summary would be: 59, 61, 65, 68, 70.

Example 28

Returning to our quiz score data. In each case, the first quartile locator is $0.25(10) = 2.5$, so the first quartile will be the 3rd data value, and the third quartile will be the 8th data value. Creating the five-number summaries:

Section and data	5-number summary
Section A: 5 5 5 5 5 5 5 5 5 5	5, 5, 5, 5, 5
Section B: 0 0 0 0 0 10 10 10 10 10	0, 0, 5, 10, 10
Section C: 4 4 4 5 5 5 5 6 6 6	4, 4, 5, 6, 6
Section D: 0 5 5 5 5 5 5 5 5 10	0, 5, 5, 5, 10

Of course, with a relatively small data set, finding a five-number summary is a bit silly, since the summary contains almost as many values as the original data.

Try it Now 8

The total cost of textbooks for the term was collected from 36 students. Find the 5 number summary of this data.

$140 $160 $160 $165 $180 $220 $235 $240 $250 $260 $280 $285
$285 $285 $290 $300 $300 $305 $310 $310 $315 $315 $320 $320
$330 $340 $345 $350 $355 $360 $360 $380 $395 $420 $460 $460

Example 29

Returning to the household income data from earlier, create the five-number summary.

Income (thousands of dollars)	Frequency
15	6
20	8
25	11
30	17
35	19
40	20
45	12
50	7

By adding the frequencies, we can see there are 100 data values represented in the table. In Example 20, we found the median was $35 thousand. We can see in the table that the minimum income is $15 thousand, and the maximum is $50 thousand.

To find Q_1, we calculate the locator: $L = 0.25(100) = 25$. This is a whole number, so Q_1 will be the mean of the 25th and 26th data values.

Counting up in the data as we did before,

There are 6 data values of $15, so   Values 1 to 6 are $15 thousand
The next 8 data values are $20, so   Values 7 to (6+8)=14 are $20 thousand
The next 11 data values are $25, so   Values 15 to (14+11)=25 are $25 thousand
The next 17 data values are $30, so   Values 26 to (25+17)=42 are $30 thousand

The 25th data value is $25 thousand, and the 26th data value is $30 thousand, so Q_1 will be the mean of these: $(25 + 30)/2 = \$27.5$ thousand.

To find Q_3, we calculate the locator: $L = 0.75(100) = 75$. This is a whole number, so Q_3 will be the mean of the 75th and 76th data values. Continuing our counting from earlier,
The next 19 data values are $35, so   Values 43 to (42+19)=61 are $35 thousand
The next 20 data values are $40, so   Values 61 to (61+20)=81 are $40 thousand

Both the 75th and 76th data values lie in this group, so Q_3 will be $40 thousand.

Putting these values together into a five-number summary, we get: 15, 27.5, 35, 40, 50

Note that the 5 number summary divides the data into four intervals, each of which will contain about 25% of the data. In the previous example, that means about 25% of households have income between $40 thousand and $50 thousand.

For visualizing data, there is a graphical representation of a 5-number summary called a **box plot**, or box and whisker graph.

Box plot
A **box plot** is a graphical representation of a five-number summary.

To create a box plot, a number line is first drawn. A box is drawn from the first quartile to the third quartile, and a line is drawn through the box at the median. "Whiskers" are extended out to the minimum and maximum values.

Example 30

The box plot below is based on the 9 female height data with 5 number summary: 59, 62, 66, 69, 72.

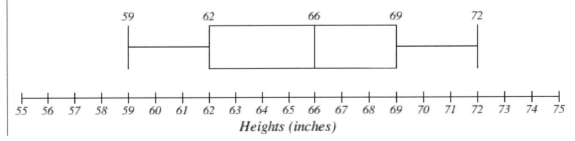

Example 31

The box plot below is based on the household income data with 5 number summary:
15, 27.5, 35, 40, 50

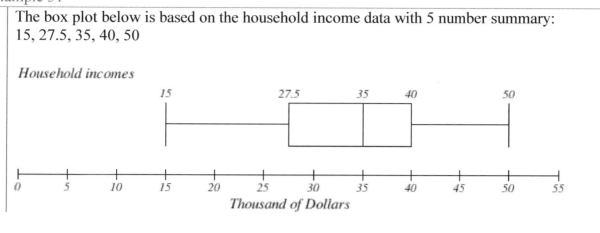

Household incomes

Try it Now 9
Create a boxplot based on the textbook price data from the last Try it Now.

Box plots are particularly useful for comparing data from two populations.

Example 32

The box plot of service times for two fast-food restaurants is shown below.

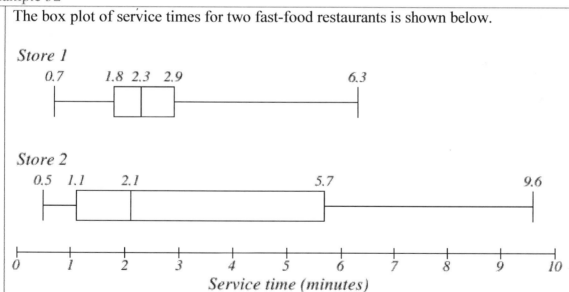

While store 2 had a slightly shorter median service time (2.1 minutes vs. 2.3 minutes), store 2 is less consistent, with a wider spread of the data.

At store 1, 75% of customers were served within 2.9 minutes, while at store 2, 75% of customers were served within 5.7 minutes.

Which store should you go to in a hurry?   That depends upon your opinions about luck –
25% of customers at store 2 had to wait between 5.7 and 9.6 minutes.

Example 33

The boxplot below is based on the birth weights of infants with severe idiopathic respiratory distress syndrome (SIRDS)[5]. The boxplot is separated to show the birth weights of infants who survived and those that did not.

Comparing the two groups, the boxplot reveals that the birth weights of the infants that died appear to be, overall, smaller than the weights of infants that survived. In fact, we can see that the median birth weight of infants that survived is the same as the third quartile of the infants that died.

Similarly, we can see that the first quartile of the survivors is larger than the median weight of those that died, meaning that over 75% of the survivors had a birth weight larger than the median birth weight of those that died.

Looking at the maximum value for those that died and the third quartile of the survivors, we can see that over 25% of the survivors had birth weights higher than the heaviest infant that died.

The box plot gives us a quick, albeit informal, way to determine that birth weight is quite likely linked to survival of infants with SIRDS.

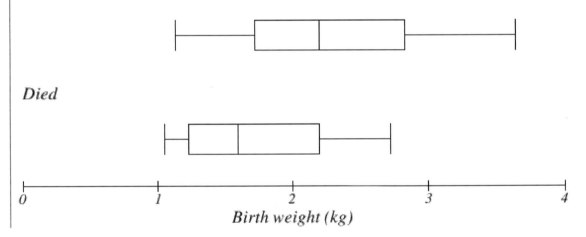

Survived

Died

Birth weight (kg)

[5] van Vliet, P.K. and Gupta, J.M. (1973) Sodium bicarbonate in idiopathic respiratory distress syndrome. *Arch. Disease in Childhood,* **48**, 249–255. As quoted on http://openlearn.open.ac.uk/mod/oucontent/view.php?id=398296&section=1.1.3

1.

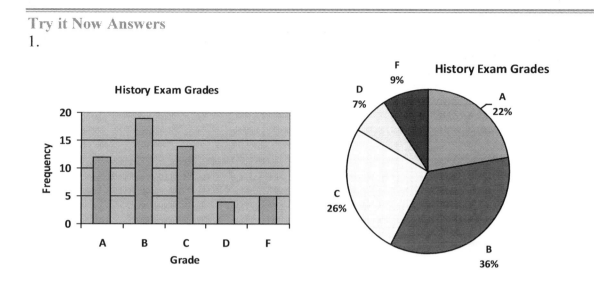

2. While the pie chart accurately depicts the relative size of the people agreeing with each candidate, the chart is confusing, since usually percents on a pie chart represent the percentage of the pie the slice represents.

3. Using a class intervals of size 55, we can group our data into six intervals:

Cost interval	Frequency
$140-194	5
$195-249	3
$250-304	9
$305-359	12
$360-414	4
$415-469	3

We can use the frequency distribution to generate the histogram

4. Adding the prices and dividing by 5 we get the mean price: $3.682

5. First we put the data in order: $3.29, $3.59, $3.75, $3.79, $3.99. Since there are an odd number of data, the median will be the middle value, $3.75.

6. There are 23 ratings.
   a. The mean is $\dfrac{1\cdot 4 + 2\cdot 8 + 3\cdot 7 + 4\cdot 3 + 5\cdot 1}{23} \approx 2.5$
   b. There are 23 data values, so the median will be the 12th data value. Ratings of 1 are the first 4 values, while a rating of 2 are the next 8 values, so the 12th value will be a rating of 2. The median is 2.
   c. The mode is the most frequent rating. The mode rating is 2.

7. Earlier we found the mean of the data was $3.682.

data value	deviation: data value - mean	deviation squared
3.29	$3.29 - 3.682 = -0.391$	0.153664
3.59	$3.59 - 3.682 = -0.092$	0.008464
3.79	$3.79 - 3.682 = 0.108$	0.011664
3.75	$3.75 - 3.682 = 0.068$	0.004624
3.99	$3.99 - 3.682 = 0.308$	0.094864

This data is from a sample, so we will add the squared deviations, divide by 4, the number of data values minus 1, and compute the square root:

$$\sqrt{\frac{0.153664 + 0.008464 + 0.011664 + 0.004624 + 0.094864}{4}} \approx \$0.261$$

8. The data is already in order, so we don't need to sort it first. The minimum value is $140 and the maximum is $460.

There are 36 data values so $n = 36$. $n/2 = 18$, which is a whole number, so the median is the mean of the 18th and 19th data values, $305 and $310. The median is $307.50.

To find the first quartile, we calculate the locator, $L = 0.25(36) = 9$. Since this is a whole number, we know Q_1 is the mean of the 9th and 10th data values, $250 and $260. $Q_1 = \$255$.

To find the third quartile, we calculate the locator, $L = 0.75(36) = 27$. Since this is a whole number, we know Q_3 is the mean of the 27th and 28th data values, $345 and $350. $Q_3 = \$347.50$.

The 5 number summary of this data is: $140, $255, $307.50, $347.50, $460

9. Boxplot of textbook costs

Exercises

Skills

1. The table below shows scores on a Math test.
   a. Complete the frequency table for the Math test scores
   b. Construct a histogram of the data
   c. Construct a pie chart of the data

80	50	50	90	70	70	100	60	70	80	70	50
90	100	80	70	30	80	80	70	100	60	60	50

2. A group of adults where asked how many cars they had in their household
   a. Complete the frequency table for the car number data
   b. Construct a histogram of the data
   c. Construct a pie chart of the data

1	4	2	2	1	2	3	3	1	4	2	2
1	2	1	3	2	2	1	2	1	1	1	2

3. A group of adults were asked how many children they have in their families. The bar graph to the right shows the number of adults who indicated each number of children.
   a. How many adults where questioned?
   b. What percentage of the adults questioned had 0 children?

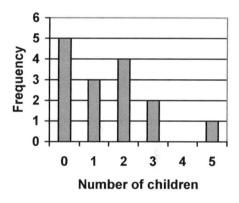

4. Jasmine was interested in how many days it would take an order from Netflix to arrive at her door.   The graph below shows the data she collected.
   a. How many movies did she order?
   b. What percentage of the movies arrived in one day?

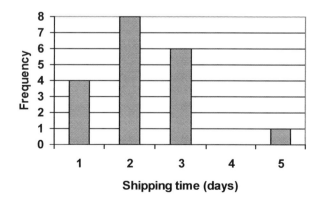

5. The bar graph below shows the *percentage* of students who received each letter grade on their last English paper. The class contains 20 students. What number of students earned an A on their paper?

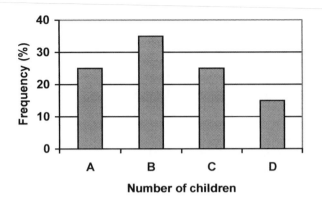

6. Kori categorized her spending for this month into four categories: Rent, Food, Fun, and Other. The percents she spent in each category are pictured here. If she spent a total of $2600 this month, how much did she spend on rent?

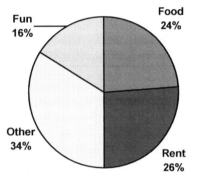

7. A group of diners were asked how much they would pay for a meal. Their responses were: $7.50, $8.25, $9.00, $8.00, $7.25, $7.50, $8.00, $7.00.
    a. Find the mean
    b. Find the median
    c. Write the 5-number summary for this data

8. You recorded the time in seconds it took for 8 participants to solve a puzzle. The times were: 15.2, 18.8, 19.3, 19.7, 20.2, 21.8, 22.1, 29.4.
    a. Find the mean
    b. Find the median
    c. Write the 5-number summary for this data

9. Refer back to the histogram from question #3.
    a. Compute the mean number of children for the group surveyed
    b. Compute the median number of children for the group surveyed
    c. Write the 5-number summary for this data.
    d. Create box plot.

10. Refer back to the histogram from question #4.
    a. Compute the mean number of shipping days
    b. Compute the median number of shipping days
    c. Write the 5-number summary for this data.
    d. Create box plot.

Concepts

11. The box plot below shows salaries for Actuaries and CPAs. Kendra makes the median salary for an Actuary. Kelsey makes the first quartile salary for a CPA. Who makes more money? How much more?

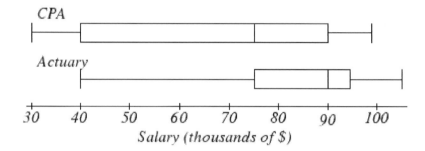

12. Referring to the boxplot above, what percentage of actuaries makes more than the median salary of a CPA?

Exploration

13. Studies are often done by pharmaceutical companies to determine the effectiveness of a treatment program. Suppose that a new AIDS antibody drug is currently under study. It is given to patients once the AIDS symptoms have revealed themselves. Of interest is the average length of time in months patients live once starting the treatment. Two researchers each follow a different set of 40 AIDS patients from the start of treatment until their deaths. The following data (in months) are collected.

    Researcher 1: 3; 4; 11; 15; 16; 17; 22; 44; 37; 16; 14; 24; 25; 15; 26; 27; 33; 29; 35; 44; 13; 21; 22; 10; 12; 8; 40; 32; 26; 27; 31; 34; 29; 17; 8; 24; 18; 47; 33; 34

    Researcher 2: 3; 14; 11; 5; 16; 17; 28; 41; 31; 18; 14; 14; 26; 25; 21; 22; 31; 2; 35; 44; 23; 21; 21; 16; 12; 18; 41; 22; 16; 25; 33; 34; 29; 13; 18; 24; 23; 42; 33; 29

    a. Create comparative histograms of the data
    b. Create comparative boxplots of the data

14. A graph appears below showing the number of adults and children who prefer each type of soda. There were 130 adults and kids surveyed. Discuss some ways in which the graph below could be improved

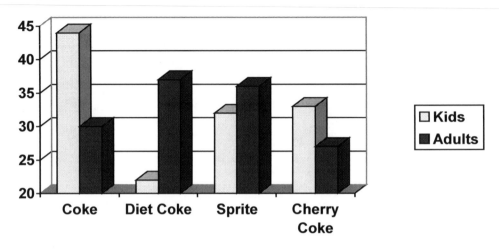

15. Make up three data sets with 5 numbers each that have:
   a. the same mean but different standard deviations.
   b. the same mean but different medians.
   c. the same median but different means.

16. A sample of 30 distance scores measured in yards has a mean of 7, a variance of 16, and a standard deviation of 4.
   a. You want to convert all your distances from yards to feet, so you multiply each score in the sample by 3. What are the new mean, median, variance, and standard deviation?
   b. You then decide that you only want to look at the distance past a certain point. Thus, after multiplying the original scores by 3, you decide to subtract 4 feet from each of the scores. Now what are the new mean, median, variance, and standard deviation?

17. In your class, design a poll on a topic of interest to you and give it to the class.
   a. Summarize the data, computing the mean and five-number summary.
   b. Create a graphical representation of the data.
   c. Write several sentences about the topic, using your computed statistics as evidence in your writing.

Probability

Introduction

The probability of a specified event is the chance or likelihood that it will occur. There are several ways of viewing probability. One would be **experimental** in nature, where we repeatedly conduct an experiment. Suppose we flipped a coin over and over and over again and it came up heads about half of the time; we would expect that in the future whenever we flipped the coin it would turn up heads about half of the time. When a weather reporter says "there is a 10% chance of rain tomorrow," she is basing that on prior evidence; that out of all days with similar weather patterns, it has rained on 1 out of 10 of those days.

Another view would be **subjective** in nature, in other words an educated guess. If someone asked you the probability that the Seattle Mariners would win their next baseball game, it would be impossible to conduct an experiment where the same two teams played each other repeatedly, each time with the same starting lineup and starting pitchers, each starting at the same time of day on the same field under the precisely the same conditions. Since there are so many variables to take into account, someone familiar with baseball and with the two teams involved might make an educated guess that there is a 75% chance they will win the game; that is, *if* the same two teams were to play each other repeatedly under identical conditions, the Mariners would win about three out of every four games. But this is just a guess, with no way to verify its accuracy, and depending upon how educated the educated guesser is, a subjective probability may not be worth very much.

We will return to the experimental and subjective probabilities from time to time, but in this course we will mostly be concerned with **theoretical** probability, which is defined as follows: Suppose there is a situation with n <u>equally likely</u> possible outcomes and that m of those n outcomes correspond to a particular event; then the **probability** of that event is defined as $\dfrac{m}{n}$.

Basic Concepts

If you roll a die, pick a card from deck of playing cards, or randomly select a person and observe their hair color, we are executing an experiment or procedure. In probability, we look at the likelihood of different outcomes. We begin with some terminology.

> **Events and Outcomes**
> The result of an experiment is called an **outcome**.
>
> An **event** is any particular outcome or group of outcomes.
>
> A **simple event** is an event that cannot be broken down further
>
> The **sample space** is the set of all possible simple events.

Example 1

If we roll a standard 6-sided die, describe the sample space and some simple events.

The sample space is the set of all possible simple events: {1,2,3,4,5,6}

Some examples of simple events:
We roll a 1
We roll a 5

Two dice   One die

Some compound events:
We roll a number bigger than 4
We roll an even number

Basic Probability

Given that all outcomes are equally likely, we can compute the probability of an event
E using this formula:

$$P(E) = \frac{\text{Number of outcomes corresponding to the event } E}{\text{Total number of equally - likely outcomes}}$$

Example 2

If we roll a 6-sided die, calculate
a) P(rolling a 1)
b) P(rolling a number bigger than 4)

Recall that the sample space is {1,2,3,4,5,6}

a) There is one outcome corresponding to "rolling a 1", so the probability is $\frac{1}{6}$

b) There are two outcomes bigger than a 4, so the probability is $\frac{2}{6} = \frac{1}{3}$

Probabilities are essentially fractions, and can be reduced to lower terms like fractions.

Example 3

Let's say you have a bag with 20 cherries, 14 sweet and 6 sour. If you pick a cherry at random, what is the probability that it will be sweet?

There are 20 possible cherries that could be picked, so the number of possible outcomes is 20. Of these 20 possible outcomes, 14 are favorable (sweet), so the probability that the cherry will be sweet is $\frac{14}{20} = \frac{7}{10}$.

There is one potential complication to this example, however. It must be assumed that the probability of picking any of the cherries is the same as the probability of picking any other. This wouldn't be true if (let us imagine) the sweet cherries are smaller than the sour ones. (The sour cherries would come to hand more readily when you sampled from the bag.) Let us keep in mind, therefore, that when we assess probabilities in terms of the ratio of favorable to all potential cases, we rely heavily on the assumption of equal probability for all outcomes.

Try it Now 1
At some random moment, you look at your clock and note the minutes reading.
a. What is probability the minutes reading is 15?
b. What is the probability the minutes reading is 15 or less?

> **Cards**
> A standard deck of 52 playing cards consists of four **suits** (hearts, spades, diamonds and clubs). Spades and clubs are black while hearts and diamonds are red. Each suit contains 13 cards, each of a different **rank**: an Ace (which in many games functions as both a low card and a high card), cards numbered 2 through 10, a Jack, a Queen and a King.

Example 4

> Compute the probability of randomly drawing one card from a deck and getting an Ace.
>
> There are 52 cards in the deck and 4 Aces so $P(Ace) = \dfrac{4}{52} = \dfrac{1}{13} \approx 0.0769$
>
> We can also think of probabilities as percents: There is a 7.69% chance that a randomly selected card will be an Ace.

Notice that the smallest possible probability is 0 – if there are no outcomes that correspond with the event. The largest possible probability is 1 – if all possible outcomes correspond with the event.

> **Certain and Impossible events**
> An impossible event has a probability of 0.
> A certain event has a probability of 1.
> The probability of any event must be $0 \le P(E) \le 1$

In the course of this chapter, *if you compute a probability and get an answer that is negative or greater than 1, you have made a mistake and should check your work.*

Working with Events

Complementary Events

Now let us examine the probability that an event does **not** happen. As in the previous section, consider the situation of rolling a six-sided die and first compute the probability of rolling a six: the answer is $P(\text{six}) = 1/6$. Now consider the probability that we do *not* roll a six: there are 5 outcomes that are not a six, so the answer is $P(\text{not a six}) = \dfrac{5}{6}$. Notice that

$$P(\text{six}) + P(\text{not a six}) = \frac{1}{6} + \frac{5}{6} = \frac{6}{6} = 1$$

This is not a coincidence. Consider a generic situation with n possible outcomes and an event E that corresponds to m of these outcomes. Then the remaining $n - m$ outcomes correspond to E not happening, thus

$$P(\text{not } E) = \frac{n-m}{n} = \frac{n}{n} - \frac{m}{n} = 1 - \frac{m}{n} = 1 - P(E)$$

> **Complement of an Event**
> The **complement** of an event is the event "E doesn't happen"
> The notation $\overline{E}$ is used for the complement of event E.
> We can compute the probability of the complement using $P(\overline{E}) = 1 - P(E)$
> Notice also that $P(E) = 1 - P(\overline{E})$

Example 5

If you pull a random card from a deck of playing cards, what is the probability it is not a heart?

There are 13 hearts in the deck, so $P(\text{heart}) = \dfrac{13}{52} = \dfrac{1}{4}$.

The probability of *not* drawing a heart is the complement:

$$P(\text{not heart}) = 1 - P(\text{heart}) = 1 - \frac{1}{4} = \frac{3}{4}$$

Probability of two independent events

Example 6

Suppose we flipped a coin and rolled a die, and wanted to know the probability of getting a head on the coin and a 6 on the die.

We could list all possible outcomes: $\{H1,H2,H3,H4,H5,H6,T1,T2,T3,T4,T5,T6\}$.

Notice there are $2 \cdot 6 = 12$ total outcomes. Out of these, only 1 is the desired outcome, so the probability is $\dfrac{1}{12}$.

The prior example was looking at two independent events.

> **Independent Events**
> Events A and B are **independent events** if the probability of Event B occurring is the same whether or not Event A occurs.

Example 7

Are these events independent?
a) A fair coin is tossed two times. The two events are (1) first toss is a head and (2) second toss is a head.

b) The two events (1) "It will rain tomorrow in Houston" and (2) "It will rain tomorrow in Galveston" (a city near Houston).

c) You draw a card from a deck, then draw a second card without replacing the first.

a) The probability that a head comes up on the second toss is 1/2 regardless of whether or not a head came up on the first toss, so these events are independent.

b) These events are not independent because it is more likely that it will rain in Galveston on days it rains in Houston than on days it does not.

c) The probability of the second card being red depends on whether the first card is red or not, so these events are not independent.

When two events are independent, the probability of both occurring is the product of the probabilities of the individual events.

> **$P(A$ and $B)$ for independent events**
> If events A and B are independent, then the probability of both A and B occurring is
>
> $P(A$ and $B) = P(A) \cdot P(B)$
>
> where $P(A$ and $B)$ is the probability of events A and B both occurring, $P(A)$ is the probability of event A occurring, and $P(B)$ is the probability of event B occurring

If you look back at the coin and die example from earlier, you can see how the number of outcomes of the first event multiplied by the number of outcomes in the second event multiplied to equal the total number of possible outcomes in the combined event.

Example 8

In your drawer you have 10 pairs of socks, 6 of which are white, and 7 tee shirts, 3 of which are white. If you randomly reach in and pull out a pair of socks and a tee shirt, what is the probability both are white?

The probability of choosing a white pair of socks is $\frac{6}{10}$.

The probability of choosing a white tee shirt is $\frac{3}{7}$.

The probability of both being white is $\frac{6}{10} \cdot \frac{3}{7} = \frac{18}{70} = \frac{9}{35}$

Try it Now 2

A card is pulled a deck of cards and noted. The card is then replaced, the deck is shuffled, and a second card is removed and noted. What is the probability that both cards are Aces?

The previous examples looked at the probability of *both* events occurring. Now we will look at the probability of *either* event occurring.

Example 9

Suppose we flipped a coin and rolled a die, and wanted to know the probability of getting a head on the coin *or* a 6 on the die.

Here, there are still 12 possible outcomes: {H1,H2,H3,H4,H5,H6,T1,T2,T3,T4,T5,T6}

By simply counting, we can see that 7 of the outcomes have a head on the coin *or* a 6 on the die *or* both – we use *or* inclusively here (these 7 outcomes are H1, H2, H3, H4, H5, H6, T6), so the probability is $\frac{7}{12}$. How could we have found this from the individual probabilities?

As we would expect, $\frac{1}{2}$ of these outcomes have a head, and $\frac{1}{6}$ of these outcomes have a 6 on the die. If we add these, $\frac{1}{2} + \frac{1}{6} = \frac{6}{12} + \frac{2}{12} = \frac{8}{12}$, which is not the correct probability. Looking at the outcomes we can see why: the outcome H6 would have been counted twice, since it contains both a head and a 6; the probability of both a head *and* rolling a 6 is $\frac{1}{12}$.

If we subtract out this double count, we have the correct probability: $\frac{8}{12} - \frac{1}{12} = \frac{7}{12}$.

> **$P(A$ or $B)$**
> The probability of either A or B occurring (or both) is
>
> $P(A$ or $B) = P(A) + P(B) - P(A$ and $B)$

Example 10

Suppose we draw one card from a standard deck. What is the probability that we get a Queen or a King?

There are 4 Queens and 4 Kings in the deck, hence 8 outcomes corresponding to a Queen or King out of 52 possible outcomes. Thus the probability of drawing a Queen or a King is:

$$P(\text{King or Queen}) = \frac{8}{52}$$

Note that in this case, there are no cards that are both a Queen and a King, so $P(\text{King and Queen}) = 0$. Using our probability rule, we could have said:

$$P(\text{King or Queen}) = P(\text{King}) + P(\text{Queen}) - P(\text{King and Queen}) = \frac{4}{52} + \frac{4}{52} - 0 = \frac{8}{52}$$

In the last example, the events were **mutually exclusive**, so $P(A$ or $B) = P(A) + P(B)$.

Example 11

Suppose we draw one card from a standard deck. What is the probability that we get a red card or a King?

Half the cards are red, so $P(\text{red}) = \dfrac{26}{52}$

There are four kings, so $P(\text{King}) = \dfrac{4}{52}$

There are two red kings, so $P(\text{Red and King}) = \dfrac{2}{52}$

We can then calculate

$$P(\text{Red or King}) = P(\text{Red}) + P(\text{King}) - P(\text{Red and King}) = \frac{26}{52} + \frac{4}{52} - \frac{2}{52} = \frac{28}{52}$$

Try it Now 3

In your drawer you have 10 pairs of socks, 6 of which are white, and 7 tee shirts, 3 of which are white. If you reach in and randomly grab a pair of socks and a tee shirt, what the probability at least one is white?

Example 12

The table below shows the number of survey subjects who have received and not received a speeding ticket in the last year, and the color of their car. Find the probability that a randomly chosen person:
a) Has a red car *and* got a speeding ticket
b) Has a red car *or* got a speeding ticket.

	Speeding ticket	No speeding ticket	Total
Red car	15	135	150
Not red car	45	470	515
Total	60	605	665

We can see that 15 people of the 665 surveyed had both a red car and got a speeding ticket, so the probability is $\dfrac{15}{665} \approx 0.0226$.

Notice that having a red car and getting a speeding ticket are not independent events, so the probability of both of them occurring is not simply the product of probabilities of each one occurring.

We could answer this question by simply adding up the numbers: 15 people with red cars and speeding tickets + 135 with red cars but no ticket + 45 with a ticket but no red car = 195 people. So the probability is $\dfrac{195}{665} \approx 0.2932$.

We also could have found this probability by:
P(had a red car) + P(got a speeding ticket) – P(had a red car and got a speeding ticket)
$$= \frac{150}{665} + \frac{60}{665} - \frac{15}{665} = \frac{195}{665}.$$

Conditional Probability

Often it is required to compute the probability of an event given that another event has occurred.

Example 13

What is the probability that two cards drawn at random from a deck of playing cards will both be aces?

It might seem that you could use the formula for the probability of two independent events and simply multiply $\dfrac{4}{52} \cdot \dfrac{4}{52} = \dfrac{1}{169}$. This would be incorrect, however, because the two events are not independent. If the first card drawn is an ace, then the probability that the second card is also an ace would be lower because there would only be three aces left in the deck.

Once the first card chosen is an ace, the probability that the second card chosen is also an ace is called the **conditional probability** of drawing an ace. In this case the "condition" is that the first card is an ace. Symbolically, we write this as:
P(ace on second draw | an ace on the first draw).

The vertical bar "|" is read as "given," so the above expression is short for "The probability that an ace is drawn on the second draw given that an ace was drawn on the first draw." What is this probability? After an ace is drawn on the first draw, there are 3 aces out of 51 total cards left. This means that the conditional probability of drawing an ace after one ace has already been drawn is $\dfrac{3}{51} = \dfrac{1}{17}$.

Thus, the probability of both cards being aces is $\dfrac{4}{52} \cdot \dfrac{3}{51} = \dfrac{12}{2652} = \dfrac{1}{221}$.

Conditional Probability
The probability the event B occurs, given that event A has happened, is represented as
$P(B \mid A)$
This is read as "the probability of B given A"

Example 14

Find the probability that a die rolled shows a 6, given that a flipped coin shows a head.

These are two independent events, so the probability of the die rolling a 6 is $\dfrac{1}{6}$, regardless of the result of the coin flip.

Example 15

The table below shows the number of survey subjects who have received and not received a speeding ticket in the last year, and the color of their car. Find the probability that a randomly chosen person:
a) Has a speeding ticket *given* they have a red car
b) Has a red car *given* they have a speeding ticket

	Speeding ticket	No speeding ticket	Total
Red car	15	135	150
Not red car	45	470	515
Total	60	605	665

a) Since we know the person has a red car, we are only considering the 150 people in the first row of the table. Of those, 15 have a speeding ticket, so

$$P(\text{ticket} \mid \text{red car}) = \frac{15}{150} = \frac{1}{10} = 0.1$$

b) Since we know the person has a speeding ticket, we are only considering the 60 people in the first column of the table. Of those, 15 have a red car, so

$$P(\text{red car} \mid \text{ticket}) = \frac{15}{60} = \frac{1}{4} = 0.25 \, .$$

Notice from the last example that P(B | A) is **not** equal to P(A | B).

These kinds of conditional probabilities are what insurance companies use to determine your insurance rates. They look at the conditional probability of you having accident, given your age, your car, your car color, your driving history, etc., and price your policy based on that likelihood.

> **Conditional Probability Formula**
> If Events A and B are not independent, then
> $P(A \text{ and } B) = P(A) \cdot P(B \mid A)$

Example 16

If you pull 2 cards out of a deck, what is the probability that both are spades?

The probability that the first card is a spade is $\frac{13}{52}$.

The probability that the second card is a spade, given the first was a spade, is $\frac{12}{51}$, since there is one less spade in the deck, and one less total cards.

The probability that both cards are spades is $\frac{13}{52} \cdot \frac{12}{51} = \frac{156}{2652} \approx 0.0588$

Example 17

If you draw two cards from a deck, what is the probability that you will get the Ace of Diamonds and a black card?

You can satisfy this condition by having Case A or Case B, as follows:
Case A) you can get the Ace of Diamonds first and then a black card or
Case B) you can get a black card first and then the Ace of Diamonds.

Let's calculate the probability of Case A. The probability that the first card is the Ace of Diamonds is $\frac{1}{52}$. The probability that the second card is black given that the first card is the Ace of Diamonds is $\frac{26}{51}$ because 26 of the remaining 51 cards are black. The probability is therefore $\frac{1}{52} \cdot \frac{26}{51} = \frac{1}{102}$.

Now for Case B: the probability that the first card is black is $\frac{26}{52} = \frac{1}{2}$. The probability that the second card is the Ace of Diamonds given that the first card is black is $\frac{1}{51}$. The probability of Case B is therefore $\frac{1}{2} \cdot \frac{1}{51} = \frac{1}{102}$, the same as the probability of Case 1.

Recall that the probability of A or B is $P(A) + P(B) - P(A \text{ and } B)$. In this problem, $P(A \text{ and } B) = 0$ since the first card cannot be the Ace of Diamonds and be a black card. Therefore, the probability of Case A or Case B is $\frac{1}{101} + \frac{1}{101} = \frac{2}{101}$. The probability that you will get the Ace of Diamonds and a black card when drawing two cards from a deck is $\frac{2}{101}$.

Try it Now 4
In your drawer you have 10 pairs of socks, 6 of which are white. If you reach in and randomly grab two pairs of socks, what is the probability that both are white?

Example 18

A home pregnancy test was given to women, then pregnancy was verified through blood tests. The following table shows the home pregnancy test results. Find
a) $P(\text{not pregnant} \mid \text{positive test result})$
b) $P(\text{positive test result} \mid \text{not pregnant})$

	Positive test	Negative test	Total
Pregnant	70	4	74
Not Pregnant	5	14	19
Total	75	18	93

a) Since we know the test result was positive, we're limited to the 75 women in the first column, of which 5 were not pregnant. $P(\text{not pregnant} \mid \text{positive test result}) = \frac{5}{75} \approx 0.067$.

b) Since we know the woman is not pregnant, we are limited to the 19 women in the second row, of which 5 had a positive test. $P(\text{positive test result} \mid \text{not pregnant}) = \frac{5}{19} \approx 0.263$

The second result is what is usually called a false positive: A positive result when the woman is not actually pregnant.

Bayes Theorem

In this section we concentrate on the more complex conditional probability problems we began looking at in the last section.

Example 19

Suppose a certain disease has an incidence rate of 0.1% (that is, it afflicts 0.1% of the population). A test has been devised to detect this disease. The test does not produce false negatives (that is, anyone who has the disease will test positive for it), but the false positive rate is 5% (that is, about 5% of people who take the test will test positive, even though they do not have the disease). Suppose a randomly selected person takes the test and tests positive. What is the probability that this person actually has the disease?

There are two ways to approach the solution to this problem. One involves an important result in probability theory called Bayes' theorem. We will discuss this theorem a bit later, but for now we will use an alternative and, we hope, much more intuitive approach.

Let's break down the information in the problem piece by piece.

Suppose a certain disease has an incidence rate of 0.1% (that is, it afflicts 0.1% of the population). The percentage 0.1% can be converted to a decimal number by moving the decimal place two places to the left, to get 0.001. In turn, 0.001 can be rewritten as a fraction: 1/1000. This tells us that about 1 in every 1000 people has the disease. (If we wanted we could write $P(\text{disease})=0.001$.)

A test has been devised to detect this disease. The test does not produce false negatives (that is, anyone who has the disease will test positive for it). This part is fairly straightforward: everyone who has the disease will test positive, or alternatively everyone who tests negative does not have the disease. (We could also say $P(\text{positive} \mid \text{disease})=1$.)

The false positive rate is 5% (that is, about 5% of people who take the test will test positive, even though they do not have the disease). This is even more straightforward. Another way of looking at it is that of every 100 people who are tested and do not have the disease, 5 will test positive even though they do not have the disease. (We could also say that $P(\text{positive} \mid \text{no disease})=0.05$.)

Suppose a randomly selected person takes the test and tests positive. What is the probability that this person actually has the disease? Here we want to compute $P(\text{disease}|\text{positive})$. We already know that $P(\text{positive}|\text{disease})=1$, but remember that conditional probabilities are not equal if the conditions are switched.

Rather than thinking in terms of all these probabilities we have developed, let's create a hypothetical situation and apply the facts as set out above. First, suppose we randomly select 1000 people and administer the test. How many do we expect to have the disease? Since about 1/1000 of all people are afflicted with the disease, 1/1000 of 1000 people is 1. (Now you know why we chose 1000.) Only 1 of 1000 test subjects actually has the disease; the other 999 do not.

We also know that 5% of all people who do not have the disease will test positive. There are 999 disease-free people, so we would expect (0.05)(999)=49.95 (so, about 50) people to test positive who do not have the disease.

Now back to the original question, computing P(disease|positive). There are 51 people who test positive in our example (the one unfortunate person who actually has the disease, plus the 50 people who tested positive but don't). Only one of these people has the disease, so

$$P(\text{disease} \mid \text{positive}) \approx \frac{1}{51} \approx 0.0196$$

or less than 2%. Does this surprise you? This means that of all people who test positive, over 98% *do not have the disease.*

The answer we got was slightly approximate, since we rounded 49.95 to 50. We could redo the problem with 100,000 test subjects, 100 of whom would have the disease and (0.05)(99,900)=4995 test positive but do not have the disease, so the exact probability of having the disease if you test positive is

$$P(\text{disease} \mid \text{positive}) \approx \frac{100}{5095} \approx 0.0196$$

which is pretty much the same answer.

But back to the surprising result. *Of all people who test positive, over 98% do not have the disease.* If your guess for the probability a person who tests positive has the disease was wildly different from the right answer (2%), don't feel bad. The exact same problem was posed to doctors and medical students at the Harvard Medical School 25 years ago and the results revealed in a 1978 *New England Journal of Medicine* article. Only about 18% of the participants got the right answer. Most of the rest thought the answer was closer to 95% (perhaps they were misled by the false positive rate of 5%).

So at least you should feel a little better that a bunch of doctors didn't get the right answer either (assuming you thought the answer was much higher). But the significance of this finding and similar results from other studies in the intervening years lies not in making math students feel better but in the possibly catastrophic consequences it might have for patient care. If a doctor thinks the chances that a positive test result nearly guarantees that a patient has a disease, they might begin an unnecessary and possibly harmful treatment regimen on a healthy patient. Or worse, as in the early days of the AIDS crisis when being HIV-positive was often equated with a death sentence, the patient might take a drastic action and commit suicide.

As we have seen in this hypothetical example, the most responsible course of action for treating a patient who tests positive would be to counsel the patient that they most likely do *not* have the disease and to order further, more reliable, tests to verify the diagnosis.

One of the reasons that the doctors and medical students in the study did so poorly is that such problems, when presented in the types of statistics courses that medical students often take, are solved by use of Bayes' theorem, which is stated as follows:

Bayes' Theorem
$$P(A\mid B) = \frac{P(A)P(B\mid A)}{P(A)P(B\mid A) + P(\overline{A})P(B\mid \overline{A})}$$

In our earlier example, this translates to

$$P(\text{disease}\mid \text{positive}) = \frac{P(\text{disease})P(\text{positive}\mid \text{disease})}{P(\text{disease})P(\text{positive}\mid \text{disease}) + P(\text{no disease})P(\text{positive}\mid \text{no disease})}$$

Plugging in the numbers gives

$$P(\text{disease}\mid \text{positive}) = \frac{(0.001)(1)}{(0.001)(1) + (0.999)(0.05)} \approx 0.0196$$

which is exactly the same answer as our original solution.

The problem is that you (or the typical medical student, or even the typical math professor) are much more likely to be able to remember the original solution than to remember Bayes' theorem. Psychologists, such as Gerd Gigerenzer, author of *Calculated Risks: How to Know When Numbers Deceive You*, have advocated that the method involved in the original solution (which Gigerenzer calls the method of "natural frequencies") be employed in place of Bayes' Theorem. Gigerenzer performed a study and found that those educated in the natural frequency method were able to recall it far longer than those who were taught Bayes' theorem. When one considers the possible life-and-death consequences associated with such calculations it seems wise to heed his advice.

Example 20

A certain disease has an incidence rate of 2%. If the false negative rate is 10% and the false positive rate is 1%, compute the probability that a person who tests positive actually has the disease.

Imagine 10,000 people who are tested. Of these 10,000, 200 will have the disease; 10% of them, or 20, will test negative and the remaining 180 will test positive. Of the 9800 who do not have the disease, 98 will test positive. So of the 278 total people who test positive, 180 will have the disease. Thus

$$P(\text{disease}\mid \text{positive}) = \frac{180}{278} \approx 0.647$$

so about 65% of the people who test positive will have the disease.

Using Bayes theorem directly would give the same result:

$$P(\text{disease}\mid \text{positive}) = \frac{(0.02)(0.90)}{(0.02)(0.90) + (0.98)(0.01)} = \frac{0.018}{0.0278} \approx 0.647$$

Try it Now 5

A certain disease has an incidence rate of 0.5%. If there are no false negatives and if the false positive rate is 3%, compute the probability that a person who tests positive actually has the disease.

Counting

Counting? You already know how to count or you wouldn't be taking a college-level math class, right? Well yes, but what we'll really be investigating here are ways of counting *efficiently*. When we get to the probability situations a bit later in this chapter we will need to count some *very* large numbers, like the number of possible winning lottery tickets. One way to do this would be to write down every possible set of numbers that might show up on a lottery ticket, but believe me: you don't want to do this.

Basic Counting

We will start, however, with some more reasonable sorts of counting problems in order to develop the ideas that we will soon need.

Example 21

Suppose at a particular restaurant you have three choices for an appetizer (soup, salad or breadsticks) and five choices for a main course (hamburger, sandwich, quiche, fajita or pizza). If you are allowed to choose exactly one item from each category for your meal, how many different meal options do you have?

Solution 1: One way to solve this problem would be to systematically list each possible meal:

soup + hamburger	soup + sandwich	soup + quiche
soup + fajita	soup + pizza	salad + hamburger
salad + sandwich	salad + quiche	salad + fajita
salad + pizza	breadsticks + hamburger	breadsticks + sandwich
breadsticks + quiche	breadsticks + fajita	breadsticks + pizza

Assuming that we did this systematically and that we neither missed any possibilities nor listed any possibility more than once, the answer would be 15. Thus you could go to the restaurant 15 nights in a row and have a different meal each night.

Solution 2: Another way to solve this problem would be to list all the possibilities in a table:

	hamburger	sandwich	quiche	fajita	pizza
soup	soup+burger				
salad	salad+burger				
bread	*etc.*				

In each of the cells in the table we could list the corresponding meal: soup + hamburger in the upper left corner, salad + hamburger below it, etc. But if we didn't really care *what* the possible meals are, only *how many* possible meals there are, we could just count the number of cells and arrive at an answer of 15, which matches our answer from the first solution. (It's always good when you solve a problem two different ways and get the same answer!)

Solution 3: We already have two perfectly good solutions. Why do we need a third? The first method was not very systematic, and we might easily have made an omission. The second method was better, but suppose that in addition to the appetizer and the main course we further complicated the problem by adding desserts to the menu: we've used the rows of the table for the appetizers and the columns for the main courses—where will the desserts go? We would need a third dimension, and since drawing 3-D tables on a 2-D page or computer screen isn't terribly easy, we need a better way in case we have three categories to choose form instead of just two.

So, back to the problem in the example. What else can we do? Let's draw a **tree diagram**:

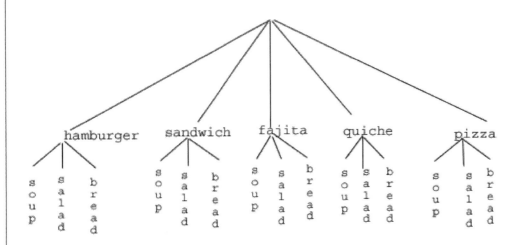

This is called a "tree" diagram because at each stage we branch out, like the branches on a tree. In this case, we first drew five branches (one for each main course) and then for each of those branches we drew three more branches (one for each appetizer). We count the number of branches at the final level and get (surprise, surprise!) 15.

If we wanted, we could instead draw three branches at the first stage for the three appetizers and then five branches (one for each main course) branching out of each of those three branches.

OK, so now we know how to count possibilities using tables and tree diagrams. These methods will continue to be useful in certain cases, but imagine a game where you have two decks of cards (with 52 cards in each deck) and you select one card from each deck. Would you really want to draw a table or tree diagram to determine the number of outcomes of this game?

Let's go back to the previous example that involved selecting a meal from three appetizers and five main courses, and look at the second solution that used a table. Notice that one way to count the number of possible meals is simply to number each of the appropriate cells in the table, as we have done above. But another way to count the number of cells in the table would be multiply the number of rows (3) by the number of columns (5) to get 15. Notice that we could have arrived at the same result without making a table at all by simply multiplying the number of choices for the appetizer (3) by the number of choices for the main course (5). We generalize this technique as the *basic counting rule*:

> **Basic Counting Rule**
> If we are asked to choose one item from each of two separate categories where there are m items in the first category and n items in the second category, then the total number of available choices is $m \cdot n$.
>
> This is sometimes called the multiplication rule for probabilities.

Example 22

There are 21 novels and 18 volumes of poetry on a reading list for a college English course. How many different ways can a student select one novel and one volume of poetry to read during the quarter?

There are 21 choices from the first category and 18 for the second, so there are $21 \cdot 18 = 378$ possibilities.

The Basic Counting Rule can be extended when there are more than two categories by applying it repeatedly, as we see in the next example.

Example 23

Suppose at a particular restaurant you have three choices for an appetizer (soup, salad or breadsticks), five choices for a main course (hamburger, sandwich, quiche, fajita or pasta) and two choices for dessert (pie or ice cream). If you are allowed to choose exactly one item from each category for your meal, how many different meal options do you have?

There are 3 choices for an appetizer, 5 for the main course and 2 for dessert, so there are $3 \cdot 5 \cdot 2 = 30$ possibilities.

Example 24

A quiz consists of 3 true-or-false questions. In how many ways can a student answer the quiz?

There are 3 questions. Each question has 2 possible answers (true or false), so the quiz may be answered in $2 \cdot 2 \cdot 2 = 8$ different ways. Recall that another way to write $2 \cdot 2 \cdot 2$ is 2^3, which is much more compact.

Try it Now 6

Suppose at a particular restaurant you have eight choices for an appetizer, eleven choices for a main course and five choices for dessert. If you are allowed to choose exactly one item from each category for your meal, how many different meal options do you have?

Permutations

In this section we will develop an even faster way to solve some of the problems we have already learned to solve by other means. Let's start with a couple examples.

Example 25

How many different ways can the letters of the word MATH be rearranged to form a four-letter code word?

This problem is a bit different. Instead of choosing one item from each of several different categories, we are repeatedly choosing items from the *same* category (the category is: the letters of the word MATH) and each time we choose an item we *do not replace* it, so there is one fewer choice at the next stage: we have 4 choices for the first letter (say we choose A), then 3 choices for the second (M, T and H; say we choose H), then 2 choices for the next letter (M and T; say we choose M) and only one choice at the last stage (T). Thus there are $4 \cdot 3 \cdot 2 \cdot 1 = 24$ ways to spell a code worth with the letters MATH.

In this example, we needed to calculate $n \cdot (n-1) \cdot (n-2) \cdots 3 \cdot 2 \cdot 1$. This calculation shows up often in mathematics, and is called the **factorial**, and is notated $n!$

> **Factorial**
> $n! = n \cdot (n-1) \cdot (n-2) \cdots 3 \cdot 2 \cdot 1$

Example 26

How many ways can five different door prizes be distributed among five people?

There are 5 choices of prize for the first person, 4 choices for the second, and so on. The number of ways the prizes can be distributed will be $5! = 5 \cdot 4 \cdot 3 \cdot 2 \cdot 1 = 120$ ways.

Now we will consider some slightly different examples.

Example 27

A charity benefit is attended by 25 people and three gift certificates are given away as door prizes: one gift certificate is in the amount of $100, the second is worth $25 and the third is worth $10. Assuming that no person receives more than one prize, how many different ways can the three gift certificates be awarded?

Using the Basic Counting Rule, there are 25 choices for the person who receives the $100 certificate, 24 remaining choices for the $25 certificate and 23 choices for the $10 certificate, so there are $25 \cdot 24 \cdot 23 = 13{,}800$ ways in which the prizes can be awarded.

Example 28

Eight sprinters have made it to the Olympic finals in the 100-meter race. In how many different ways can the gold, silver and bronze medals be awarded?

Using the Basic Counting Rule, there are 8 choices for the gold medal winner, 7 remaining choices for the silver, and 6 for the bronze, so there are $8 \cdot 7 \cdot 6 = 336$ ways the three medals can be awarded to the 8 runners.

Note that in these preceding examples, the gift certificates and the Olympic medals were awarded *without replacement*; that is, once we have chosen a winner of the first door prize or the gold medal, they are not eligible for the other prizes. Thus, at each succeeding stage of the solution there is one fewer choice (25, then 24, then 23 in the first example; 8, then 7, then 6 in the second). Contrast this with the situation of a multiple choice test, where there might be five possible answers — A, B, C, D or E — for each question on the test.

Note also that *the order of selection was important* in each example: for the three door prizes, being chosen first means that you receive substantially more money; in the Olympics example, coming in first means that you get the gold medal instead of the silver or bronze. In each case, if we had chosen the same three people in a different order there might have been a different person who received the $100 prize, or a different goldmedalist. (Contrast this with the situation where we might draw three names out of a hat to each receive a $10 gift certificate; in this case the order of selection is *not* important since each of the three people receive the same prize. Situations where the order is *not* important will be discussed in the next section.)

We can generalize the situation in the two examples above to any problem *without replacement* where the *order of selection is important*. If we are arranging in order *r* items out of *n* possibilities (instead of 3 out of 25 or 3 out of 8 as in the previous examples), the number of possible arrangements will be given by

$n \cdot (n - 1) \cdot (n - 2) \cdots (n - r + 1)$

If you don't see why $(n - r + 1)$ is the right number to use for the last factor, just think back to the first example in this section, where we calculated $25 \cdot 24 \cdot 23$ to get 13,800. In this case $n = 25$ and $r = 3$, so $n - r + 1 = 25 - 3 + 1 = 23$, which is exactly the right number for the final factor.

Now, why would we want to use this complicated formula when it's actually easier to use the Basic Counting Rule, as we did in the first two examples? Well, we won't actually use this formula all that often, we only developed it so that we could attach a special notation and a special definition to this situation where we are choosing *r* items out of *n* possibilities *without replacement* and where the *order of selection is important*. In this situation we write:

> **Permutations**
> $$_nP_r = n \cdot (n-1) \cdot (n-2) \cdots (n-r+1)$$
>
> We say that there are $_nP_r$ **permutations** of size r that may be selected from among n choices *without replacement* when *order matters*.
>
> It turns out that we can express this result more simply using factorials.
>
> $$_nP_r = \frac{n!}{(n-r)!}$$

In practicality, we usually use technology rather than factorials or repeated multiplication to compute permutations.

Example 29

I have nine paintings and have room to display only four of them at a time on my wall. How many different ways could I do this?

Since we are choosing 4 paintings out of 9 *without replacement* where the *order of selection is important* there are $_9P_4 = 9 \cdot 8 \cdot 7 \cdot 6 = 3{,}024$ permutations.

Example 30

How many ways can a four-person executive committee (president, vice-president, secretary, treasurer) be selected from a 16-member board of directors of a non-profit organization?

We want to choose 4 people out of 16 without replacement and where the order of selection is important. So the answer is $_{16}P_4 = 16 \cdot 15 \cdot 14 \cdot 13 = 43{,}680$.

Try it Now 7
How many 5 character passwords can be made using the letters A through Z
a. if repeats are allowed
b. if no repeats are allowed

Combinations

In the previous section we considered the situation where we chose r items out of n possibilities *without replacement* and where the *order of selection was important*. We now consider a similar situation in which the order of selection is *not* important.

Example 31

A charity benefit is attended by 25 people at which three $50 gift certificates are given away as door prizes. Assuming no person receives more than one prize, how many different ways can the gift certificates be awarded?

Using the Basic Counting Rule, there are 25 choices for the first person, 24 remaining choices for the second person and 23 for the third, so there are $25 \cdot 24 \cdot 23 = 13{,}800$ ways to choose three people. Suppose for a moment that Abe is chosen first, Bea second and Cindy third; this is one of the 13,800 possible outcomes. Another way to award the prizes would be to choose Abe first, Cindy second and Bea third; this is another of the 13,800 possible outcomes. But either way Abe, Bea and Cindy each get $50, so it doesn't really matter the order in which we select them. In how many different orders can Abe, Bea and Cindy be selected? It turns out there are 6:

ABC   ACB   BAC   BCA   CAB   CBA

How can we be sure that we have counted them all? We are really just choosing 3 people out of 3, so there are $3 \cdot 2 \cdot 1 = 6$ ways to do this; we didn't really need to list them all, we can just use permutations!

So, out of the 13,800 ways to select 3 people out of 25, six of them involve Abe, Bea and Cindy. The same argument works for any other group of three people (say Abe, Bea and David or Frank, Gloria and Hildy) so each three-person group is counted *six times*. Thus the 13,800 figure is six times too big. The number of distinct three-person groups will be $13{,}800/6 = 2300$.

We can generalize the situation in this example above to any problem of choosing a collection of items *without replacement* where the *order of selection is **not** important*. If we are choosing r items out of n possibilities (instead of 3 out of 25 as in the previous examples), the number of possible choices will be given by $\dfrac{{}_nP_r}{{}_rP_r}$, and we could use this formula for computation. However this situation arises so frequently that we attach a special notation and a special definition to this situation where we are choosing r items out of n possibilities *without replacement* where the *order of selection is **not** important*.

> **Combinations**
>
> $${}_nC_r = \frac{{}_nP_r}{{}_rP_r}$$
>
> We say that there are ${}_nC_r$ **combinations** of size r that may be selected from among n choices *without replacement* where *order doesn't matter*.
>
> We can also write the combinations formula in terms of factorials:
>
> $${}_nC_r = \frac{n!}{(n-r)!\,r!}$$

Example 32

A group of four students is to be chosen from a 35-member class to represent the class on the student council. How many ways can this be done?

Since we are choosing 4 people out of 35 *without replacement* where the *order of selection is* **not** *important* there are $_{35}C_4 = \dfrac{35 \cdot 34 \cdot 33 \cdot 32}{4 \cdot 3 \cdot 2 \cdot 1} = 52{,}360$ combinations.

Try it Now 8

The United States Senate Appropriations Committee consists of 29 members; the Defense Subcommittee of the Appropriations Committee consists of 19 members. Disregarding party affiliation or any special seats on the Subcommittee, how many different 19-member subcommittees may be chosen from among the 29 Senators on the Appropriations Committee?

In the preceding Try it Now problem we assumed that the 19 members of the Defense Subcommittee were chosen without regard to party affiliation. In reality this would never happen: if Republicans are in the majority they would never let a majority of Democrats sit on (and thus control) any subcommittee. (The same of course would be true if the Democrats were in control.) So let's consider the problem again, in a slightly more complicated form:

Example 33

The United States Senate Appropriations Committee consists of 29 members, 15 Republicans and 14 Democrats. The Defense Subcommittee consists of 19 members, 10 Republicans and 9 Democrats. How many different ways can the members of the Defense Subcommittee be chosen from among the 29 Senators on the Appropriations Committee?

In this case we need to choose 10 of the 15 Republicans and 9 of the 14 Democrats. There are $_{15}C_{10} = 3003$ ways to choose the 10 Republicans and $_{14}C_9 = 2002$ ways to choose the 9 Democrats. But now what? How do we finish the problem?

Suppose we listed all of the possible 10-member Republican groups on 3003 slips of red paper and all of the possible 9-member Democratic groups on 2002 slips of blue paper. How many ways can we choose one red slip and one blue slip? This is a job for the Basic Counting Rule! We are simply making one choice from the first category and one choice from the second category, just like in the restaurant menu problems from earlier.

There must be $3003 \cdot 2002 = 6{,}012{,}006$ possible ways of selecting the members of the Defense Subcommittee.

Probability using Permutations and Combinations

We can use permutations and combinations to help us answer more complex probability questions

Example 34

A 4 digit PIN number is selected. What is the probability that there are no repeated digits?

There are 10 possible values for each digit of the PIN (namely: 0, 1, 2, 3, 4, 5, 6, 7, 8, 9), so there are $10 \cdot 10 \cdot 10 \cdot 10 = 10^4 = 10000$ total possible PIN numbers.

To have no repeated digits, all four digits would have to be different, which is selecting without replacement. We could either compute $10 \cdot 9 \cdot 8 \cdot 7$, or notice that this is the same as the permutation $_{10}P_4 = 5040$.

The probability of no repeated digits is the number of 4 digit PIN numbers with no repeated digits divided by the total number of 4 digit PIN numbers. This probability is

$$\frac{_{10}P_4}{10^4} = \frac{5040}{10000} = 0.504$$

Example 35

In a certain state's lottery, 48 balls numbered 1 through 48 are placed in a machine and six of them are drawn at random. If the six numbers drawn match the numbers that a player had chosen, the player wins $1,000,000. In this lottery, the order the numbers are drawn in doesn't matter. Compute the probability that you win the million-dollar prize if you purchase a single lottery ticket.

In order to compute the probability, we need to count the total number of ways six numbers can be drawn, and the number of ways the six numbers on the player's ticket could match the six numbers drawn from the machine. Since there is no stipulation that the numbers be in any particular order, the number of possible outcomes of the lottery drawing is $_{48}C_6 = 12,271,512$. Of these possible outcomes, only one would match all six numbers on the player's ticket, so the probability of winning the grand prize is:

$$\frac{_6C_6}{_{48}C_6} = \frac{1}{12271512} \approx 0.0000000815$$

Example 36

In the state lottery from the previous example, if five of the six numbers drawn match the numbers that a player has chosen, the player wins a second prize of $1,000. Compute the probability that you win the second prize if you purchase a single lottery ticket.

As above, the number of possible outcomes of the lottery drawing is $_{48}C_6 = 12,271,512$. In order to win the second prize, five of the six numbers on the ticket must match five of the six winning numbers; in other words, we must have chosen five of the six winning numbers and one of the 42 losing numbers. The number of ways to choose 5 out of the 6 winning numbers is given by $_6C_5 = 6$ and the number of ways to choose 1 out of the 42 losing numbers is given by $_{42}C_1 = 42$. Thus the number of favorable outcomes is then given by the Basic Counting Rule: $_6C_5 \cdot {}_{42}C_1 = 6 \cdot 42 = 252$. So the probability of winning the second prize is.

$$\frac{\left(_6C_5\right)\left(_{42}C_1\right)}{_{48}C_6} = \frac{252}{12271512} \approx 0.0000205$$

Try it Now 9

A multiple-choice question on an economics quiz contains 10 questions with five possible answers each. Compute the probability of randomly guessing the answers and getting 9 questions correct.

Example 37

Compute the probability of randomly drawing five cards from a deck and getting exactly one Ace.

In many card games (such as poker) the order in which the cards are drawn is not important (since the player may rearrange the cards in his hand any way he chooses); in the problems that follow, we will assume that this is the case unless otherwise stated. Thus we use combinations to compute the possible number of 5-card hands, $_{52}C_5$. This number will go in the denominator of our probability formula, since it is the number of possible outcomes.

For the numerator, we need the number of ways to draw one Ace and four other cards (none of them Aces) from the deck. Since there are four Aces and we want exactly one of them, there will be $_4C_1$ ways to select one Ace; since there are 48 non-Aces and we want 4 of them, there will be $_{48}C_4$ ways to select the four non-Aces. Now we use the Basic Counting Rule to calculate that there will be $_4C_1 \cdot {}_{48}C_4$ ways to choose one ace and four non-Aces.

Putting this all together, we have

$$P(\text{one Ace}) = \frac{\left(_4C_1\right)\left(_{48}C_4\right)}{_{52}C_5} = \frac{778320}{2598960} \approx 0.299$$

Example 38

Compute the probability of randomly drawing five cards from a deck and getting exactly two Aces.

The solution is similar to the previous example, except now we are choosing 2 Aces out of 4 and 3 non-Aces out of 48; the denominator remains the same:

$$P(\text{two Aces}) = \frac{\left(_4C_2\right)\left(_{48}C_3\right)}{_{52}C_5} = \frac{103776}{2598960} \approx 0.0399$$

It is useful to note that these card problems are remarkably similar to the lottery problems discussed earlier.

Try it Now 10
Compute the probability of randomly drawing five cards from a deck of cards and getting three Aces and two Kings.

Birthday Problem

Let's take a pause to consider a famous problem in probability theory:

> Suppose you have a room full of 30 people. What is the probability that there is at least one shared birthday?

Take a guess at the answer to the above problem. Was your guess fairly low, like around 10%? That seems to be the intuitive answer (30/365, perhaps?). Let's see if we should listen to our intuition. Let's start with a simpler problem, however.

Example 39

Suppose three people are in a room. What is the probability that there is at least one shared birthday among these three people?

There are a lot of ways there could be at least one shared birthday. Fortunately there is an easier way. We ask ourselves "What is the alternative to having at least one shared birthday?" In this case, the alternative is that there are **no** shared birthdays. In other words, the alternative to "at least one" is having **none**. In other words, since this is a complementary event,

P(at least one) = 1 – P(none)

We will start, then, by computing the probability that there is no shared birthday. Let's imagine that you are one of these three people. Your birthday can be anything without conflict, so there are 365 choices out of 365 for your birthday. What is the probability that the second person does not share your birthday? There are 365 days in the year (let's ignore leap years) and removing your birthday from contention, there are 364 choices that will guarantee that you do not share a birthday with this person, so the probability that the second person does not share your birthday is 364/365. Now we move to the third person. What is the probability that this third person does not have the same birthday as either you or the second person? There are 363 days that will not duplicate your birthday or the second person's, so the probability that the third person does not share a birthday with the first two is 363/365.

We want the second person not to share a birthday with you *and* the third person not to share a birthday with the first two people, so we use the multiplication rule:

$$P(\text{no shared birthday}) = \frac{365}{365} \cdot \frac{364}{365} \cdot \frac{363}{365} \approx 0.9918$$

and then subtract from 1 to get

$$P(\text{shared birthday}) = 1 - P(\text{no shared birthday}) = 1 - 0.9918 = 0.0082.$$

This is a pretty small number, so maybe it makes sense that the answer to our original problem will be small. Let's make our group a bit bigger.

Example 40

Suppose five people are in a room. What is the probability that there is at least one shared birthday among these five people?

Continuing the pattern of the previous example, the answer should be

$$P(\text{shared birthday}) = 1 - \frac{365}{365} \cdot \frac{364}{365} \cdot \frac{363}{365} \cdot \frac{362}{365} \cdot \frac{361}{365} \approx 0.0271$$

Note that we could rewrite this more compactly as

$$P(\text{shared birthday}) = 1 - \frac{{}_{365}P_5}{365^5} \approx 0.0271$$

which makes it a bit easier to type into a calculator or computer, and which suggests a nice formula as we continue to expand the population of our group.

Example 41

Suppose 30 people are in a room. What is the probability that there is at least one shared birthday among these 30 people?

Here we can calculate

$$P(\text{shared birthday}) = 1 - \frac{{}_{365}P_{30}}{365^{30}} \approx 0.706$$

which gives us the surprising result that when you are in a room with 30 people there is a 70% chance that there will be at least one shared birthday!

If you like to bet, and if you can convince 30 people to reveal their birthdays, you might be able to win some money by betting a friend that there will be at least two people with the same birthday in the room anytime you are in a room of 30 or more people. (Of course, you would need to make sure your friend hasn't studied probability!) You wouldn't be guaranteed to win, but you should win more than half the time.

This is one of many results in probability theory that is counterintuitive; that is, it goes against our gut instincts. If you still don't believe the math, you can carry out a simulation.

Just so you won't have to go around rounding up groups of 30 people, someone has kindly developed a Java applet so that you can conduct a computer simulation. Go to this web page: http://www-stat.stanford.edu/~susan/surprise/Birthday.html, and once the applet has loaded, select 30 birthdays and then keep clicking Start and Reset. If you keep track of the number of times that there is a repeated birthday, you should get a repeated birthday about 7 out of every 10 times you run the simulation.

Try it Now 11
Suppose 10 people are in a room. What is the probability that there is at least one shared birthday among these 10 people?

Expected Value

Expected value is perhaps the most useful probability concept we will discuss. It has many applications, from insurance policies to making financial decisions, and it's one thing that the casinos and government agencies that run gambling operations and lotteries hope most people never learn about.

Example 42

[1]In the casino game roulette, a wheel with 38 spaces (18 red, 18 black, and 2 green) is spun. In one possible bet, the player bets $1 on a single number. If that number is spun on the wheel, then they receive $36 (their original $1 + $35). Otherwise, they lose their $1. On average, how much money should a player expect to win or lose if they play this game repeatedly?

Suppose you bet $1 on each of the 38 spaces on the wheel, for a total of $38 bet. When the winning number is spun, you are paid $36 on that number. While you won on that one number, overall you've lost $2. On a per-space basis, you have "won" -$2/$38 ≈ -$0.053. In other words, on average you lose 5.3 cents per space you bet on.

We call this average gain or loss the expected value of playing roulette. Notice that no one ever loses exactly 5.3 cents: most people (in fact, about 37 out of every 38) lose $1 and a very few people (about 1 person out of every 38) gain $35 (the $36 they win minus the $1 they spent to play the game).

There is another way to compute expected value without imagining what would happen if we play every possible space. There are 38 possible outcomes when the wheel spins, so the probability of winning is $\frac{1}{38}$. The complement, the probability of losing, is $\frac{37}{38}$.

---

[1] Photo CC-BY-SA http://www.flickr.com/photos/stoneflower/

Summarizing these along with the values, we get this table:

Outcome	Probability of outcome
$35	$\dfrac{1}{38}$
-$1	$\dfrac{37}{38}$

Notice that if we multiply each outcome by its corresponding probability we get $\$35 \cdot \dfrac{1}{38} = 0.9211$ and $-\$1 \cdot \dfrac{37}{38} = -0.9737$, and if we add these numbers we get $0.9211 + (-0.9737) \approx -0.053$, which is the expected value we computed above.

Expected Value

Expected Value is the average gain or loss of an event if the procedure is repeated many times.

We can compute the expected value by multiplying each outcome by the probability of that outcome, then adding up the products.

Try it Now 12

You purchase a raffle ticket to help out a charity. The raffle ticket costs $5. The charity is selling 2000 tickets. One of them will be drawn and the person holding the ticket will be given a prize worth $4000. Compute the expected value for this raffle.

Example 43

In a certain state's lottery, 48 balls numbered 1 through 48 are placed in a machine and six of them are drawn at random. If the six numbers drawn match the numbers that a player had chosen, the player wins $1,000,000. If they match 5 numbers, then win $1,000. It costs $1 to buy a ticket. Find the expected value.

Earlier, we calculated the probability of matching all 6 numbers and the probability of matching 5 numbers:

$$\frac{_6C_6}{_{48}C_6} = \frac{1}{12271512} \approx 0.0000000815 \text{ for all 6 numbers,}$$

$$\frac{(_6C_5)(_{42}C_1)}{_{48}C_6} = \frac{252}{12271512} \approx 0.0000205 \text{ for 5 numbers.}$$

Our probabilities and outcome values are:

Outcome	Probability of outcome
$999,999	$\dfrac{1}{12271512}$
$999	$\dfrac{252}{12271512}$
-$1	$1 - \dfrac{253}{12271512} = \dfrac{12271259}{12271512}$

The expected value, then is:

$$(\$999,999)\cdot\frac{1}{12271512} + (\$999)\cdot\frac{252}{12271512} + (-\$1)\cdot\frac{12271259}{12271512} \approx -\$0.898$$

On average, one can expect to lose about 90 cents on a lottery ticket. Of course, most players will lose $1.

In general, if the expected value of a game is negative, it is not a good idea to play the game, since on average you will lose money. It would be better to play a game with a positive expected value (good luck trying to find one!), although keep in mind that even if the *average* winnings are positive it could be the case that most people lose money and one very fortunate individual wins a great deal of money. If the expected value of a game is 0, we call it a **fair game**, since neither side has an advantage.

Not surprisingly, the expected value for casino games is negative for the player, which is positive for the casino. It must be positive or they would go out of business. Players just need to keep in mind that when they play a game repeatedly, their expected value is negative. That is fine so long as you enjoy playing the game and think it is worth the cost. But it would be wrong to expect to come out ahead.

Try it Now 13
A friend offers to play a game, in which you roll 3 standard 6-sided dice. If all the dice roll different values, you give him $1. If any two dice match values, you get $2. What is the expected value of this game? Would you play?

Expected value also has applications outside of gambling. Expected value is very common in making insurance decisions.

Example 44

A 40-year-old man in the U.S. has a 0.242% risk of dying during the next year[2]. An insurance company charges $275 for a life-insurance policy that pays a $100,000 death benefit. What is the expected value for the person buying the insurance?

The probabilities and outcomes are

Outcome	Probability of outcome
$100,000 - $275 = $99,725	0.00242
-$275	1 - 0.00242 = 0.99758

The expected value is ($99,725)(0.00242) + (-$275)(0.99758) = -$33.

Not surprisingly, the expected value is negative; the insurance company can only afford to offer policies if they, on average, make money on each policy. They can afford to pay out the occasional benefit because they offer enough policies that those benefit payouts are balanced by the rest of the insured people.

For people buying the insurance, there is a negative expected value, but there is a security that comes from insurance that is worth that cost.

Try it Now Answers

1. There are 60 possible readings, from 00 to 59. a. $\dfrac{1}{60}$ b. $\dfrac{16}{60}$ (counting 00 through 15)

2. Since the second draw is made after replacing the first card, these events are independent. The probability of an ace on each draw is $\dfrac{4}{52} = \dfrac{1}{13}$, so the probability of an Ace on both draws is $\dfrac{1}{13} \cdot \dfrac{1}{13} = \dfrac{1}{169}$

3. P(white sock and white tee) = $\dfrac{6}{10} \cdot \dfrac{3}{7} = \dfrac{9}{35}$

   P(white sock or white tee) = $\dfrac{6}{10} + \dfrac{3}{7} - \dfrac{9}{35} = \dfrac{27}{35}$

4. a. $\dfrac{6}{10} \cdot \dfrac{5}{9} = \dfrac{30}{90} = \dfrac{1}{3}$

---

[2] According to the estimator at http://www.numericalexample.com/index.php?view=article&id=91

5. Out of 100,000 people, 500 would have the disease. Of those, all 500 would test positive. Of the 99,500 without the disease, 2,985 would falsely test positive and the other 96,515 would test negative.

$$P(\text{disease} \mid \text{positive}) = \frac{500}{500 + 2985} = \frac{500}{3485} \approx 14.3\%$$

6. $8 \cdot 11 \cdot 5 = 440$ menu combinations

7. There are 26 characters. a. $26^5 = 11{,}881{,}376$.   b. $_{26}P_5 = 26 \cdot 25 \cdot 24 \cdot 23 \cdot 22 = 7{,}893{,}600$

8. Order does not matter. $_{29}C_{19} = 20{,}030{,}010$ possible subcommittees

9. There are $5^{10} = 9{,}765{,}625$ different ways the exam can be answered. There are 9 possible locations for the one missed question, and in each of those locations there are 4 wrong answers, so there are 36 ways the test could be answered with one wrong answer.

$$P(\text{9 answers correct}) = \frac{36}{5^{10}} \approx 0.0000037 \text{ chance}$$

10. $P(\text{three Aces and two Kings}) = \dfrac{\left(_4C_3\right)\left(_4C_2\right)}{_{52}C_5} = \dfrac{24}{2598960} \approx 0.0000092$

11. $P(\text{shared birthday}) = 1 - \dfrac{_{365}P_{10}}{365^{10}} \approx 0.117$

12. $(\$3{,}995) \cdot \dfrac{1}{2000} + (-\$5) \cdot \dfrac{1999}{2000} \approx -\$3.00$

13. Suppose you roll the first die. The probability the second will be different is $\dfrac{5}{6}$. The probability that the third roll is different than the previous two is $\dfrac{4}{6}$, so the probability that the three dice are different is $\dfrac{5}{6} \cdot \dfrac{4}{6} = \dfrac{20}{36}$. The probability that two dice will match is the complement, $1 - \dfrac{20}{36} = \dfrac{16}{36}$.

The expected value is: $(\$2) \cdot \dfrac{16}{36} + (-\$1) \cdot \dfrac{20}{36} = \dfrac{12}{36} \approx \$0.33$. Yes, it is in your advantage to play. On average, you'd win $0.33 per play.

Exercises

1. A ball is drawn randomly from a jar that contains 6 red balls, 2 white balls, and 5 yellow balls. Find the probability of the given event.
    a. A red ball is drawn
    b. A white ball is drawn

2. Suppose you write each letter of the alphabet on a different slip of paper and put the slips into a hat. What is the probability of drawing one slip of paper from the hat at random and getting:
    a. A consonant
    b. A vowel

3. A group of people were asked if they had run a red light in the last year. 150 responded "yes", and 185 responded "no". Find the probability that if a person is chosen at random, they have run a red light in the last year.

4. In a survey, 205 people indicated they prefer cats, 160 indicated they prefer dots, and 40 indicated they don't enjoy either pet. Find the probability that if a person is chosen at random, they prefer cats.

5. Compute the probability of tossing a six-sided die (with sides numbered 1 through 6) and getting a 5.

6. Compute the probability of tossing a six-sided die and getting a 7.

7. Giving a test to a group of students, the grades and gender are summarized below. If one student was chosen at random, find the probability that the student was female.

	A	B	C	Total
Male	8	18	13	39
Female	10	4	12	26
Total	18	22	25	65

8. The table below shows the number of credit cards owned by a group of individuals. If one person was chosen at random, find the probability that the person had no credit cards.

	Zero	One	Two or more	Total
Male	9	5	19	33
Female	18	10	20	48
Total	27	15	39	81

9. Compute the probability of tossing a six-sided die and getting an even number.

10. Compute the probability of tossing a six-sided die and getting a number less than 3.

11. If you pick one card at random from a standard deck of cards, what is the probability it will be a King?

12. If you pick one card at random from a standard deck of cards, what is the probability it will be a Diamond?

13. Compute the probability of rolling a 12-sided die and getting a number other than 8.

14. If you pick one card at random from a standard deck of cards, what is the probability it is not the Ace of Spades?

15. Referring to the grade table from question #7, what is the probability that a student chosen at random did NOT earn a C?

16. Referring to the credit card table from question #8, what is the probability that a person chosen at random has at least one credit card?

17. A six-sided die is rolled twice. What is the probability of showing a 6 on both rolls?

18. A fair coin is flipped twice. What is the probability of showing heads on both flips?

19. A die is rolled twice. What is the probability of showing a 5 on the first roll and an even number on the second roll?

20. Suppose that 21% of people own dogs. If you pick two people at random, what is the probability that they both own a dog?

21. Suppose a jar contains 17 red marbles and 32 blue marbles. If you reach in the jar and pull out 2 marbles at random, find the probability that both are red.

22. Suppose you write each letter of the alphabet on a different slip of paper and put the slips into a hat. If you pull out two slips at random, find the probability that both are vowels.

23. Bert and Ernie each have a well-shuffled standard deck of 52 cards. They each draw one card from their own deck. Compute the probability that:
    a. Bert and Ernie both draw an Ace.
    b. Bert draws an Ace but Ernie does not.
    c. neither Bert nor Ernie draws an Ace.
    d. Bert and Ernie both draw a heart.
    e. Bert gets a card that is not a Jack and Ernie draws a card that is not a heart.

24. Bert has a well-shuffled standard deck of 52 cards, from which he draws one card; Ernie has a 12-sided die, which he rolls at the same time Bert draws a card. Compute the probability that:
    a. Bert gets a Jack and Ernie rolls a five.
    b. Bert gets a heart and Ernie rolls a number less than six.
    c. Bert gets a face card (Jack, Queen or King) and Ernie rolls an even number.
    d. Bert gets a red card and Ernie rolls a fifteen.
    e. Bert gets a card that is not a Jack and Ernie rolls a number that is not twelve.

25. Compute the probability of drawing a King from a deck of cards and then drawing a Queen.

26. Compute the probability of drawing two spades from a deck of cards.

27. A math class consists of 25 students, 14 female and 11 male. Two students are selected at random to participate in a probability experiment. Compute the probability that
    a. a male is selected, then a female.
    b. a female is selected, then a male.
    c. two males are selected.
    d. two females are selected.
    e. no males are selected.

28. A math class consists of 25 students, 14 female and 11 male. Three students are selected at random to participate in a probability experiment. Compute the probability that
    a. a male is selected, then two females.
    b. a female is selected, then two males.
    c. two females are selected, then one male.
    d. three males are selected.
    e. three females are selected.

29. Giving a test to a group of students, the grades and gender are summarized below. If one student was chosen at random, find the probability that the student was female and earned an A.

	A	B	C	Total
Male	8	18	13	39
Female	10	4	12	26
Total	18	22	25	65

30. The table below shows the number of credit cards owned by a group of individuals. If one person was chosen at random, find the probability that the person was male and had two or more credit cards.

	Zero	One	Two or more	Total
Male	9	5	19	33
Female	18	10	20	48
Total	27	15	39	81

31. A jar contains 6 red marbles numbered 1 to 6 and 8 blue marbles numbered 1 to 8. A marble is drawn at random from the jar. Find the probability the marble is red or odd-numbered.

32. A jar contains 4 red marbles numbered 1 to 4 and 10 blue marbles numbered 1 to 10. A marble is drawn at random from the jar. Find the probability the marble is blue or even-numbered.

33. Referring to the table from #29, find the probability that a student chosen at random is female or earned a B.

34. Referring to the table from #30, find the probability that a person chosen at random is male or has no credit cards.

35. Compute the probability of drawing the King of hearts or a Queen from a deck of cards.

36. Compute the probability of drawing a King or a heart from a deck of cards.

37. A jar contains 5 red marbles numbered 1 to 5 and 8 blue marbles numbered 1 to 8. A marble is drawn at random from the jar. Find the probability the marble is
   a. Even-numbered given that the marble is red.
   b. Red given that the marble is even-numbered.

38. A jar contains 4 red marbles numbered 1 to 4 and 8 blue marbles numbered 1 to 8. A marble is drawn at random from the jar. Find the probability the marble is
   a. Odd-numbered given that the marble is blue.
   b. Blue given that the marble is odd-numbered.

39. Compute the probability of flipping a coin and getting heads, given that the previous flip was tails.

40. Find the probability of rolling a "1" on a fair die, given that the last 3 rolls were all ones.

41. Suppose a math class contains 25 students, 14 females (three of whom speak French) and 11 males (two of whom speak French). Compute the probability that a randomly selected student speaks French, given that the student is female.

42. Suppose a math class contains 25 students, 14 females (three of whom speak French) and 11 males (two of whom speak French). Compute the probability that a randomly selected student is male, given that the student speaks French.

43. A certain virus infects one in every 400 people. A test used to detect the virus in a person is positive 90% of the time if the person has the virus and 10% of the time if the person does not have the virus. Let A be the event "the person is infected" and B be the event "the person tests positive".
   a. Find the probability that a person has the virus given that they have tested positive, i.e. find P(A | B).
   b. Find the probability that a person does not have the virus given that they test negative, i.e. find P(not A | not B).

44. A certain virus infects one in every 2000 people. A test used to detect the virus in a person is positive 96% of the time if the person has the virus and 4% of the time if the person does not have the virus. Let A be the event "the person is infected" and B be the event "the person tests positive".
   a. Find the probability that a person has the virus given that they have tested positive, i.e. find P(A | B).
   b. Find the probability that a person does not have the virus given that they test negative, i.e. find P(not A | not B).

45. A certain disease has an incidence rate of 0.3%. If the false negative rate is 6% and the false positive rate is 4%, compute the probability that a person who tests positive actually has the disease.

46. A certain disease has an incidence rate of 0.1%. If the false negative rate is 8% and the false positive rate is 3%, compute the probability that a person who tests positive actually has the disease.

47. A certain group of symptom-free women between the ages of 40 and 50 are randomly selected to participate in mammography screening. The incidence rate of breast cancer among such women is 0.8%. The false negative rate for the mammogram is 10%. The false positive rate is 7%. If a the mammogram results for a particular woman are positive (indicating that she has breast cancer), what is the probability that she actually has breast cancer?

48. About 0.01% of men with no known risk behavior are infected with HIV. The false negative rate for the standard HIV test 0.01% and the false positive rate is also 0.01%. If a randomly selected man with no known risk behavior tests positive for HIV, what is the probability that he is actually infected with HIV?

49. A boy owns 2 pairs of pants, 3 shirts, 8 ties, and 2 jackets. How many different outfits can he wear to school if he must wear one of each item?

50. At a restaurant you can choose from 3 appetizers, 8 entrees, and 2 desserts. How many different three-course meals can you have?

51. How many three-letter "words" can be made from 4 letters "FGHI" if
    a. repetition of letters is allowed
    b. repetition of letters is not allowed

52. How many four-letter "words" can be made from 6 letters "AEBWDP" if
    a. repetition of letters is allowed
    b. repetition of letters is not allowed

53. All of the license plates in a particular state feature three letters followed by three digits (e.g. ABC 123). How many different license plate numbers are available to the state's Department of Motor Vehicles?

54. A computer password must be eight characters long. How many passwords are possible if only the 26 letters of the alphabet are allowed?

55. A pianist plans to play 4 pieces at a recital. In how many ways can she arrange these pieces in the program?

56. In how many ways can first, second, and third prizes be awarded in a contest with 210 contestants?

57. Seven Olympic sprinters are eligible to compete in the 4 x 100 m relay race for the USA Olympic team. How many four-person relay teams can be selected from among the seven athletes?

58. A computer user has downloaded 25 songs using an online file-sharing program and wants to create a CD-R with ten songs to use in his portable CD player. If the order that the songs are placed on the CD-R is important to him, how many different CD-Rs could he make from the 25 songs available to him?

59. In western music, an octave is divided into 12 pitches. For the film *Close Encounters of the Third Kind*, director Steven Spielberg asked composer John Williams to write a five-note theme, which aliens would use to communicate with people on Earth. Disregarding rhythm and octave changes, how many five-note themes are possible if no note is repeated?

60. In the early twentieth century, proponents of the Second Viennese School of musical composition (including Arnold Schönberg, Anton Webern and Alban Berg) devised the twelve-tone technique, which utilized a tone row consisting of all 12 pitches from the chromatic scale in any order, but with not pitches repeated in the row. Disregarding rhythm and octave changes, how many tone rows are possible?

61. In how many ways can 4 pizza toppings be chosen from 12 available toppings?

62. At a baby shower 17 guests are in attendance and 5 of them are randomly selected to receive a door prize. If all 5 prizes are identical, in how many ways can the prizes be awarded?

63. In the 6/50 lottery game, a player picks six numbers from 1 to 50. How many different choices does the player have if order doesn't matter?

64. In a lottery daily game, a player picks three numbers from 0 to 9. How many different choices does the player have if order doesn't matter?

65. A jury pool consists of 27 people. How many different ways can 11 people be chosen to serve on a jury and one additional person be chosen to serve as the jury foreman?

66. The United States Senate Committee on Commerce, Science, and Transportation consists of 23 members, 12 Republicans and 11 Democrats. The Surface Transportation and Merchant Marine Subcommittee consists of 8 Republicans and 7 Democrats. How many ways can members of the Subcommittee be chosen from the Committee?

67. You own 16 CDs. You want to randomly arrange 5 of them in a CD rack. What is the probability that the rack ends up in alphabetical order?

68. A jury pool consists of 27 people, 14 men and 13 women. Compute the probability that a randomly selected jury of 12 people is all male.

69. In a lottery game, a player picks six numbers from 1 to 48. If 5 of the 6 numbers match those drawn, they player wins second prize. What is the probability of winning this prize?

70. In a lottery game, a player picks six numbers from 1 to 48. If 4 of the 6 numbers match those drawn, they player wins third prize. What is the probability of winning this prize?

71. Compute the probability that a 5-card poker hand is dealt to you that contains all hearts.

72. Compute the probability that a 5-card poker hand is dealt to you that contains four Aces.

73. A bag contains 3 gold marbles, 6 silver marbles, and 28 black marbles. Someone offers to play this game: You randomly select on marble from the bag. If it is gold, you win $3. If it is silver, you win $2. If it is black, you lose $1. What is your expected value if you play this game?

74. A friend devises a game that is played by rolling a single six-sided die once. If you roll a 6, he pays you $3; if you roll a 5, he pays you nothing; if you roll a number less than 5, you pay him $1. Compute the expected value for this game. Should you play this game?

75. In a lottery game, a player picks six numbers from 1 to 23. If the player matches all six numbers, they win 30,000 dollars. Otherwise, they lose $1. Find the expected value of this game.

76. A game is played by picking two cards from a deck. If they are the same value, then you win $5, otherwise you lose $1. What is the expected value of this game?

77. A company estimates that 0.7% of their products will fail after the original warranty period but within 2 years of the purchase, with a replacement cost of $350. If they offer a 2 year extended warranty for $48, what is the company's expected value of each warranty sold?

78. An insurance company estimates the probability of an earthquake in the next year to be 0.0013. The average damage done by an earthquake it estimates to be $60,000. If the company offers earthquake insurance for $100, what is their expected value of the policy?

Exploration

Some of these questions were adapted from puzzles at mindyourdecisions.com.

79. A small college has been accused of gender bias in its admissions to graduate programs.
    a. Out of 500 men who applied, 255 were accepted. Out of 700 women who applied, 240 were accepted. Find the acceptance rate for each gender. Does this suggest bias?
    b. The college then looked at each of the two departments with graduate programs, and found the data below. Compute the acceptance rate within each department by gender. Does this suggest bias?

Department	Men		Women	
	Applied	Admitted	Applied	Admitted
Dept A	400	240	100	90
Dept B	100	15	600	150

    c. Looking at our results from Parts *a* and *b*, what can you conclude? Is there gender bias in this college's admissions? If so, in which direction?

80. A bet on "black" in Roulette has a probability of 18/38 of winning. If you win, you double your money. You can bet anywhere from $1 to $100 on each spin.
    a. Suppose you have $10, and are going to play until you go broke or have $20. What is your best strategy for playing?
    b. Suppose you have $10, and are going to play until you go broke or have $30. What is your best strategy for playing?

81. Your friend proposes a game: You flip a coin. If it's heads, you win $1. If it's tails, you lose $1. However, you are worried the coin might not be fair coin. How could you change the game to make the game fair, without replacing the coin?

82. Fifty people are in a line. The first person in the line to have a birthday matching someone in front of them will win a prize. Of course, this means the first person in the line has no chance of winning. Which person has the highest likelihood of winning?

83. Three people put their names in a hat, then each draw a name, as part of a randomized gift exchange. What is the probability that no one draws their own name? What about with four people?

84. How many different "words" can be formed by using all the letters of each of the following words exactly once?
    a. "ALICE"
    b. "APPLE"

85. How many different "words" can be formed by using all the letters of each of the following words exactly once?
    a. "TRUMPS"
    b. "TEETER"

86. The *Monty Hall problem* is named for the host of the game show *Let's make a Deal*. In this game, there would be three doors, behind one of which there was a prize. The contestant was asked to choose one of the doors. Monty Hall would then open one of the other doors to show there was no prize there. The contestant was then asked if they wanted to stay with their original door, or switch to the other unopened door. Is it better to stay or switch, or does it matter?

87. Suppose you have two coins, where one is a fair coin, and the other coin comes up heads 70% of the time. What is the probability you have the fair coin given each of the following outcomes from a series of flips?
    a. 5 Heads and 0 Tails
    b. 8 Heads and 3 Tails
    c. 10 Heads and 10 Tails
    d. 3 Heads and 8 Tails

88. Suppose you have six coins, where five are fair coins, and one coin comes up heads 80% of the time. What is the probability you have a fair coin given each of the following outcomes from a series of flips?
    a. 5 Heads and 0 Tails
    b. 8 Heads and 3 Tails
    c. 10 Heads and 10 Tails
    d. 3 Heads and 8 Tails

89. In this problem, we will explore probabilities from a series of events.
    a. If you flip 20 coins, how many would you *expect* to come up "heads", on average? Would you expect *every* flip of 20 coins to come up with exactly that many heads?
    b. If you were to flip 20 coins, what would you consider a "usual" result? An "unusual" result?
    c. Flip 20 coins (or one coin 20 times) and record how many come up "heads". Repeat this experiment 9 more times. Collect the data from the entire class.
    d. When flipping 20 coins, what is the theoretic probability of flipping 20 heads?
    e. Based on the class's experimental data, what appears to be the probability of flipping 10 heads out of 20 coins?
    f. The formula $_nC_x p^x(1-p)^{n-x}$ will compute the probability of an event with probability p occurring x times out of n, such as flipping x heads out of n coins where the probability of heads is $p = \frac{1}{2}$. Use this to compute the theoretic probability of flipping 10 heads out of 20 coins.
    g. If you were to flip 20 coins, based on the class's experimental data, what range of values would you consider a "usual" result? What is the combined probability of these results? What would you consider an "unusual" result? What is the combined probability of these results?
    h. We'll now consider a simplification of a case from the 1960s. In the area, about 26% of the jury eligible population was black. In the court case, there were 100 men on the juror panel, of which 8 were black. Does this provide evidence of racial bias in jury selection?

Sets

It is natural for us to classify items into groups, or sets, and consider how those sets overlap with each other. We can use these sets understand relationships between groups, and to analyze survey data.

Basics

An art collector might own a collection of paintings, while a music lover might keep a collection of CDs. Any collection of items can form a **set**.

> **Set**
> A **set** is a collection of distinct objects, called **elements** of the set
>
> A set can be defined by describing the contents, or by listing the elements of the set, enclosed in curly brackets.

Example 1

Some examples of sets defined by describing the contents:
a) The set of all even numbers
b) The set of all books written about travel to Chile

Some examples of sets defined by listing the elements of the set:
a) {1, 3, 9, 12}
b) {red, orange, yellow, green, blue, indigo, purple}

A set simply specifies the contents; order is not important. The set represented by {1, 2, 3} is equivalent to the set {3, 1, 2}.

> **Notation**
> Commonly, we will use a variable to represent a set, to make it easier to refer to that set later.
>
> The symbol $\in$ means "is an element of".
>
> A set that contains no elements, { }, is called the **empty set** and is notated $\emptyset$

Example 2

Let $A = \{1, 2, 3, 4\}$

To notate that 2 is element of the set, we'd write $2 \in A$

Sometimes a collection might not contain all the elements of a set. For example, Chris owns three Madonna albums. While Chris's collection is a set, we can also say it is a **subset** of the larger set of all Madonna albums.

Subset

A **subset** of a set A is another set that contains only elements from the set A, but may not contain all the elements of A.

If B is a subset of A, we write $B \subseteq A$

A **proper subset** is a subset that is not identical to the original set – it contains fewer elements.

If B is a proper subset of A, we write $B \subset A$

Example 3

Consider these three sets

A = the set of all even numbers $B = \{2, 4, 6\}$ $C = \{2, 3, 4, 6\}$

Here $B \subset A$ since every element of B is also an even number, so is an element of A.

More formally, we could say $B \subset A$ since if $x \in B$, then $x \in A$.

It is also true that $B \subset C$.

C is not a subset of A, since C contains an element, 3, that is not contained in A

Example 4

Suppose a set contains the plays "Much Ado About Nothing", "MacBeth", and "A Midsummer's Night Dream". What is a larger set this might be a subset of?

There are many possible answers here. One would be the set of plays by Shakespeare. This is also a subset of the set of all plays ever written. It is also a subset of all British literature.

Try it Now 1

The set $A = \{1, 3, 5\}$. What is a larger set this might be a subset of?

Union, Intersection, and Complement

Commonly sets interact. For example, you and a new roommate decide to have a house party, and you both invite your circle of friends. At this party, two sets are being combined, though it might turn out that there are some friends that were in both sets.

Union, Intersection, and Complement
The **union** of two sets contains all the elements contained in either set (or both sets).
The union is notated $A \cup B$.
More formally, $x \in A \cup B$ if $x \in A$ or $x \in B$ (or both)

The **intersection** of two sets contains only the elements that are in both sets.
The intersection is notated $A \cap B$.
More formally, $x \in A \cap B$ if $x \in A$ and $x \in B$

The **complement** of a set A contains everything that is *not* in the set A.
The complement is notated A', or A^c, or sometimes $\sim A$.

Example 5

Consider the sets: A = {red, green, blue} B = {red, yellow, orange}
C = {red, orange, yellow, green, blue, purple}

a) Find $A \cup B$

The union contains all the elements in either set: $A \cup B$ = {red, green, blue, yellow, orange}
Notice we only list red once.

b) Find $A \cap B$

The intersection contains all the elements in both sets: $A \cap B$ = {red}

c) Find $A^c \cap C$

Here we're looking for all the elements that are *not* in set A and are also in C.
$A^c \cap C$ = {orange, yellow, purple}

Try it Now 2
Using the sets from the previous example, find $A \cup C$ and $B^c \cap A$

Notice that in the example above, it would be hard to just ask for A^c, since everything from the color fuchsia to puppies and peanut butter are included in the complement of the set. For this reason, complements are usually only used with intersections, or when we have a universal set in place.

> **Universal Set**
>
> A **universal set** is a set that contains all the elements we are interested in. This would have to be defined by the context.
>
> A complement is relative to the universal set, so A^c contains all the elements in the universal set that are not in A.

Example 6

a) If we were discussing searching for books, the universal set might be all the books in the library.
b) If we were grouping your Facebook friends, the universal set would be all your Facebook friends.
c) If you were working with sets of numbers, the universal set might be all whole numbers, all integers, or all real numbers

Example 7

Suppose the universal set is U = all whole numbers from 1 to 9. If $A = \{1, 2, 4\}$, then

$A^c = \{3, 5, 6, 7, 8, 9\}$.

As we saw earlier with the expression $A^c \cap C$, set operations can be grouped together. Grouping symbols can be used like they are with arithmetic – to force an order of operations.

Example 8

Suppose $H = \{$cat, dog, rabbit, mouse$\}$, $F = \{$dog, cow, duck, pig, rabbit$\}$
$W = \{$duck, rabbit, deer, frog, mouse$\}$

a) Find $(H \cap F) \cup W$

We start with the intersection: $H \cap F = \{$dog, rabbit$\}$
Now we union that result with W: $(H \cap F) \cup W = \{$dog, duck, rabbit, deer, frog, mouse$\}$

b) Find $H \cap (F \cup W)$

We start with the union: $F \cup W = \{$dog, cow, rabbit, duck, pig, deer, frog, mouse$\}$
Now we intersect that result with H: $H \cap (F \cup W) = \{$dog, rabbit, mouse$\}$

c) Find $(H \cap F)^c \cap W$

We start with the intersection: $H \cap F = \{$dog, rabbit$\}$
Now we want to find the elements of W that are *not* in $H \cap F$
$(H \cap F)^c \cap W = \{$duck, deer, frog, mouse$\}$

Venn Diagrams

To visualize the interaction of sets, John Venn in 1880 thought to use overlapping circles, building on a similar idea used by Leonhard Euler in the 18th century. These illustrations now called **Venn Diagrams**.

> **Venn Diagram**
> A Venn diagram represents each set by a circle, usually drawn inside of a containing box representing the universal set. Overlapping areas indicate elements common to both sets.

Basic Venn diagrams can illustrate the interaction of two or three sets.

Example 9

Create Venn diagrams to illustrate $A \cup B$, $A \cap B$, and $A^c \cap B$

$A \cup B$ contains all elements in *either* set.

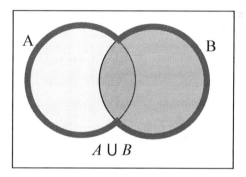

$A \cap B$ contains only those elements in both sets – in the overlap of the circles.

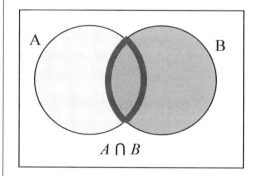

A^c will contain all elements *not* in the set A. $A^c \cap B$ will contain the elements in set B that are not in set A.

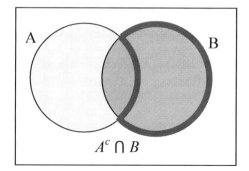

Example 10

Use a Venn diagram to illustrate $(H \cap F)^c \cap W$

We'll start by identifying everything in the set $H \cap F$

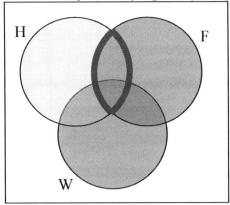

Now, $(H \cap F)^c \cap W$ will contain everything *not* in the set identified above that is also in set W.

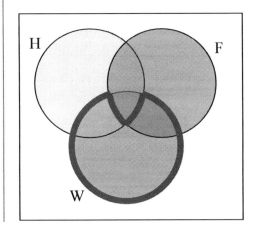

Example 11

Create an expression to represent the outlined part of the Venn diagram shown.

The elements in the outlined set *are* in sets H and F, but are not in set W. So we could represent this set as $H \cap F \cap W^c$

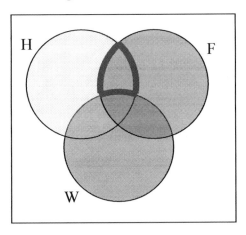

Try it Now 3
Create an expression to represent the outlined portion of the Venn diagram shown

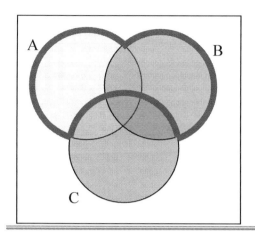

Cardinality

Often times we are interested in the number of items in a set or subset. This is called the cardinality of the set.

> **Cardinality**
> The number of elements in a set is the cardinality of that set.
>
> The cardinality of the set A is often notated as $|A|$ or n(A)

Example 12

Let $A = \{1, 2, 3, 4, 5, 6\}$ and $B = \{2, 4, 6, 8\}$.
What is the cardinality of B? $A \cup B$, $A \cap B$?

The cardinality of B is 4, since there are 4 elements in the set.
The cardinality of $A \cup B$ is 7, since $A \cup B = \{1, 2, 3, 4, 5, 6, 8\}$, which contains 7 elements.
The cardinality of $A \cap B$ is 3, since $A \cap B = \{2, 4, 6\}$, which contains 3 elements.

Example 13

What is the cardinality of P = the set of English names for the months of the year?

The cardinality of this set is 12, since there are 12 months in the year.

Sometimes we may be interested in the cardinality of the union or intersection of sets, but not know the actual elements of each set. This is common in surveying.

Example 14

A survey asks 200 people "What beverage do you drink in the morning", and offers choices:
- Tea only
- Coffee only
- Both coffee and tea

Suppose 20 report tea only, 80 report coffee only, 40 report both. How many people drink tea in the morning? How many people drink neither tea or coffee?

This question can most easily be answered by creating a Venn diagram. We can see that we can find the people who drink tea by adding those who drink only tea to those who drink both: 60 people.

We can also see that those who drink neither are those not contained in the any of the three other groupings, so we can count those by subtracting from the cardinality of the universal set, 200. 200 – 20 – 80 – 40 = 60 people who drink neither.

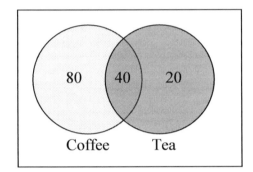

Example 15

A survey asks: Which online services have you used in the last month:
- Twitter
- Facebook
- Have used both

The results show 40% of those surveyed have used Twitter, 70% have used Facebook, and 20% have used both. How many people have used neither Twitter or Facebook?

Let T be the set of all people who have used Twitter, and F be the set of all people who have used Facebook. Notice that while the cardinality of F is 70% and the cardinality of T is 40%, the cardinality of $F \cup T$ is not simply 70% + 40%, since that would count those who use both services twice. To find the cardinality of $F \cup T$, we can add the cardinality of F and the cardinality of T, then subtract those in intersection that we've counted twice. In symbols,
$$n(F \cup T) = n(F) + n(T) - n(F \cap T)$$
$$n(F \cup T) = 70\% + 40\% - 20\% = 90\%$$

Now, to find how many people have not used either service, we're looking for the cardinality of $(F \cup T)^c$. Since the universal set contains 100% of people and the cardinality of $F \cup T = 90\%$, the cardinality of $(F \cup T)^c$ must be the other 10%.

The previous example illustrated two important properties

Cardinality properties

$n(A \cup B) = n(A) + n(B) - n(A \cap B)$

$n(A^c) = n(U) - n(A)$

Notice that the first property can also be written in an equivalent form by solving for the cardinality of the intersection:

$n(A \cap B) = n(A) + n(B) - n(A \cup B)$

Example 16

Fifty students were surveyed, and asked if they were taking a social science (SS), humanities (HM) or a natural science (NS) course the next quarter.

21 were taking a SS course	26 were taking a HM course
19 were taking a NS course	9 were taking SS and HM
7 were taking SS and NS	10 were taking HM and NS
3 were taking all three	7 were taking none

How many students are only taking a SS course?

It might help to look at a Venn diagram.
From the given data, we know that there are
3 students in region *e* and
7 students in region *h*.

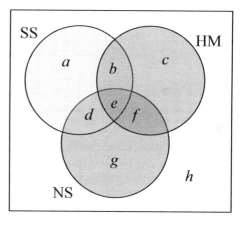

Since 7 students were taking a SS and NS course, we know that $n(d) + n(e) = 7$. Since we know there are 3 students in region 3, there must be $7 - 3 = 4$ students in region *d*.

Similarly, since there are 10 students taking HM and NS, which includes regions *e* and *f*, there must be $10 - 3 = 7$ students in region *f*.

Since 9 students were taking SS and HM, there must be $9 - 3 = 6$ students in region *b*.

Now, we know that 21 students were taking a SS course. This includes students from regions *a*, *b*, *d*, and *e*. Since we know the number of students in all but region *a*, we can determine that $21 - 6 - 4 - 3 = 8$ students are in region *a*.

8 students are taking only a SS course.

Try it Now 4

One hundred fifty people were surveyed and asked if they believed in UFOs, ghosts, and Bigfoot.

> 43 believed in UFOs   44 believed in ghosts
> 25 believed in Bigfoot   10 believed in UFOs and ghosts
> 8 believed in ghosts and Bigfoot   5 believed in UFOs and Bigfoot
> 2 believed in all three

How many people surveyed believed in at least one of these things?

Try it Now Answers

1. There are several answers: The set of all odd numbers less than 10. The set of all odd numbers. The set of all integers. The set of all real numbers.

2. $A \cup C$ = {red, orange, yellow, green, blue purple}
$B^c \cap A$ = {green, blue}

3. $A \cup B \cap C^c$

4. Starting with the intersection of all three circles, we work our way out. Since 10 people believe in UFOs and Ghosts, and 2 believe in all three, that leaves 8 that believe in only UFOs and Ghosts. We work our way out, filling in all the regions. Once we have, we can add up all those regions, getting 91 people in the union of all three sets. This leaves 150 − 91 = 59 who believe in none.

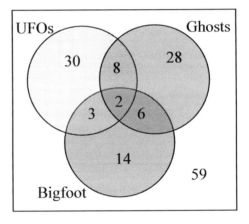

Exercises

1. List out the elements of the set "The letters of the word Mississipi"

2. List out the elements of the set "Months of the year"

3. Write a verbal description of the set {3, 6, 9}

4. Write a verbal description of the set {a, i, e, o, u}

5. Is {1, 3, 5} a subset of the set of odd integers?

6. Is {A, B, C} a subset of the set of letters of the alphabet?

For problems 7-12, consider the sets below, and indicate if each statement is true or false.
$A = \{1, 2, 3, 4, 5\}$ $B = \{1, 3, 5\}$ $C = \{4, 6\}$ $U = \{$numbers from 0 to 10$\}$

7. $3 \in B$ 8. $5 \in C$ 9. $B \subset A$ 10. $C \subset A$ 11. $C \subset B$ 12. $C \subset D$

Using the sets from above, and treating U as the Universal set, find each of the following:

13. $A \cup B$ 14. $A \cup C$ 15. $A \cap C$ 16. $B \cap C$ 17. A^c 18. B^c

Let $D = \{b, a, c, k\}$, $E = \{t, a, s, k\}$, $F = \{b, a, t, h\}$. Using these sets, find the following:

19. $D^c \cap E$ 20. $F^c \cap D$ 21. $(D \cap E) \cup F$ 22. $D \cap (E \cup F)$

23. $(F \cap E)^c \cap D$ 24. $(D \cup E)^c \cap F$

Create a Venn diagram to illustrate each of the following:

25. $(F \cap E) \cup D$ 26. $(D \cup E)^c \cap F$

27. $(F^c \cap E^c) \cap D$ 28. $(D \cup E) \cup F$

Write an expression for the shaded region.

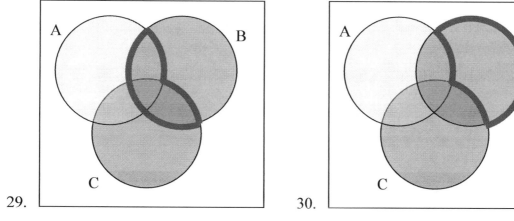

29.   30.

330

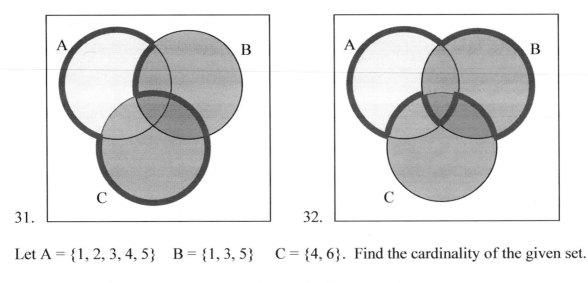

31.   32.

Let A = {1, 2, 3, 4, 5}   B = {1, 3, 5}   C = {4, 6}. Find the cardinality of the given set.

33. n(A)   34. n(B)   35. n($A \cup C$)   36. n($A \cap C$)

The Venn diagram here shows the cardinality of each set. Use this in 37-40 to find the cardinality of given set.

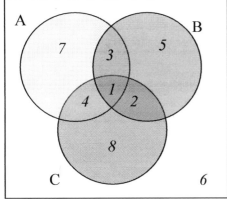

37. n($A \cap C$)   38. n($B \cup C$)   39. n($A \cap B \cap C^c$)   40. n($A \cap B^c \cap C$)

41.   If n(G) = 20, n(H) = 30, n($G \cap H$) = 5, find n($G \cup H$)

42.   If n(G) = 5, n(H) = 8, n($G \cap H$) = 4, find n($G \cup H$)

43.   A survey was given asking whether they watch movies at home from Netflix, Redbox, or a video store. Use the results to determine how many people use Redbox.

        52 only use Netflix   62 only use Redbox
        24 only use a video store   16 use only a video store and Redbox
        48 use only Netflix and Redbox   30 use only a video store and Netflix
        10 use all three   25 use none of these

44. A survey asked buyers whether color, size, or brand influenced their choice of cell phone. The results are below. How many people were influenced by brand?

5 only said color	8 only said size
16 only said brand	20 said only color and size
42 said only color and brand	53 said only size and brand
102 said all three	20 said none of these

45. Use the given information to complete a Venn diagram, then determine: a) how many students have seen exactly one of these movies, and b) how many had seen only *Star Wars*.

18 had seen *The Matrix* (*M*)	24 had seen *Star Wars (SW)*
20 had seen *Lord of the Rings (LotR)*	10 had seen *M* and *SW*
14 had seen *LotR* and *SW*	12 had seen *M* and *LotR*
6 had seen all three	

46. A survey asked people what alternative transportation modes they use. Using the data to complete a Venn diagram, then determine: a) what percent of people only ride the bus, and b) how many people don't use any alternate transportation.

30% use the bus	20% ride a bicycle
25% walk	5% use the bus and ride a bicycle
10% ride a bicycle and walk	12% use the bus and walk
2% use all three	

Historical Counting Systems

Introduction and Basic Number and Counting Systems

Introduction

As we begin our journey through the history of mathematics, one question to be asked is "Where do we start?" Depending on how you view mathematics or numbers, you could choose any of a number of launching points from which to begin. Howard Eves suggests the following list of possibilities.[1]

Where to start the study of the history of mathematics…
- At the first logical geometric "proofs" traditionally credited to Thales of Miletus (600 BCE).
- With the formulation of methods of measurement made by the Egyptians and Mesopotamians/Babylonians.
- Where prehistoric peoples made efforts to organize the concepts of size, shape, and number.
- In pre–human times in the very simple number sense and pattern recognition that can be displayed by certain animals, birds, etc.
- Even before that in the amazing relationships of numbers and shapes found in plants.
- With the spiral nebulae, the natural course of planets, and other universe phenomena.

We can choose no starting point at all and instead agree that mathematics has *always* existed and has simply been waiting in the wings for humans to discover. Each of these positions can be defended to some degree and which one you adopt (if any) largely depends on your philosophical ideas about mathematics and numbers.

Nevertheless, we need a starting point. Without passing judgment on the validity of any of these particular possibilities, we will choose as our starting point the emergence of the idea of number and the process of counting as our launching pad. This is done primarily as a practical matter given the nature of this course. In the following chapter, we will try to focus on two main ideas. The first will be an examination of basic number and counting systems and the symbols that we use for numbers. We will look at our own modern (Western) number system as well those of a couple of selected civilizations to see the differences and diversity that is possible when humans start counting. The second idea we will look at will be base systems. By comparing our own base-ten (decimal) system with other bases, we will quickly become aware that the system that we are so used to, when slightly changed, will challenge our notions about numbers and what symbols for those numbers actually mean.

Recognition of More vs. Less

The idea of number and the process of counting goes back far beyond history began to be recorded. There is some archeological evidence that suggests that humans were counting as far back as 50,000 years ago.[2] However, we do not really know how this process started or developed over time. The best we can do is to make a good guess as to how things progressed. It is probably not hard to believe that even the earliest humans had some sense of *more* and *less*. Even some small animals have been shown to have such a sense. For example, one naturalist tells of how he would secretly remove one egg each day from a plover's nest. The mother was diligent in laying an extra egg every day to make up for the missing egg. Some research has shown that hens can be trained to distinguish between even and odd numbers of pieces of food.[3] With these sorts of findings in mind, it is not hard to conceive that early humans had (at least) a similar sense of more and less. However, our conjectures about how and when these ideas emerged among humans are simply that; educated guesses based on our own assumptions of what might or could have been.

The Need for Simple Counting

As societies and humankind evolved, simply having a sense of more or less, even or odd, etc., would prove to be insufficient to meet the needs of everyday living. As tribes and groups formed, it became important to be able to know how many members were in the group, and perhaps how many were in the enemy's camp. Certainly it was important for them to know if the flock of sheep or other possessed animals were increasing or decreasing in size. "Just how many of them do we have, anyway?" is a question that we do not have a hard time imagining them asking themselves (or each other).

In order to count items such as animals, it is often conjectured that one of the earliest methods of doing so would be with "tally sticks." These are objects used to track the numbers of items to be counted. With this method, each "stick" (or pebble, or whatever counting device being used) represents one animal or object. This method uses the idea of **one to one correspondence**. In a one to one correspondence, items that are being counted are uniquely linked with some counting tool.

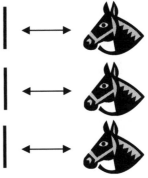

In the picture to the right, you see each stick corresponding to one horse. By examining the collection of sticks in hand one knows how many animals should be present. You can imagine the usefulness of such a system, at least for smaller numbers of items to keep track of. If a herder wanted to "count off" his animals to make sure they were all present, he could mentally (or methodically) assign each stick to one animal and continue to do so until he was satisfied that all were accounted for.

Of course, in our modern system, we have replaced the sticks with more abstract objects. In particular, the top stick is replaced with our symbol "1," the second stick gets replaced by a "2" and the third stick is represented by the symbol "3," but we are getting ahead of ourselves here. These modern symbols took many centuries to emerge.

Another possible way of employing the "tally stick" counting method is by making marks or cutting notches into pieces of wood, or even tying knots in string (as we shall see later). In 1937, Karl Absolom discovered a wolf bone that goes back possibly 30,000 years. It is believed to be a counting device.[4] Another example of this kind of tool is the Ishango Bone, discovered in 1960 at Ishango, and shown below.[5] It is reported to be between six and nine thousand years old and shows what appear to be markings used to do counting of some sort.

The markings on rows (a) and (b) each add up to 60. Row (b) contains the prime numbers between 10 and 20. Row (c) seems to illustrate for the method of doubling and multiplication used by the Egyptians. It is believed that this may also represent a lunar phase counter.

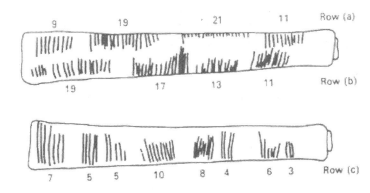

Spoken Words

As methods for counting developed, and as language progressed as well, it is natural to expect that spoken words for numbers would appear. Unfortunately, the developments of these words, especially those corresponding to the numbers from one through ten, are not easy to trace. Past ten, however, we do see some patterns:

Eleven comes from "ein lifon," meaning "one left over."
Twelve comes from "twe lif," meaning "two left over."
Thirteen comes from "Three and ten" as do fourteen through nineteen.
Twenty appears to come from "twe–tig" which means "two tens."
Hundred probably comes from a term meaning "ten times."

Written Numbers

When we speak of "written" numbers, we have to be careful because this could mean a variety of things. It is important to keep in mind that modern paper is only a little more than 100 years old, so "writing" in times past often took on forms that might look quite unfamiliar to us today.

As we saw earlier, some might consider wooden sticks with notches carved in them as writing as these are means of recording information on a medium that can be "read" by others. Of course, the symbols used (simple notches) certainly did not leave a lot of flexibility for communicating a wide variety of ideas or information.

Other mediums on which "writing" may have taken place include carvings in stone or clay tablets, rag paper made by hand (12th century in Europe, but earlier in China), papyrus (invented by the Egyptians and used up until the Greeks), and parchments from animal skins. And these are just a few of the many possibilities.

These are just a few examples of early methods of counting and simple symbols for representing numbers. Extensive books, articles and research have been done on this topic and could provide enough information to fill this entire course if we allowed it to. The range and diversity of creative thought that has been used in the past to describe numbers and to count objects and people is staggering. Unfortunately, we don't have time to examine them all, but it is fun and interesting to look at one system in more detail to see just how ingenious people have been.

The Number and Counting System of the Inca Civilization

Background

There is generally a lack of books and research material concerning the historical foundations of the Americas. Most of the "important" information available concentrates on the eastern hemisphere, with Europe as the central focus. The reasons for this may be twofold: first, it is thought that there was a lack of specialized mathematics in the American regions; second, many of the secrets of ancient mathematics in the Americas have been closely guarded.[6] The Peruvian system does not seem to be an exception here. Two researchers, Leland Locke and Erland Nordenskiold, have carried out research that has attempted to discover what mathematical knowledge was known by the Incas and how they used the Peruvian quipu, a counting system using cords and knots, in their mathematics. These researchers have come to certain beliefs about the quipu that we will summarize here.

Counting Boards

It should be noted that the Incas did not have a complicated system of computation. Where other peoples in the regions, such as the Mayans, were doing computations related to their rituals and calendars, the Incas seem to have been more concerned with the simpler task of record–keeping. To do this, they used what are called the "quipu" to record quantities of items. (We will describe them in more detail in a moment.) However, they first often needed to do computations whose results would be recorded on quipu. To do these computations, they would sometimes use a counting board constructed with a slab of stone. In the slab were cut rectangular and square compartments so that an octagonal (eight–sided) region was

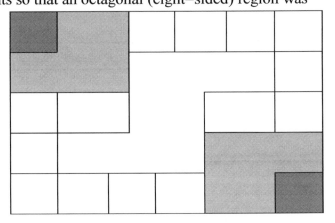

left in the middle. Two opposite corner rectangles were raised. Another two sections were mounted on the original surface of the slab so that there were actually three levels available. In the figure shown, the darkest shaded corner regions represent the highest, third level. The lighter shaded regions surrounding the corners are the second highest levels, while the clear white rectangles are the compartments cut into the stone slab.

Pebbles were used to keep accounts and their positions within the various levels and compartments gave totals. For example, a pebble in a smaller (white) compartment represented one unit. Note that there are 12 such squares around the outer edge of the figure. If a pebble was put into one of the two (white) larger, rectangular compartments, its value was doubled. When a pebble was put in the octagonal region in the middle of the slab, its value was tripled. If a pebble was placed on the second (shaded) level, its value was multiplied by six. And finally, if a pebble was found on one of the two highest corner levels, its value was multiplied by twelve. Different objects could be counted at the same time by representing different objects by different colored pebbles.

Example 1

Suppose you have the following counting board with two different kind of pebbles places as illustrated. Let the solid black pebble represent a dog and the striped pebble represent a cat. How many dogs are being represented?

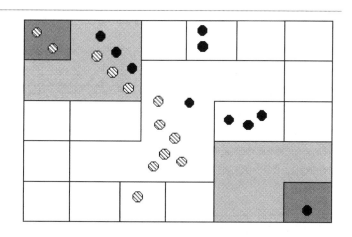

There are two black pebbles in the outer square regions…these represent 2 dogs.

There are three black pebbles in the larger (white) rectangular compartments. These represent 6 dogs.

There is one black pebble in the middle region…this represents 3 dogs.

There are three black pebbles on the second level…these represent 18 dogs.

Finally, there is one black pebble on the highest corner level…this represents 12 dogs. We then have a total of 2+6+3+18+12 = 41 dogs.

Try it Now 1

How many cats are represented on this board?

The Quipu

This kind of board was good for doing quick computations, but it did not provide a good way to keep a permanent recording of quantities or computations. For this purpose, they used the quipu. The quipu is a collection of cords with knots in them. These cords and knots are carefully arranged so that the position and type of cord or knot gives specific information on how to decipher the cord.

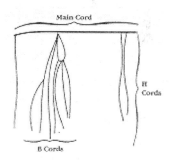

A quipu is made up of a main cord which has other cords (branches) tied to it. See pictures to the right.[7]

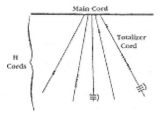

Locke called the branches H cords. They are attached to the main cord. B cords, in turn, were attached to the H cords. Most of these cords would have knots on them. Rarely are knots found on the main cord, however, and tend to be mainly on the H and B cords. A quipu might also have a "totalizer" cord that summarizes all of the information on the cord group in one place.

Locke points out that there are three types of knots, each representing a different value, depending on the kind of knot used and its position on the cord. The Incas, like us, had a decimal (base–ten) system, so each kind of knot had a specific decimal value. The Single knot, pictured in

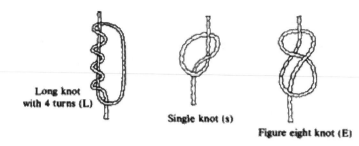

Long knot with 4 turns (L)

Single knot (s)

Figure eight knot (E)

the middle of the diagram[8] was used to denote tens, hundreds, thousands, and ten thousands. They would be on the upper levels of the H cords. The figure–eight knot on the end was used to denote the integer "one." Every other integer from 2 to 9 was represented with a long knot, shown on the left of the figure. (Sometimes long knots were used to represents tens and hundreds.) Note that the long knot has several turns in it...the number of turns indicates which integer is being represented. The units (ones) were placed closest to the bottom of the cord, then tens right above them, then the hundreds, and so on.

In order to make reading these pictures easier, we will adopt a convention that is consistent. For the long knot with turns in it (representing the numbers 2 through 9), we will use the following notation:

$$\equiv\!\rangle$$

The four horizontal bars represent four turns and the curved arc on the right links the four turns together. This would represent the number 4.

We will represent the single knot with a large dot (•) and we will represent the figure eight knot with a sideways eight (∞).

Example 2

What number is represented on the cord shown?

On the cord, we see a long knot with four turns in it...this represents four in the ones place. Then 5 single knots appear in the tens position immediately above that, which represents 5 tens, or 50. Finally, 4 single knots are tied in the hundreds, representing four 4 hundreds, or 400. Thus, the total shown on this cord is 454.

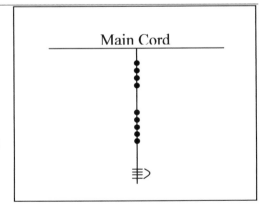

Main Cord

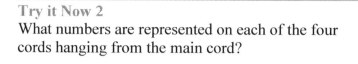

Try it Now 2
What numbers are represented on each of the four cords hanging from the main cord?

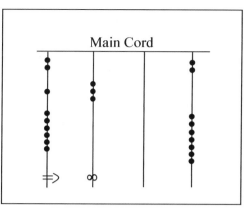

The colors of the cords had meaning and could distinguish one object from another. One color could represent llamas, while a different color might represent sheep, for example. When all the colors available were exhausted, they would have to be re-used. Because of this, the ability to read the quipu became a complicated task and specially trained individuals did this job. They were called Quipucamayoc, which means keeper of the quipus. They would build, guard, and decipher quipus.

As you can see from this photograph of an actual quipu, they could get quite complex.

There were various purposes for the quipu. Some believe that they were used to keep an account of their traditions and history, using knots to record history rather than some other formal system of writing. One writer has even suggested that the quipu replaced writing as it formed a role in the Incan postal system.[9] Another proposed use of the quipu is as a translation tool. After the conquest of the Incas by the Spaniards and subsequent "conversion" to Catholicism, an Inca

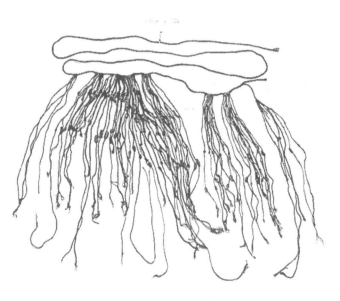

supposedly could use the quipu to confess their sins to a priest. Yet another proposed use of the quipu was to record numbers related to magic and astronomy, although this is not a widely accepted interpretation.

The mysteries of the quipu have not been fully explored yet. Recently, Ascher and Ascher have published a book, *The Code of the Quipu: A Study in Media, Mathematics, and Culture*, which is "an extensive elaboration of the logical-numerical system of the quipu."[10] For more information on the quipu, you may want to check out the following Internet link:

http://www.anthropology.wisc.edu/salomon/Chaysimire/khipus.php

We are so used to seeing the symbols 1, 2, 3, 4, etc. that it may be somewhat surprising to see such a creative and innovative way to compute and record numbers. Unfortunately, as we proceed through our mathematical education in grade and high school, we receive very little information about the wide range of number systems that have existed and which still exist all over the world. That's not to say our own system is not important or efficient. The fact that it has survived for hundreds of years and shows no sign of going away any time soon suggests that we may have finally found a system that works well and may not need further improvement, but only time will tell that whether or not that conjecture is valid or not. We now turn to a brief historical look at how our current system developed over history.

The Hindu – Arabic Number System

The Evolution of a System

Our own number system, composed of the ten symbols {0,1,2,3,4,5,6,7,8,9} is called the *Hindu–Arabic system*. This is a base–ten (decimal) system since place values increase by powers of ten. Furthermore, this system is positional, which means that the position of a symbol has bearing on the value of that symbol within the number. For example, the position of the symbol 3 in the number 435,681 gives it a value much greater than the value of the symbol 8 in that same number. We'll explore base systems more thoroughly later. The development of these ten symbols and their use in a positional system comes to us primarily from India.[11]

It was not until the 15th century that the symbols that we are familiar with today first took form in Europe. However, the history of these numbers and their development goes back hundreds of years. One important source of information on this topic is the writer al–Biruni, whose picture is shown here.[12] Al–Biruni, who was born in modern day Uzbekistan, had visited India on several occasions and made comments on the Indian number system. When we look at the origins of the numbers that al–Biruni encountered, we have to go back to the third century B.C.E. to explore their origins. It is then that the Brahmi numerals were being used.

The Brahmi numerals were more complicated than those used in our own modern system. They had separate symbols for the numbers 1 through 9, as well as distinct symbols for 10, 100, 1000,…, also for 20, 30, 40,…, and others for 200, 300, 400, …, 900. The Brahmi symbols for 1, 2, and 3 are shown below.[13]

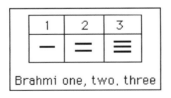

Brahmi one, two, three

These numerals were used all the way up to the 4th century C.E., with variations through time and geographic location. For example, in the first century C.E., one particular set of Brahmi numerals took on the following form[14]:

1	2	3	4	5	6	7	8	9
—	=	≡	+	ђ	ụ	？	↜	?

From the 4th century on, you can actually trace several different paths that the Brahmi numerals took to get to different points and incarnations. One of those paths led to our current numeral system, and went through what are called the Gupta numerals. The Gupta numerals were prominent during a time ruled by the Gupta dynasty and were spread throughout that empire as they conquered lands during the 4th through 6th centuries. They have the following form[15]:

1	2	3	4	5	6	7	8	9
—	=	≡	ϥ	Ь	ऊ	η	ς	ɔ

How the numbers got to their Gupta form is open to considerable debate. Many possible hypotheses have been offered, most of which boil down to two basic types[16]. The first type of hypothesis states that the numerals came from the initial letters of the names of the numbers. This is not uncommon…the Greek numerals developed in this manner. The second type of hypothesis states that they were derived from some earlier number system. However, there are other hypotheses that are offered, one of which is by the researcher Ifrah. His theory is that there were originally nine numerals, each represented by a corresponding number of vertical lines. One possibility is this:[17]

1	2	3	4	5	6	7	8	9
I	II	III	II II	III II	IIII II	IIII III	IIII IIII	III III III

Because these symbols would have taken a lot of time to write, they eventually evolved into cursive symbols that could be written more quickly. If we compare these to the Gupta numerals above, we can try to see how that evolutionary process might have taken place, but our imagination would be just about all we would have to depend upon since we do not know exactly how the process unfolded.

The Gupta numerals eventually evolved into another form of numerals called the Nagari numerals, and these continued to evolve until the 11th century, at which time they looked like this:[18]

1	2	3	4	5	6	7	8	9	0
٩	२	३	8	५	६	७	८	९	०

Note that by this time, the symbol for 0 has appeared! The Mayans in the Americas had a symbol for zero long before this, however, as we shall see later in the chapter.

These numerals were adopted by the Arabs, most likely in the eighth century during Islamic incursions into the northern part of India.[19] It is believed that the Arabs were instrumental in spreading them to other parts of the world, including Spain (see below).

Other examples of variations up to the eleventh century include:

Devangari, eighth century[20]:

West Arab Gobar, tenth century[21]:

Spain, 976 B.C.E.[22]:

Finally, one more graphic[23] shows various forms of these numerals as they developed and eventually converged to the 15th century in Europe.

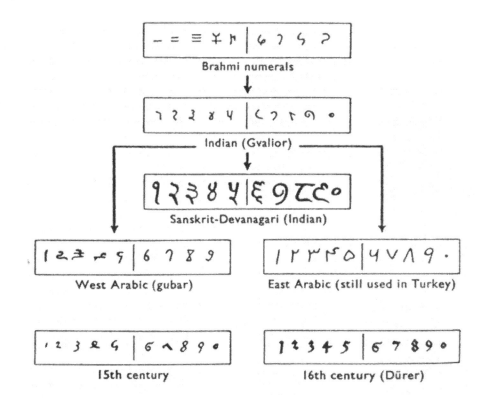

Brahmi numerals

Indian (Gvalior)

Sanskrit-Devanagari (Indian)

West Arabic (gubar)   East Arabic (still used in Turkey)

15th century   16th century (Dürer)

The Positional System

More important than the form of the number symbols is the development of the place value system. Although it is in slight dispute, the earliest known document in which the Indian system displays a positional system dates back to 346 C.E. However, some evidence suggests that they may have actually developed a positional system as far back as the first century C.E.

The Indians were not the first to use a positional system. The Babylonians (as we will see in Chapter 3) used a positional system with 60 as their base. However, there is not much evidence that the Babylonian system had much impact on later numeral systems, except with the Greeks. Also, the Chinese had a base–10 system, probably derived from the use of a counting board[24]. Some believe that the positional system used in India was derived from the Chinese system.

Wherever it may have originated, it appears that around 600 C.E., the Indians abandoned the use of symbols for numbers higher than nine and began to use our familiar system where the position of the symbol determines its overall value.[25] Numerous documents from the seventh century demonstrate the use of this positional system.

Interestingly, the earliest dated inscriptions using the system with a symbol for zero come from Cambodia. In 683, the 605th year of the Saka era is written with three digits and a dot in the middle. The 608th year uses three digits with a modern 0 in the middle.[26] The dot as a symbol for zero also appears in a Chinese work (*Chiu–chih li*). The author of this document gives a strikingly clear description of how the Indian system works:

> *Using the [Indian] numerals, multiplication and division are carried out. Each numeral is written in one stroke. When a number is counted to ten, it is advanced into the higher place. In each vacant place a dot is always put. Thus the numeral is always denoted in each place. Accordingly there can be no error in determining the place. With the numerals, calculations is easy…"[27]*

Transmission to Europe

It is not completely known how the system got transmitted to Europe. Traders and travelers of the Mediterranean coast may have carried it there. It is found in a tenth–century Spanish manuscript and may have been introduced to Spain by the Arabs, who invaded the region in 711 C.E. and were there until 1492.

In many societies, a division formed between those who used numbers and calculation for practical, every day business and those who used them for ritualistic purposes or for state business.[28] The former might often use older systems while the latter were inclined to use the newer, more elite written numbers. Competition between the two groups arose and continued for quite some time.

In a 14th century manuscript of Boethius' *The Consolations of Philosophy*, there appears a well–known drawing of two mathematicians. One is a merchant and is using an abacus (the "abacist"). The other is a Pythagorean philosopher (the "algorist") using his "sacred" numbers. They are in a competition that is being judged by the goddess of number. By 1500 C.E., however, the newer symbols and system had won out and has persevered until today. The Seattle Times recently reported that the Hindu–Arabic numeral system has been included in the book *The Greatest Inventions of the Past 2000 Years*.[29]

One question to answer is *why* the Indians would develop such a positional notation. Unfortunately, an answer to that question is not currently known. Some suggest that the system has its origins with the Chinese counting boards. These boards were portable and it is thought that Chinese travelers who passed through India took their boards with them and ignited an idea in Indian mathematics.[30] Others, such as G. G. Joseph propose that it is the Indian fascination with very large numbers that drove them to develop a system whereby these kinds of big numbers could easily be written down. In this theory, the system developed entirely within the Indian mathematical framework without considerable influence from other civilizations.

The Development and Use of Different Number Bases

Introduction and Basics

During the previous discussions, we have been referring to positional base systems. In this section of the chapter, we will explore exactly what a base system is and what it means if a system is "positional." We will do so by first looking at our own familiar, base-ten system and then deepen our exploration by looking at other possible base systems. In the next part of this section, we will journey back to Mayan civilization and look at their unique base system, which is based on the number 20 rather than the number 10.

A base system is a structure within which we count. The easiest way to describe a base system is to think about our own base–ten system. The base–ten system, which we call the "decimal" system, requires a total of ten different symbols/digits to write any number. They are, of course, 0, 1, 2, ….. 9.

The decimal system is also an example of a *positional* base system, which simply means that the position of a digit gives its place value. Not all civilizations had a positional system even though they did have a base with which they worked.

In our base–ten system, a number like 5,783,216 has meaning to us because we are familiar with the system and its places. As we know, there are six ones, since there is a 6 in the ones place. Likewise, there are seven "hundred thousands," since the 7 resides in that place. Each digit has a value that is explicitly determined by its position within the number. We make a distinction between digit, which is just a symbol such as 5, and a number, which is made up of one or more digits. We can take this number and assign each of its digits a value. One way to do this is with a table, which follows:

5,000,000	$= 5 \times 1,000,000$	$= 5 \times 10^6$	Five million
+700,000	$= 7 \times 100,000$	$= 7 \times 10^5$	Seven hundred thousand
+80,000	$= 8 \times 10,000$	$= 8 \times 10^4$	Eighty thousand
+3,000	$= 3 \times 1000$	$= 3 \times 10^3$	Three thousand
+200	$= 2 \times 100$	$= 2 \times 10^2$	Two hundred
+10	$= 1 \times 10$	$= 1 \times 10^1$	Ten
+6	$= 6 \times 1$	$= 6 \times 10^0$	Six
5,783,216	Five million, seven hundred eighty-three thousand, two hundred sixteen		

From the third column in the table we can see that each place is simply a multiple of ten. Of course, this makes sense given that our base is ten. The digits that are multiplying each place simply tell us how many of that place we have. We are restricted to having at most 9 in any one place before we have to "carry" over to the next place. We cannot, for example, have 11 in the hundreds place. Instead, we would carry 1 to the thousands place and retain 1 in the hundreds place. This comes as no surprise to us since we readily see that 11 hundreds is the same as one thousand, one hundred. Carrying is a pretty typical occurrence in a base system.

However, base-ten is not the only option we have. Practically any positive integer greater than or equal to 2 can be used as a base for a number system. Such systems can work just like the decimal system except the number of symbols will be different and each position will depend on the base itself.

Other Bases

For example, let's suppose we adopt a base–five system. The only modern digits we would need for this system are 0,1,2,3 and 4. What are the place values in such a system? To answer that, we start with the ones place, as most base systems do. However, if we were to count in this system, we could only get to four (4) before we had to jump up to the next place. Our base is 5, after all! What is that next place that we would jump to? It would not be tens, since we are no longer in base–ten. We're in a different numerical world. As the base–ten system progresses from 10^0 to 10^1, so the base–five system moves from 5^0 to $5^1 = 5$. Thus, we move from the ones to the fives.

After the fives, we would move to the 5^2 place, or the twenty fives. Note that in base–ten, we would have gone from the tens to the hundreds, which is, of course, 10^2.

Let's take an example and build a table. Consider the number 30412 in base five. We will write this as 30412_5, where the subscript 5 is not part of the number but indicates the base we're using. First off, note that this is NOT the number "thirty thousand, four hundred twelve." We must be careful not to impose the base-ten system on this number. Here's what our table might look like. We will use it to convert this number to our more familiar base–ten system.

	Base 5	This column coverts to base–ten	In Base–Ten
	3×5^4	$= 3 \times 625$	$= 1875$
+	0×5^3	$= 0 \times 125$	$= 0$
+	4×5^2	$= 4 \times 25$	$= 100$
+	1×5^1	$= 1 \times 5$	$= 5$
+	2×5^0	$= 2 \times 1$	$= 2$
		Total	1982

As you can see, the number 30412_5 is equivalent to 1,982 in base–ten. We will say $30412_5 = 1982_{10}$. All of this may seem strange to you, but that's only because you are so used to the only system that you've ever seen.

Example 3

Convert 6234_7 to a base 10 number.

We first note that we are given a base-7 number that we are to convert. Thus, our places will start at the ones (7^0), and then move up to the 7's, 49's (7^2), etc. Here's the breakdown:

	Base 7	Convert	Base 10
	$= 6 \times 7^3$	$= 6 \times 343$	$= 2058$
+	$= 2 \times 7^2$	$= 2 \times 49$	$= 98$
+	$= 3 \times 7$	$= 3 \times 7$	$= 21$
+	$= 4 \times 1$	$= 4 \times 1$	$= 4$
		Total	2181

Thus $6234_7 = 2181_{10}$

Try it Now 3
Convert 41065_7 to a base 10 number.

Converting from Base 10 to Other Bases

Converting from an unfamiliar base to the familiar decimal system is not that difficult once you get the hang of it. It's only a matter of identifying each place and then multiplying each digit by the appropriate power. However, going the other direction can be a little trickier. Suppose you have a base–ten number and you want to convert to base–five. Let's start with some simple examples before we get to a more complicated one.

Example 4

Convert twelve to a base–five number.

We can probably easily see that we can rewrite this number as follows:
$12 = (2 \times 5) + (2 \times 1)$

Hence, we have two fives and 2 ones. Hence, in base–five we would write twelve as 22_5. Thus, $12_{10} = 22_5$.

Example 5

Convert sixty–nine to a base–five number.

We can see now that we have more than 25, so we rewrite sixty–nine as follows:
$69 = (2 \times 25) + (3 \times 5) + (4 \times 1)$

Here, we have two twenty–fives, 3 fives, and 4 ones. Hence, in base five we have 234. Thus, $69_{10} = 234_5$.

Example 6

Convert the base–seven number 3261_7 to base 10.

The powers of 7 are:
$7^0 = 1$
$7^1 = 7$
$7^2 = 49$
$7^3 = 343$
Etc…

$3261_7 = (3 \times 343) + (2 \times 49) + (6 \times 7) + (1 \times 1) = 1170_{10}$.
Thus $3261_7 = 1170_{10}$.

Try it Now 4
Convert 143 to base 5

Try it Now 5
Convert the base–three number 21021_3 to base 10.

In general, when converting from base–ten to some other base, it is often helpful to determine the highest power of the base that will divide into the given number at least once. In the last example, $5^2 = 25$ is the largest power of five that is present in 69, so that was our starting point. If we had moved to $5^3 = 125$, then 125 would not divide into 69 at least once.

> **Converting from Base 10 to Base _b_**
> 1. Find the highest power of the base _b_ that will divide into the given number at least once and then divide.
> 2. Write down the whole number part, then use the remainder from division in the next step.
> 3. Repeat step two, dividing by the next highest power of the base _b_, writing down the whole number part (including 0), and using the remainder in the next step.
> 4. Continue until the remainder is smaller than the base. This last remainder will be in the "ones" place.
> 5. Collect all your whole number parts to get your number in base _b_ notation.

Example 7

Convert the base–ten number 348 to base–five.

The powers of five are:
$5^0 = 1$
$5^1 = 5$
$5^2 = 25$
$5^3 = 125$
$5^4 = 625$
Etc…

Since 348 is smaller than 625, but bigger than 125, we see that $5^3 = 125$ is the highest power of five present in 348. So we divide 125 into 348 to see how many of them there are:
$348 \div 125 = 2$ with remainder 98

We write down the whole part, 2, and continue with the remainder. There are 98 left over, so we see how many 25's (the next smallest power of five) there are in the remainder:
$98 \div 25 = 3$ with remainder 23

We write down the whole part, 2, and continue with the remainder. There are 23 left over, so we look at the next place, the 5's:
$23 \div 5 = 4$ with remainder 3

This leaves us with 3, which is less than our base, so this number will be in the "ones" place. We are ready to assemble our base–five number:
$348 = (2 \times 5^3) + (3 \times 5^2) + (4 \times 5^1) + (3 \times 1)$

Hence, our base–five number is 2343. We'll say that $348_{10} = 2343_5$.

Example 8

Convert the base–ten number 4,509 to base–seven.

The powers of 7 are:

$7^0 = 1$
$7^1 = 7$
$7^2 = 49$
$7^3 = 343$
$7^4 = 2401$
$7^5 = 16807$
Etc…

The highest power of 7 that will divide into 4,509 is $7^4 = 2401$.
With division, we see that it will go in 1 time with a remainder of 2108. So we have 1 in the 7^4 place.

The next power down is $7^3 = 343$, which goes into 2108 six times with a new remainder of 50. So we have 6 in the 7^3 place.

The next power down is $7^2 = 49$, which goes into 50 once with a new remainder of 1. So there is a 1 in the 7^2 place.

The next power down is 7^1 but there was only a remainder of 1, so that means there is a 0 in the 7's place and we still have 1 as a remainder.

That, of course, means that we have 1 in the ones place.

Putting all of this together means that $4,509_{10} = 16101_7$.

$$4,509 \div 7^4 = \quad 1 \text{ R } 2108$$
$$2108 \div 7^3 = \quad 6 \text{ R } 50$$
$$50 \div 7^2 = \quad 1 \text{ R } 1$$
$$1 \div 7^1 = \quad 0 \text{ R } 1$$
$$1 \div 7^0 = \quad 1$$

$$4,509_{10} = 16101_7$$

Try it Now 6
Convert 657_{10} to a base 4 number.

Try it Now 7
Convert 8377_{10} to a base 8 number.

Another Method For Converting From Base 10 to Other Bases

As you read the solution to this last example and attempted the "Try it Now" problems, you may have had to repeatedly stop and think about what was going on. The fact that you are probably struggling to follow the explanation and reproduce the process yourself is mostly due to the fact that the non-decimal systems are so unfamiliar to you. In fact, the only system that you are probably comfortable with is the decimal system.

As budding mathematicians, you should always be asking questions like "How could I simplify this process?" In general, that is one of the main things that mathematicians do…they look for ways to take complicated situations and make them easier or more familiar. In this section we will attempt to do that.

To do so, we will start by looking at our own decimal system. What we do may seem obvious and maybe even intuitive but that's the point. We want to find a process that we readily recognize works and makes sense to us in a familiar system and then use it to extend our results to a different, unfamiliar system.

Let's start with the decimal number, 4863_{10}. We will convert this number to base 10. Yeah, I know it's already in base 10, but if you carefully follow what we're doing, you'll see it makes things work out very nicely with other bases later on. We first note that the highest power of 10 that will divide into 4863 at least once is $10^3 = 1000$. *In general, this is the first step in our new process; we find the highest power that a given base that will divide at least once into our given number.*

We now divide 1000 into 4863:

$$4863 \div 1000 = 4.863$$

This says that there are four thousands in 4863 (obviously). However, it also says that there are 0.863 thousands in 4863. This fractional part is our remainder and will be converted to lower powers of our base (10). If we take that decimal and multiply by 10 (since that's the base we're in) we get the following:

$$0.863 \times 10 = 8.63$$

Why multiply by 10 at this point? We need to recognize here that 0.863 thousands is the same as 8.63 hundreds. Think about that until it sinks in.

$$(0.863)(1000) = 863$$
$$(8.63)(100) = 863$$

These two statements are equivalent. So, what we are really doing here by multiplying by 10 is rephrasing or converting from one place (thousands) to the next place down (hundreds).

$$0.863 \times 10 \Rightarrow 8.63$$
$$\text{(Parts of Thousands)} \times 10 \Rightarrow \text{Hundreds}$$

What we have now is 8 hundreds and a remainder of 0.63 hundreds, which is the same as 6.3 tens. We can do this again with the 0.63 that remains after this first step.

$$0.63 \times 10 \Rightarrow 6.3$$
$$\text{Hundreds} \times 10 \Rightarrow \text{Tens}$$

So we have six tens and 0.3 tens, which is the same as 3 ones, our last place value.

Now here's the punch line. Let's put all of the together in one place:

$$4863 \div 1000 = \enspace ④.863$$
$$0.863 \times 10 = \enspace ⑧.63$$
$$0.63 \times 10 = \enspace ⑥.3$$
$$0.3 \times 10 = \enspace ③.0$$

Note that in each step, the remainder is carried down to the next step and multiplied by 10, the base. Also, at each step, the whole number part, which is circled, gives the digit that belongs in that particular place.   What is amazing is that this works for any base! So, to convert from a base 10 number to some other base, b, we have the following steps we can follow:

Converting from Base 10 to Base b: Another method
1. Find the highest power of the base b that will divide into the given number at least once and then divide.
2. Keep the whole number part, and multiply the fractional part by the base b.
3. Repeat step two, keeping the whole number part (including 0), carrying the fractional part to the next step until only a whole number result is obtained.
4. Collect all your whole number parts to get your number in base b notation.

We will illustrate this procedure with some examples.

Example 9

Convert the base 10 number, 348_{10}, to base 5.

This is actually a conversion that we have done in a previous example. The powers of five are:

$5^0 = 1$
$5^1 = 5$
$5^2 = 25$
$5^3 = 125$
$5^4 = 625$
Etc…

The highest power of five that will go into 348 at least once is 5^3.

We divide by 125 and then proceed.

$$348 \div 5^3 = ②.784$$

$$0.784 \times 5 = ③.92$$

$$0.92 \times 5 = ④0.6$$

$$0.6 \times 5 = ③.0$$

By keeping all the whole number parts, from top bottom, gives 2343 as our base 5 number. Thus, $2343_5 = 348_{10}$.

We can compare our result with what we saw earlier, or simply check with our calculator, and find that these two numbers really are equivalent to each other.

Example 10

Convert the base 10 number, 3007_{10}, to base 5.

The highest power of 5 that divides at least once into 3007 is $5^4 = 625$. Thus, we have:

$$3007 \div 625 = ④.8112$$
$$0.8112 \times 5 = ④.056$$
$$0.056 \times 5 = ⓪.28$$
$$0.28 \times 5 = ①0.4$$
$$0.4 \times 5 = ②0.0$$

This gives us that $3007_{10} = 44012_5$. Notice that in the third line that multiplying by 5 gave us 0 for our whole number part. We don't discard that! The zero tells us that a zero in that place. That is, there are no 5^2's in this number.

This last example shows the importance of using a calculator in certain situations and taking care to avoid clearing the calculator's memory or display until you get to the very end of the process.

Example 11

Convert the base 10 number, 63201_{10}, to base 7.

The powers of 7 are:

$7^0 = 1$
$7^1 = 7$
$7^2 = 49$
$7^3 = 343$
$7^4 = 2401$
$7^5 = 16807$
etc…

The highest power of 7 that will divide at least once into 63201 is 7^5. When we do the initial division on a calculator, we get the following:

$63201 \div 7^5 = 3.760397453$

The decimal part actually fills up the calculators display and we don't know if it terminates at some point or perhaps even repeats down the road. So if we clear our calculator at this point, we will introduce error that is likely to keep this process from ever ending. To avoid this problem, we leave the result in the calculator and simply subtract 3 from this to get the fractional part all by itself. DO NOT ROUND OFF! Subtraction and then multiplication by seven gives:

$63201 \div 7^5 = ③.760397453$
$0.760397453 \times 7 = ⑤.322782174$
$0.322782174 \times 7 = ②.259475219$
$0.259475219 \times 7 = ①.816326531$
$0.816326531 \times 7 = ⑤.714285714$
$0.714285714 \times 7 = ⑤.000000000$

Yes, believe it or not, that last product is exactly 5, *as long as you don't clear anything out on your calculator*. This gives us our final result: $63201_{10} = 352155_7$.

If we round, even to two decimal places in each step, clearing our calculator out at each step along the way, we will get a series of numbers that do not terminate, but begin repeating themselves endlessly. (Try it!) We end up with something that doesn't make any sense, at least not in this context. So be careful to use your calculator cautiously on these conversion problems.

Also, remember that if your first division is by 7^5, then you expect to have 6 digits in the final answer, corresponding to the places for 7^5, 7^4, and so on down to 7^0. If you find yourself with more than 6 digits due to rounding errors, you know something went wrong.

Try it Now 8
Convert the base 10 number, 9352_{10}, to base 5.

Try it Now 9
Convert the base 10 number, 1500, to base 3.

Be careful not to clear your calculator on this one. Also, if you're not careful in each step, you may not get all of the digits you're looking for, so move slowly and with caution.

The Mayan Numeral System

Background

As you might imagine, the development of a base system is an important step in making the counting process more efficient. Our own base–ten system probably arose from the fact that we have 10 fingers (including thumbs) on two hands. This is a natural development. However, other civilizations have had a variety of bases other than ten. For example, the Natives of Queensland used a base–two system, counting as follows: "one, two, two and one, two two's, much." Some Modern South American Tribes have a base–five system counting in this way: "one, two, three, four, hand, hand and one, hand and two," and so on. The Babylonians used a base–sixty (sexigesimal) system. In this chapter, we wrap up with a specific example of a civilization that actually used a base system other than 10.

The Mayan civilization is generally dated from 1500 B.C.E to 1700 C.E. The Yucatan Peninsula (see map[31]) in Mexico was the scene for the development of one of the most advanced civilizations of the ancient world. The Mayans had a sophisticated ritual system that was overseen by a priestly class. This class of priests developed a philosophy with time as divine and eternal.[32] The calendar, and calculations related to it, were thus very important to the ritual life of the priestly class, and hence the Mayan people. In fact, much of what we know about this culture comes from their calendar records and astronomy data. Another important source of information on the Mayans is the writings of Father Diego de Landa, who went to Mexico as a missionary in 1549.

1. Chichen Itza
2. Uxmal
3. Tulum
4. Palenque
5. Bonampak, Yaxchilan
6. Tikal
7. Altun Ha
8. Copán

There were two numeral systems developed by the Mayans – one for the common people and one for the priests. Not only did these two systems use different symbols, they also used different base systems. For the priests, the number system was governed by ritual. The days

of the year were thought to be gods, so the formal symbols for the days were decorated heads,[33] like the sample to the left[34] Since the basic calendar was based on 360 days, the priestly numeral system used a mixed base system employing multiples of 20 and 360. This makes for a confusing system, the details of which we will skip.

The Mayan Number System

Instead, we will focus on the numeration system of the "common" people, which used a more consistent base system. As we stated earlier, the Mayans used a base–20 system, called the "vigesimal" system. Like our system, it is positional, meaning that the position of a numeric symbol indicates its place value. In the following table you can see the place value in its vertical format.[35]

Powers	Base–Ten Value	Place Name
20^7	12,800,000,000	Hablat
20^6	64,000,000	Alau
20^5	3,200,000	Kinchil
20^4	160,000	Cabal
20^3	8,000	Pic
20^2	400	*Bak*
20^1	20	Kal
20^0	1	Hun

In order to write numbers down, there were only three symbols needed in this system. A horizontal bar represented the quantity 5, a dot represented the quantity 1, and a special symbol (thought to be a shell) represented zero. The Mayan system may have been the first to make use of zero as a placeholder/number. The first 20 numbers are shown in the table to the right.[36]

Unlike our system, where the ones place starts on the right and then moves to the left, the Mayan systems places the ones on the <u>bottom</u> of a vertical orientation and moves up as the place value increases.

When numbers are written in vertical form, there should never be more than four dots in a single place. When writing Mayan numbers, every group of five dots becomes one bar. Also, there should never be more than three bars in a single place…four bars would be converted to one dot in the next place up. It's the same as 10 getting converted to a 1 in the next place up when we carry during addition.

Number	Vertical Form	Number	Vertical Form
0	⬭	10	═
1	○	11	═ with ○
2	○ ○	12	═ with ○ ○
3	○ ○ ○	13	═ with ○ ○ ○
4	○ ○ ○ ○	14	═ with ○ ○ ○ ○
5	—	15	≡
6	— with ○	16	≡ with ○
7	— with ○ ○	17	≡ with ○ ○
8	— with ○ ○ ○	18	≡ with ○ ○ ○
9	— with ○ ○ ○ ○	19	≡ with ○ ○ ○ ○

Example 12

What is the value of this number, which is shown in vertical form?

○ ○ ○

○ ○ ○
=====

Starting from the bottom, we have the ones place. There are two bars and three dots in this place. Since each bar is worth 5, we have 13 ones when we count the three dots in the ones place. Looking to the place value above it (the twenties places), we see there are three dots so we have three twenties.

○ ○ ○   ◄——— 20's

○ ○ ○   ◄——— 1's
=====

Hence we can write this number in base–ten as:

$$\left(3\times 20^1\right)+\left(13\times 20^0\right)=\left(3\times 20\right)+\left(13\times 1\right)$$
$$=60+13$$
$$=73$$

Example 13

What is the value of the following Mayan number?

This number has 11 in the ones place, zero in the 20's place, and 18 in the $20^2=400$'s place. Hence, the value of this number in base–ten is:

$$18\times 400 + 0\times 20 + 11\times 1 = 7211.$$

Try it Now 10

Convert the Mayan number below to base 10.

● ● ●

● ● ●
=====

● ●

Example 14

Convert the base 10 number 3575_{10} to Mayan numerals.

This problem is done in two stages. First we need to convert to a base 20 number. We will do so using the method provided in the last section of the text. The second step is to convert that number to Mayan symbols.

The highest power of 20 that will divide into 3575 is $20^2 = 400$, so we start by dividing that and then proceed from there:

$$3575 \div 400 = 8.9375$$
$$0.9375 \times 20 = 18.75$$
$$0.75 \times 20 = 15.0$$

This means that $3575_{10} = 8,18,15_{20}$

The second step is to convert this to Mayan notation. This number indicates that we have 15 in the ones position. That's three bars at the bottom of the number. We also have 18 in the 20's place, so that's three bars and three dots in the second position. Finally, we have 8 in the 400's place, so that's one bar and three dots on the top. We get the following

Note that in the previous example a new notation was used when we wrote $8,18,15_{20}$. The commas between the three numbers 8, 18, and 15 are now separating place values for us so that we can keep them separate from each other. This use of the comma is slightly different than how they're used in the decimal system. When we write a number in base 10, such as 7,567,323, the commas are used primarily as an aide to read the number easily but they do not separate single place values from each other. We will need this notation whenever the base we use is larger than 10.

Writing numbers with bases bigger than 10
When the base of a number is larger than 10, separate each "digit" with a comma to make the separation of digits clear.

For example, in base 20, to write the number corresponding to $17 \times 20^2 + 6 \times 20^1 + 13 \times 20^0$, we'd write $17,6,13_{20}$.

Try it Now 11

Convert the base 10 number 10553_{10} to Mayan numerals.

Try it Now 12

Convert the base 10 number 5617_{10} to Mayan numerals.

Adding Mayan Numbers

When adding Mayan numbers together, we'll adopt a scheme that the Mayans probably did not use but which will make life a little easier for us.

Example 15

Add, in Mayan, the numbers 37 and 29:

First draw a box around each of the vertical places. This will help keep the place values from being mixed up.

Next, put all of the symbols from both numbers into a single set of places (boxes), and to the right of this new number draw a set of empty boxes where you will place the final sum:

You are now ready to start carrying. Begin with the place that has the lowest value, just as you do with Arabic numbers. Start at the bottom place, where each dot is worth 1. There are six dots, but a maximum of four are allowed in any one place; once you get to five dots, you must convert to a bar. Since five dots make one bar, we draw a bar through five of the dots, leaving us with one dot which is under the four-dot limit. Put this dot into the bottom place of the empty set of boxes you just drew:

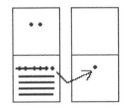

Now look at the bars in the bottom place. There are five, and the maximum number the place can hold is three. *Four bars are equal to one dot in the next highest place.*

Whenever we have four bars in a single place we will automatically convert that to a *dot* in the next place up. We draw a circle around four of the bars and an arrow up to the dots' section of the higher place. At the end of that arrow, draw a new dot. That dot represents 20 just the same as the other dots in that place. Not counting the circled bars in the bottom place, there is one bar left. One bar is under the three-bar limit; put it under the dot in the set of empty places to the right.

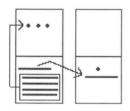

Now there are only three dots in the next highest place, so draw them in the corresponding empty box.

We can see here that we have 3 twenties (60), and 6 ones, for a total of 66. We check and note that 37 + 29 = 66, so we have done this addition correctly. Is it easier to just do it in base–ten? Probably, but that's only because it's more familiar to you. Your task here is to try to learn a new base system and how addition can be done in slightly different ways than what you have seen in the past. Note, however, that the concept of carrying is still used, just as it is in our own addition algorithm.

Try it Now 13
Try adding 174 and 78 in Mayan by first converting to Mayan numbers and then working entirely within that system. Do not add in base–ten (decimal) until the very end when you *check* your work.

Conclusion

In this chapter, we have briefly sketched the development of numbers and our counting system, with the emphasis on the "brief" part. There are numerous sources of information and research that fill many volumes of books on this topic. Unfortunately, we cannot begin to come close to covering all of the information that is out there.

We have only scratched the surface of the wealth of research and information that exists on the development of numbers and counting throughout human history. What is important to

360

note is that the system that we use every day is a product of thousands of years of progress and development. It represents contributions by many civilizations and cultures. It does not come down to us from the sky, a gift from the gods. It is not the creation of a textbook publisher. It is indeed as human as we are, as is the rest of mathematics. Behind every symbol, formula and rule there is a human face to be found, or at least sought.

Furthermore, we hope that you now have a basic appreciation for just how interesting and diverse number systems can get. Also, we're pretty sure that you have also begun to recognize that we take our own number system for granted so much that when we try to adapt to other systems or bases, we find ourselves truly having to concentrate and think about what is going on.

Try it Now Answers

1. $1+6\times3+3\times6+2\times12 = 61$ cats.

2. From left to right:
   Cord 1 = 2,162
   Cord 2 = 301
   Cord 3 = 0
   Cord 4 = 2,070

3. $41065_7 = 9994_{10}$
4. $143_{10} = 1033_5$
5. $21021_3 = 196_{10}$
6. $657_{10} = 22101_4$
7. $8377_{10} = 20271_8$
8. $9352_{10} = 244402_5$
9. $1500_{10} = 2001120_3$
10. 1562
11. $10553_{10} = 1,6,7,13_{20}$

12. $5617_{10} = 14,0,17_{20}$. Note that there is a zero in the 20's place, so you'll need to use the appropriate zero symbol in between the ones and 400's places.

13. A sample solution is shown.

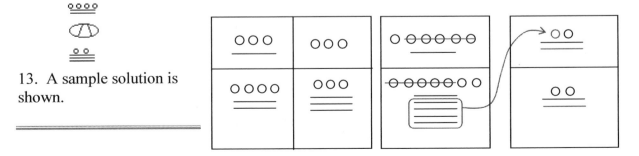

Exercises

Skills

Counting Board And Quipu

1. In the following Peruvian counting board, determine how many of each item is represented. Please show all of your calculations along with some kind of explanation of how you got your answer. Note the key at the bottom of the drawing.

● = jars   ◉ = baskets   ◍ = tools

2. Draw a quipu with a main cord that has branches (H cords) that show each of the following numbers on them. (You should produce <u>one</u> drawing for this problem with the cord for part **a** on the left and moving to the right for parts **b** through **d**.)
    a. 232   b. 5065
    c. 23,451   d. 3002

Basic Base Conversions

3.   423 in base 5 to base 10   4. 3044 in base 5 to base 10

5.   387 in base 10 to base 5   6. 2546 in base 10 to base 5

7.   110101 in base 2 to base 10   8. 11010001 in base 2 to base 10

9.   100 in base 10 to base 2   10. 2933 in base 10 to base 2

11. Convert 653 in base 7 to base 10.   12. Convert 653 in base 10 to base 7

13. 3412 in base 5 to base 2   14. 10011011 in base 2 to base 5
(Hint: convert first to base 10 then to the final desired base)

The Caidoz System

Suppose you were to discover an ancient base–12 system made up twelve symbols. Let's call this base system the Caidoz system. Here are the symbols for each of the numbers 0 through 12:

0 = ♈	6 = ♎
1 = ♉	7 = ♏
2 = ♊	8 = ♐
3 = ♋	9 = ♑
4 = ♌	10 = ♒
5 = ♍	11 = ♓

Convert each of the following numbers in Caidoz to base 10

15. ♏ ♉ ♒

16. ♐ ♋ ♈ ♓

17. ♎ ♌ ♊

18. ♍ ♉ ♒ ♈

Convert the following base 10 numbers to Caidoz, using the symbols shown above.

19. 175

20. 3030

21. 10,000

22. 5507

Mayan Conversions

Convert the following numbers to Mayan notation. Show your calculations used to get your answers.

23. 135

24. 234

25. 360

26. 1,215

27. 10,500

28. 1,100,000

Convert the following Mayan numbers to decimal (base–10) numbers. Show all calculations.

29.

30.

31.

32.

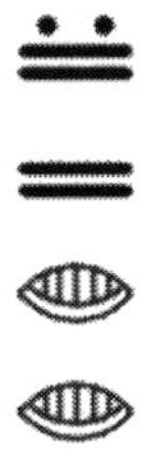

James Bidwell has suggested that Mayan addition was done by "simply combining bars and dots and carrying to the next higher place." He goes on to say, "After the combining of dots and bars, the second step is to exchange every five dots for one bar in the same position." After converting the following base 10 numbers into vertical Maya notation (in base 20, of course), perform the indicated addition:

33.   32 + 11

34.   82 + 15

35.   35 + 148

36.   2412 + 5000

37.   450 + 844

38.   10,000 + 20,000

39.   4,500 + 3,500

40.   130,000 + 30,000

41.   Use the fact that the Mayans had a base-20 number system to complete the following multiplication table. The table entries should be in Mayan notation. Remember: Their zero looked like this… ⬭. *Xerox and then cut out the table below, fill it in, and paste it onto your homework assignment if you do not want to duplicate the table with a ruler.*

(To think about but not write up: Bidwell claims that only these entries are needed for "Mayan multiplication." What does he mean?)

×	•	••	•••	••••	—	=	≡
•							
••							
•••							
••••							
—							
=							
≡							

Binary and Hexadecimal Conversions

*Modern computers operate in a world of "on" and "off" electronic switches, so use a **binary** counting system – base 2, consisting of only two digits: 0 and 1.*

Convert the following binary numbers to decimal (base–10) numbers.

42.   1001

43.   1101

44.   110010

45.   101110

Convert the following base-10 numbers to binary

46.   7

47.   12

48.   36

49.   27

Four binary digits together can represent any base-10 number from 0 to 15. To create a more human-readable representation of binary-coded numbers, hexadecimal numbers, base 16, are commonly used. Instead of using the $8,13,12_{16}$ notation used earlier, the letter A is used to represent the digit 10, B for 11, up to F for 15, so $8,13,12_{16}$ would be written as 8DC.

Convert the following hexadecimal numbers to decimal (base–10) numbers.

50.   C3

51.   4D

52.   3A6

53.   BC2

Convert the following base-10 numbers to hexadecimal

54.   152

55.   176

56.   2034

57.   8263

Exploration

58.   What are the advantages and disadvantages of bases other than ten.

59.   Supposed you are charged with creating a base–15 number system. What symbols would you use for your system and why? Explain with at least two specific examples how you would convert between your base–15 system and the decimal system.

60.   Describe an interesting aspect of Mayan civilization that we did not discuss in class. Your findings must come from some source such as an encyclopedia article, or internet site and you must provide reference(s) of the materials you used (either the publishing information or Internet address).

61. For a Papuan tribe in southeast New Guinea, it was necessary to translate the bible passage John 5:5 "And a certain man was there, which had an infirmity 30 and 8 years" into "A man lay ill one man, both hands, five and three years." Based on your own understanding of bases systems (and some common sense), furnish an explanation of the translation. Please use complete sentences to do so. (Hint: To do this problem, I am asking you to think about how base systems work, where they come from, and how they are used. You won't necessarily find an "answer" in readings or such…you'll have to think it through and come up with a reasonable response. Just make sure that you clearly explain why the passage was translated the way that it was.)

62. The Mayan calendar was largely discussed leading up to December 2012. Research how the Mayan calendar works, and how the counts are related to the number based they use.

Endnotes

[1] Eves, Howard; An Introduction to the History of Mathematics, p. 9.

[2] Eves, p. 9.

[3] McLeish, John; The Story of Numbers – How Mathematics Has Shaped Civilization, p. 7.

[4] Bunt, Lucas; Jones, Phillip; Bedient, Jack; The Historical Roots of Elementary Mathematics, p. 2.

[5] http://www.math.buffalo.edu/mad/Ancient-Africa/mad_zaire-uganda.html

[6] Diana, Lind Mae; The Peruvian Quipu in *Mathematics Teacher,* Issue 60 (Oct., 1967), p. 623–28.

[7] Diana, Lind Mae; The Peruvian Quipu in *Mathematics Teacher,* Issue 60 (Oct., 1967), p. 623–28.

[8] http://wiscinfo.doit.wisc.edu/chaysimire/titulo2/khipus/what.htm

[9] Diana, Lind Mae; The Peruvian Quipu in *Mathematics Teacher,* Issue 60 (Oct., 1967), p. 623–28.

[10] http://www.cs.uidaho.edu/~casey931/seminar/quipu.html

[11] http://www-groups.dcs.st-and.ac.uk/~history/HistTopics/Indian_numerals.html

[12] http://www-groups.dcs.st-and.ac.uk/~history/Mathematicians/Al-Biruni.html

[13] http://www-groups.dcs.st-and.ac.uk/~history/HistTopics/Indian_numerals.html

[14] http://www-groups.dcs.st-and.ac.uk/~history/HistTopics/Indian_numerals.html

[15] Ibid

[16] Ibid

[17] Ibid

[18] Ibid

[19] Katz, page 230

[20] Burton, David M., *History of Mathematics, An Introduction*, p. 254–255

[21] Ibid

[22] Ibid

[23] Katz, page 231.

[24] Ibid, page 230

[25] Ibid, page 231.

[26] Ibid, page 232.

[27] Ibid, page 232.

[28] McLeish, p. 18

[29] http://seattletimes.nwsource.com/news/health-science/html98/invs_20000201.html, Seattle Times, Feb. 1, 2000

[30] Ibid, page 232.

[31] http://www.gorp.com/gorp/location/latamer/map_maya.htm

[32] Bidwell, James; Mayan Arithmetic in *Mathematics Teacher*, Issue 74 (Nov., 1967), p. 762–68.

[33] http://www.ukans.edu/~lctls/Mayan/numbers.html

[34] http://www.ukans.edu/~lctls/Mayan/numbers.html

[35] Bidwell

[36] http://www.vpds.wsu.edu/fair_95/gym/UM001.html

[37] http://forum.swarthmore.edu/k12/mayan.math/mayan2.html

Fractals

Fractals are mathematical sets, usually obtained through recursion, that exhibit interesting dimensional properties. We'll explore what that sentence means through the rest of the chapter. For now, we can begin with the idea of self-similarity, a characteristic of most fractals.

> **Self-similarity**
> A shape is **self-similar** when it looks essentially the same from a distance as it does closer up.

Self-similarity can often be found in nature. In the Romanesco broccoli pictured below[1], if we zoom in on part of the image, the piece remaining looks similar to the whole.

Likewise, in the fern frond below[2], one piece of the frond looks similar to the whole.

Similarly, if we zoom in on the coastline of Portugal[3], each zoom reveals previously hidden detail, and the coastline, while not identical to the view from further way, does exhibit similar characteristics.

---

[1] http://en.wikipedia.org/wiki/File:Cauliflower_Fractal_AVM.JPG
[2] http://www.flickr.com/photos/cjewel/3261398909/
[3] Openstreetmap.org, CC-BY-SA

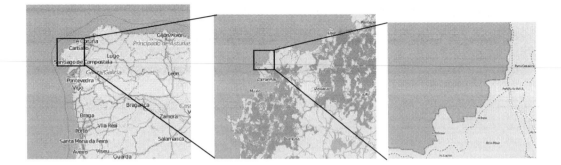

Iterated Fractals

This self-similar behavior can be replicated through recursion: repeating a process over and over.

Example 1

Suppose that we start with a filled-in triangle. We connect the midpoints of each side and remove the middle triangle. We then repeat this process.

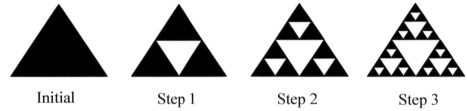

| Initial | Step 1 | Step 2 | Step 3 |

If we repeat this process, the shape that emerges is called the Sierpinski gasket. Notice that it exhibits self-similarity – any piece of the gasket will look identical to the whole. In fact, we can say that the Sierpinski gasket contains three copies of itself, each half as tall and wide as the original. Of course, each of those copies also contains three copies of itself.

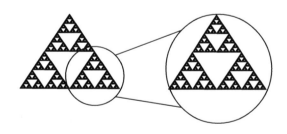

We can construct other fractals using a similar approach. To formalize this a bit, we're going to introduce the idea of initiators and generators.

> **Initiators and Generators**
> An **initiator** is a starting shape
> A **generator** is an arranged collection of scaled copies of the initiator

To generate fractals from initiators and generators, we follow a simple rule:

> **Fractal Generation Rule**
> At each step, replace every copy of the initiator with a scaled copy of the generator, rotating as necessary

This process is easiest to understand through example.

Example 2

Use the initiator and generator shown to create the iterated fractal.

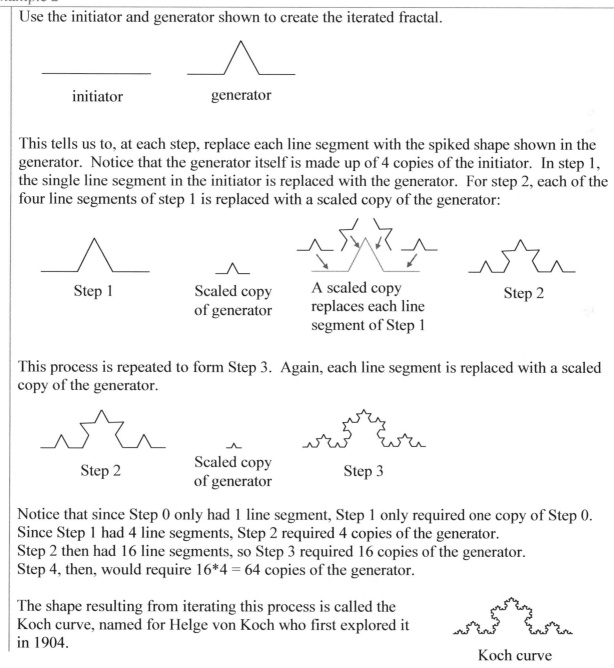

This tells us to, at each step, replace each line segment with the spiked shape shown in the generator. Notice that the generator itself is made up of 4 copies of the initiator. In step 1, the single line segment in the initiator is replaced with the generator. For step 2, each of the four line segments of step 1 is replaced with a scaled copy of the generator:

This process is repeated to form Step 3. Again, each line segment is replaced with a scaled copy of the generator.

Notice that since Step 0 only had 1 line segment, Step 1 only required one copy of Step 0.
Since Step 1 had 4 line segments, Step 2 required 4 copies of the generator.
Step 2 then had 16 line segments, so Step 3 required 16 copies of the generator.
Step 4, then, would require 16*4 = 64 copies of the generator.

The shape resulting from iterating this process is called the Koch curve, named for Helge von Koch who first explored it in 1904.

370

Notice that the Sierpinski gasket can also be described using the initiator-generator approach

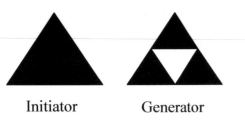

Initiator   Generator

Example 3

Use the initiator and generator below, however only iterate on the "branches." Sketch several steps of the iteration.

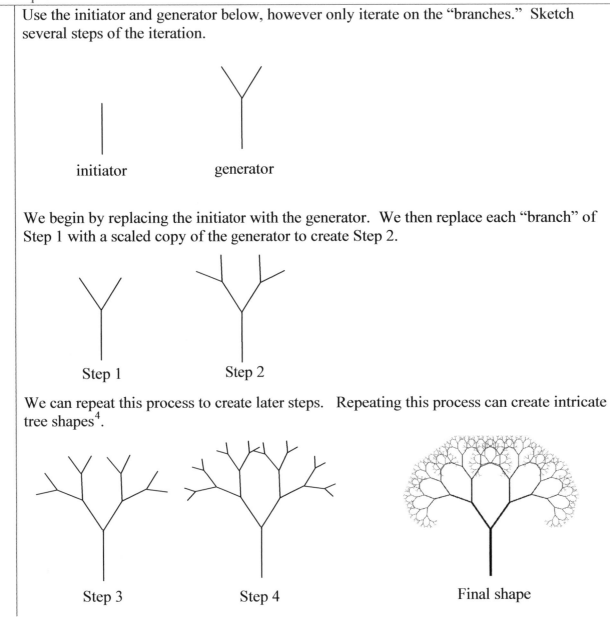

initiator   generator

We begin by replacing the initiator with the generator. We then replace each "branch" of Step 1 with a scaled copy of the generator to create Step 2.

Step 1   Step 2

We can repeat this process to create later steps. Repeating this process can create intricate tree shapes[4].

Step 3   Step 4   Final shape

---

[4] http://www.flickr.com/photos/visualarts/5436068969/

Use the initiator and generator shown to produce the next two stages

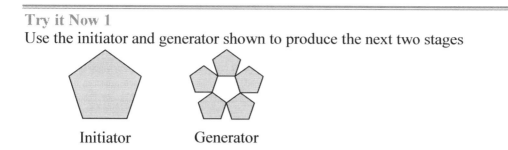

Initiator   Generator

Using iteration processes like those above can create a variety of beautiful images evocative of nature[56].

More natural shapes can be created by adding in randomness to the steps.

Example 4

Create a variation on the Sierpinski gasket by randomly skewing the corner points each time an iteration is made.

Suppose we start with the triangle below. We begin, as before, by removing the middle triangle.   We then add in some randomness.

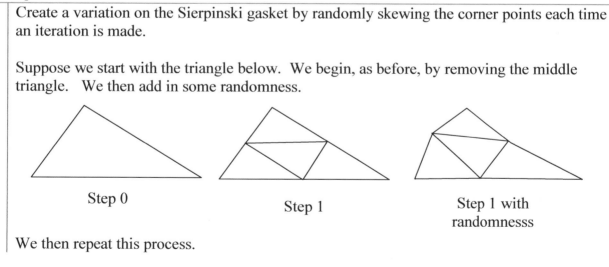

Step 0   Step 1   Step 1 with
                                              randomnesss

We then repeat this process.

[5] http://en.wikipedia.org/wiki/File:Fractal_tree_%28Plate_b_-_2%29.jpg
[6] http://en.wikipedia.org/wiki/File:Barnsley_Fern_fractals_-_4_states.PNG

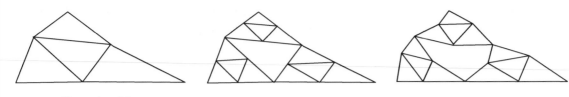

Step 1 with randomnesss

Step 2

Step 2 with randomnesss

Continuing this process can create mountain-like structures.

The landscape to the right[7] was created using fractals, then colored and textured.

Fractal Dimension

In addition to visual self-similarity, fractals exhibit other interesting properties. For example, notice that each step of the Sierpinski gasket iteration removes one quarter of the remaining area. If this process is continued indefinitely, we would end up essentially removing all the area, meaning we started with a 2-dimensional area, and somehow end up with something less than that, but seemingly more than just a 1-dimensional line.

To explore this idea, we need to discuss dimension. Something like a line is 1-dimensional; it only has length. Any curve is 1-dimensional. Things like boxes and circles are 2-dimensional, since they have length and width, describing an area. Objects like boxes and cylinders have length, width, and height, describing a volume, and are 3-dimensional.

1-dimensional   2-dimensional   3-dimensional

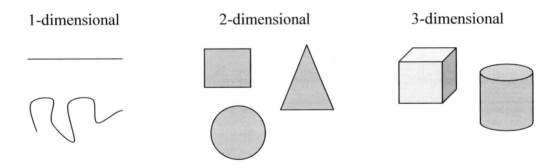

Certain rules apply for scaling objects, related to their dimension.

If I had a line with length 1, and wanted scale its length by 2, I would need two copies of the original line. If I had a line of length 1, and wanted to scale its length by 3, I would need three copies of the original.

---

[7] http://en.wikipedia.org/wiki/File:FractalLandscape.jpg

If I had a rectangle with length 2 and height 1, and wanted to scale its length and width by 2, I would need four copies of the original rectangle. If I wanted to scale the length and width by 3, I would need nine copies of the original rectangle.

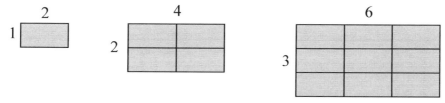

If I had a cubical box with sides of length 1, and wanted to scale its length and width by 2, I would need eight copies of the original cube. If I wanted to scale the length and width by 3, I would need 27 copies of the original cube.

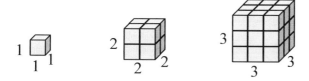

Notice that in the 1-dimensional case, copies needed = scale.
In the 2-dimensional case, copies needed = scale^2.
In the 3-dimensional case, copies needed = scale^3.

From these examples, we might infer a pattern.

Scaling-Dimension Relation
To scale a D-dimensional shape by a scaling factor S, the number of copies C of the original shape needed will be given by:

$$\text{Copies} = \text{Scale}^{\text{Dimension}}, \text{ or } C = S^D$$

Example 5

Use the scaling-dimension relation to determine the dimension of the Sierpinski gasket.

Suppose we define the original gasket to have side length 1. The larger gasket shown is twice as wide and twice as tall, so has been scaled by a factor of 2.

Notice that to construct the larger gasket, 3 copies of the original gasket are needed.

Using the scaling-dimension relation $C = S^D$, we obtain the equation $3 = 2^D$.

Since $2^1 = 2$ and $2^2 = 4$, we can immediately see that D is somewhere between 1 and 2; the gasket is more than a 1-dimensional shape, but we've taken away so much area its now less than 2-dimensional.

Solving the equation $3 = 2^D$ requires logarithms. If you studied logarithms earlier, you may recall how to solve this equation (if not, just skip to the box below and use that formula):

$3 = 2^D$	Take the logarithm of both sides
$\log(3) = \log(2^D)$	Use the exponent property of logs
$\log(3) = D\log(2)$	Divide by $\log(2)$
$D = \dfrac{\log(3)}{\log(2)} \approx 1.585$	The dimension of the gasket is about 1.585

Scaling-Dimension Relation, to find Dimension
To find the dimension D of a fractal, determine the scaling factor S and the number of copies C of the original shape needed, then use the formula
$$D = \frac{\log(C)}{\log(S)}$$

Try it Now 2
Determine the fractal dimension of the fractal produced using the initiator and generator

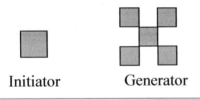

Initiator   Generator

We will now turn our attention to another type of fractal, defined by a different type of recursion. To understand this type, we are first going to need to discuss complex numbers.

Complex Numbers[8]

The numbers you are most familiar with are called **real numbers**. These include numbers like 4, 275, -200, 10.7, ½, π, and so forth. All these real numbers can be plotted on a number line. For example, if we wanted to show the number 3, we plot a point:

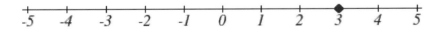

To solve certain problems like $x^2 = -4$, it became necessary to introduce **imaginary numbers**.

Imaginary Number i
The imaginary number i is defined to be $i = \sqrt{-1}$.
Any real multiple of i, like $5i$, is also an imaginary number.

Example 6

Simplify $\sqrt{-9}$.

We can separate $\sqrt{-9}$ as $\sqrt{9}\sqrt{-1}$. We can take the square root of 9, and write the square root of -1 as i.

$$\sqrt{-9} = \sqrt{9}\sqrt{-1} = 3i$$

A complex number is the sum of a real number and an imaginary number.

Complex Number
A **complex number** is a number $z = a + bi$, where a and b are real numbers
a is the real part of the complex number
b is the imaginary part of the complex number

To plot a complex number like $3 - 4i$, we need more than just a number line since there are two components to the number. To plot this number, we need two number lines, crossed to form a complex plane.

Complex Plane
In the **complex plane**, the horizontal axis is the
real axis and the vertical axis is the imaginary axis.

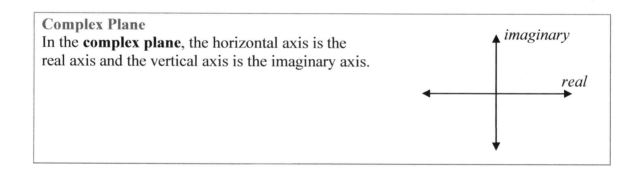

Example 7

Plot the number $3 - 4i$ on the complex plane.

The real part of this number is 3, and the imaginary part is -4. To plot this, we draw a point 3 units to the right of the origin in the horizontal direction and 4 units down in the vertical direction.

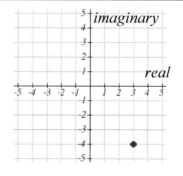

Because this is analogous to the Cartesian coordinate system for plotting points, we can think about plotting our complex number $z = a + bi$ as if we were plotting the point (a, b) in Cartesian coordinates. Sometimes people write complex numbers as $z = x + yi$ to highlight this relation.

Arithmetic on Complex Numbers

Before we dive into the more complicated uses of complex numbers, let's make sure we remember the basic arithmetic involved. To add or subtract complex numbers, we simply add the like terms, combining the real parts and combining the imaginary parts.

Example 8

Add $3 - 4i$ and $2 + 5i$.

Adding $(3 - 4i) + (2 + 5i)$, we add the real parts and the imaginary parts
$3 + 2 - 4i + 5i$
$5 + i$

Try it Now 3
Subtract $2 + 5i$ from $3 - 4i$

When we add complex numbers, we can visualize the addition as a shift, or translation, of a point in the complex plane.

Example 9

Visualize the addition $3 - 4i$ and $-1 + 5i$.

The initial point is $3 - 4i$. When we add $-1 + 3i$, we add -1 to the real part, moving the point 1 units to the left, and we add 5 to the imaginary part, moving the point 5 units vertically. This shifts the point $3 - 4i$ to $2 + 1i$

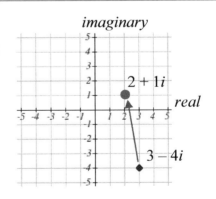

We can also multiply complex numbers by a real number, or multiply two complex numbers.

Example 10

Multiply: $4(2+5i)$.

To multiply the complex number by a real number, we simply distribute as we would when multiplying polynomials.

$4(2+5i)$ Distribute
$=4\cdot2+4\cdot5i$ Simplify
$=8+20i$

Example 11

Multiply: $(2+5i)(4+i)$.

$(2+5i)(4+i)$ Expand
$=8+20i+2i+5i^2$ Since $i=\sqrt{-1}$, $i^2=-1$
$=8+20i+2i+5(-1)$ Simplify
$=3+22i$

Try it Now 4
Multiply $3-4i$ and $2+3i$.

To understand the effect of multiplication visually, we'll explore three examples.

Example 12

Visualize the product $2(1+2i)$

Multiplying we'd get
$2\cdot1+2\cdot2i$
$=2+4i$

Notice both the real and imaginary parts have been scaled by 2. Visually, this will stretch the point outwards, away from the origin.

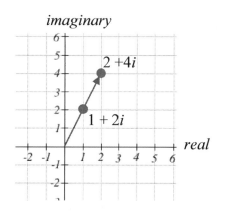

Example 13

Visualize the product $i(1+2i)$

Multiplying, we'd get
$i \cdot 1 + i \cdot 2i$

$= i + 2i^2$

$= i + 2(-1)$

$= -2 + i$

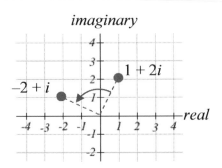

In this case, the distance from the origin has not changed, but the point has been rotated about the origin, 90° counter-clockwise.

Example 14

Visualize the result of multiplying $1+2i$ by $1+i$. Then show the result of multiplying by $1+i$ again.

Multiplying $1+2i$ by $1+i$,
$(1+2i)(1+i)$

$= 1 + i + 2i + 2i^2$

$= 1 + 3i + 2(-1)$

$= -1 + 3i$

Multiplying by $1 + i$ again,
$(-1 + 3i)(1 + i)$

$= -1 - i + 3i + 3i^2$

$= -1 + 2i + 3(-1)$

$= -4 + 2i$

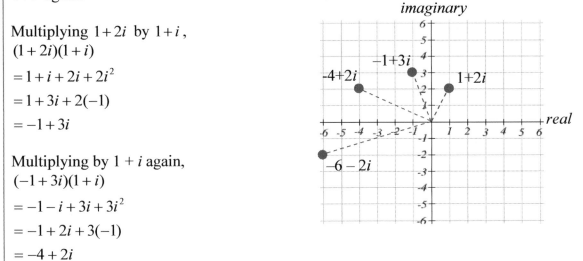

If we multiplied by $1+i$ again, we'd get –6–2i. Plotting these numbers in the complex plane, you may notice that each point gets both further from the origin, and rotates counterclockwise, in this case by 45°.

In general, multiplication by a complex number can be thought of as a **scaling**, changing the distance from the origin, combined with a **rotation** about the origin.

Complex Recursive Sequences

We will now explore recursively defined sequences of complex numbers.

> **Recursive Sequence**
> A **recursive relationship** is a formula which relates the next value, z_{n+1}, in a sequence to the previous value, z_n. In addition to the formula, we need an initial value, z_0.
>
> The sequence of values produced is the recursive sequence.

Example 15

Given the recursive relationship $z_{n+1} = z_n + 2, \quad z_0 = 4$, generate several terms of the recursive sequence.

We are given the starting value, $z_0 = 4$. The recursive formula holds for any value of n, so if $n = 0$, then $z_{n+1} = z_n + 2$ would tell us $z_{0+1} = z_0 + 2$, or more simply, $z_1 = z_0 + 2$.

Notice this defines z_1 in terms of the known z_0, so we can compute the value:
$z_1 = z_0 + 2 = 4 + 2 = 6$.

Now letting $n = 1$, the formula tells us $z_{1+1} = z_1 + 2$, or $z_2 = z_1 + 2$. Again, the formula gives the next value in the sequence in terms of the previous value.
$z_2 = z_1 + 2 = 6 + 2 = 8$

Continuing,
$z_3 = z_2 + 2 = 8 + 2 = 10$
$z_4 = z_3 + 2 = 10 + 2 = 12$

The previous example generated a basic linear sequence of real numbers. The same process can be used with complex numbers.

Example 16

Given the recursive relationship $z_{n+1} = z_n \cdot i + (1 - i), \quad z_0 = 4$, generate several terms of the recursive sequence.

We are given $z_0 = 4$. Using the recursive formula:
$z_1 = z_0 \cdot i + (1 - i) = 4 \cdot i + (1 - i) = 1 + 3i$
$z_2 = z_1 \cdot i + (1 - i) = (1 + 3i) \cdot i + (1 - i) = i + 3i^2 + (1 - i) = i - 3 + (1 - i) = -2$
$z_3 = z_2 \cdot i + (1 - i) = (-2) \cdot i + (1 - i) = -2i + (1 - i) = 1 - 3i$
$z_4 = z_3 \cdot i + (1 - i) = (1 - 3i) \cdot i + (1 - i) = i - 3i^2 + (1 - i) = i + 3 + (1 - i) = 4$
$z_5 = z_4 \cdot i + (1 - i) = 4 \cdot i + (1 - i) = 1 + 3i$

Notice this sequence is exhibiting an interesting pattern – it began to repeat itself.

Mandelbrot Set

The Mandelbrot Set is a set of numbers defined based on recursive sequences

> **Mandelbrot Set**
>
> For any complex number c, define the sequence $z_{n+1} = z_n^2 + c, \quad z_0 = 0$
>
> If this sequence always stays close to the origin (within 2 units), then the number c is part of the **Mandelbrot Set**. If the sequence gets far from the origin, then the number c is not part of the set.

Example 17

Determine if $c = 1 + i$ is part of the Mandelbrot set.

We start with $z_0 = 0$. We continue, omitting some detail of the calculations

$z_1 = z_0^2 + 1 + i = 0 + 1 + i = 1 + i$

$z_2 = z_1^2 + 1 + i = (1 + i)^2 + 1 + i = 1 + 3i$

$z_3 = z_2^2 + 1 + i = (1 + 3i)^2 + 1 + i = -7 + 7i$

$z_4 = z_3^2 + 1 + i = (-7 + 7i)^2 + 1 + i = 1 - 97i$

We can already see that these values are getting quite large. It does not appear that $c = 1 + i$ is part of the Mandelbrot set.

Example 18

Determine if $c = 0.5i$ is part of the Mandelbrot set.

We start with $z_0 = 0$. We continue, omitting some detail of the calculations

$z_1 = z_0^2 + 0.5i = 0 + 0.5i = 0.5i$

$z_2 = z_1^2 + 0.5i = (0.5i)^2 + 0.5i = -0.25 + 0.5i$

$z_3 = z_2^2 + 0.5i = (-0.25 + 0.5i)^2 + 0.5i = -0.1875 + 0.25i$

$z_4 = z_3^2 + 0.5i = (-0.1875 + 0.25i)^2 + 0.5i = -0.02734 + 0.40625i$

While not definitive with this few iterations, it does appear that this value is remaining small, suggesting that $0.5i$ is part of the Mandelbrot set.

Try it Now 5

Determine if $c = 0.4 + 0.3i$ is part of the Mandelbrot set.

If all complex numbers are tested, and we plot each number that is in the Mandelbrot set on the complex plane, we obtain the shape to the right[9].

The boundary of this shape exhibits quasi-self-similarity, in that portions look very similar to the whole.

In addition to coloring the Mandelbrot set itself black, it is common to the color the points in the complex plane surrounding the set. To create a meaningful coloring, often people count the number of iterations of the recursive sequence that are required for a point to get further than 2 units away from the origin. For example, using $c = 1 + i$ above, the sequence was distance 2 from the origin after only two recursions.

For some other numbers, it may take tens or hundreds of iterations for the sequence to get far from the origin. Numbers that get big fast are colored one shade, while colors that are slow to grow are colored another shade. For example, in the image below[10], light blue is used for numbers that get large quickly, while darker shades are used for numbers that grow more slowly. Greens, reds, and purples can be seen when we zoom in – those are used for numbers that grow very slowly.

The Mandelbrot set, for having such a simple definition, exhibits immense complexity. Zooming in on other portions of the set yields fascinating swirling shapes.

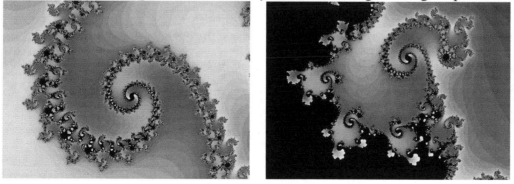

---

[9] http://en.wikipedia.org/wiki/File:Mandelset_hires.png
[10] This series was generated using Scott's Mandelbrot Set Explorer

382

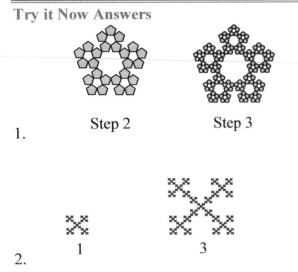

Step 2 Step 3

1.

2.

      1   3

Scaling the fractal by a factor of 3 requires 5 copies of the original. $D = \dfrac{\log(5)}{\log(3)} \approx 1.465$

3. $(3 - 4i) - (2 + 5i) = 1 - 9i$

4. Multiply $(3 - 4i)(2 + 3i) = 6 + 9i - 8i - 12i^2 = 6 + i - 12(-1) = 18 + i$

5. $z_1 = z_0^2 + 0.4 + 0.3i = 0 + 0.4 + 0.3i = 0.4 + 0.3i$

$z_2 = z_1^2 + 0.4 + 0.3i = (0.4 + 0.3i)^2 + 0.4 + 0.3i =$

$z_3 = z_2^2 + 0.5i = (-0.25 + 0.5i)^2 + 0.5i = -0.1875 + 0.25i$

$z_4 = z_3^2 + 0.5i = (-0.1875 + 0.25i)^2 + 0.5i = -0.02734 + 0.40625i$

Additional Resources

A much more extensive coverage of fractals can be found on the Fractal Geometry site. This site includes links to several Java software programs for exploring fractals.

The Mandelbrot Explorer site, provides more details on the Mandelbrot set, including a nice visualization of Mandelbrot sequences.

Exercises

Iterated Fractals

Using the initiator and generator shown, draw the next two stages of the iterated fractal.

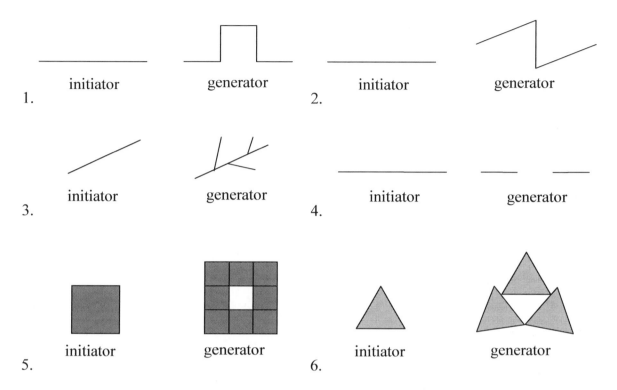

7. Create your own version of Sierpinski gasket with added randomness.

8. Create a version of the branching tree fractal from example #3 with added randomness.

Fractal Dimension

9. Determine the fractal dimension of the Koch curve.

10. Determine the fractal dimension of the curve generated in exercise #1

11. Determine the fractal dimension of the Sierpinski carpet generated in exercise #5

12. Determine the fractal dimension of the Cantor set generated in exercise #4

Complex Numbers

13. Plot each number in the complex plane: a) 4   b) $-3i$ c) $-2 + 3i$ d) $2 + i$

14. Plot each number in the complex plane: a) -2 b) $4i$ c) $1 + 2i$ d) $-1 - i$

15. Compute: a) $(2+3i)+(3-4i)$ b) $(3-5i)-(-2-i)$

16. Compute: a) $(1-i)+(2+4i)$ b) $(-2-3i)-(4-2i)$

17. Multiply: a) $3(2+4i)$ b) $(2i)(-1-5i)$ c) $(2-4i)(1+3i)$

18. Multiply: a) $2(-1+3i)$ b) $(3i)(2-6i)$ c) $(1-i)(2+5i)$

19. Plot the number $2+3i$. Does multiplying by $1-i$ move the point closer to or further from the origin? Does it rotate the point, and if so which direction?

20. Plot the number $2+3i$. Does multiplying by $0.75+0.5i$ move the point closer to or further from the origin? Does it rotate the point, and if so which direction?

Recursive Sequences

21. Given the recursive relationship $z_{n+1} = iz_n + 1$, $z_0 = 2$, generate the next 3 terms of the recursive sequence.

22. Given the recursive relationship $z_{n+1} = 2z_n + i$, $z_0 = 3-2i$, generate the next 3 terms of the recursive sequence.

23. Using $c = -0.25$, calculate the first 4 terms of the Mandelbrot sequence.

24. Using $c = 1 - i$, calculate the first 4 terms of the Mandelbrot sequence.

For a given value of c, the Mandelbrot sequence can be described as *escaping* (growing large), a *attracted* (it approaches a fixed value), or *periodic* (it jumps between several fixed values). A periodic cycle is typically described the number if values it jumps between; a 2-cycle jumps between 2 values, and a 4-cycle jumps between 4 values.

For questions 25 – 30, you'll want to use a calculator that can compute with complex numbers, or use <u>an online calculator</u> which can compute a Mandelbrot sequence. For each value of c, examine the Mandelbrot sequence and determine if the value appears to be escaping, attracted, or periodic?

25. $c = -0.5 + 0.25i$.   26. $c = 0.25 + 0.25i$.

27. $c = -1.2$.   28. $c = i$.

29. $c = 0.5 + 0.25i$.   30. $c = -0.5 + 0.5i$.

31. $c = -0.12 + 0.75i$.   32. $c = -0.5 + 0.5i$.

Exploration

The Julia Set for c is another fractal, related to the Mandelbrot set. The Julia Set for c uses the recursive sequence: $z_{n+1} = z_n^2 + c, \quad z_0 = d$, where c is constant for any particular Julia set, and d is the number being tested. A value d is part of the Julia Set for c if the sequence does not grow large.

For example, the Julia Set for -2 would be defined by $z_{n+1} = z_n^2 - 2, \quad z_0 = d$. We then pick values for d, and test each to determine if it is part of the Julia Set for -2. If so, we color black the point in the complex plane corresponding with the number d. If not, we can color the point d based on how fast it grows, like we did with the Mandelbrot Set.

For questions 33-34, you will probably want to use the online calculator again.

33. Determine which of these numbers are in the Julia Set at $c = -0.12i + 0.75i$
    a) $0.25i$ b) 0.1 c) $0.25 + 0.25i$

34. Determine which of these numbers are in the Julia Set at $c = -0.75$
    a) $0.5i$ b) 1 c) $0.5 - 0.25i$

You can find many images online of various Julia Sets[11].

35. Explain why no point with initial distance from the origin greater than 2 will be part of the Mandelbrot sequence

---

[11] For example, http://www.jcu.edu/math/faculty/spitz/juliaset/juliaset.htm,

Cryptography

When people need to secretly store or communicate messages, they turn to cryptography. Cryptography involves using techniques to obscure a message so outsiders cannot read the message. It is typically split into two steps: encryption, in which the message is obscured, and decryption, in which the original message is recovered from the obscured form.

Substitution Ciphers

One simple encryption method is called a **substitution cipher**.

Substitution Cipher
A substitution cipher replaces each letter in the message with a different letter, following some established mapping.

A simple example of a substitution cipher is called the **Caesar cipher**, sometimes called a shift cipher. In this approach, each letter is replaced with a letter some fixed number of positions later in the alphabet. For example, if we use a shift of 3, then the letter A would be replaced with D, the letter 3 positions later in the alphabet. The entire mapping would look like: [1]

Original: ABCDEFGHIJKLMNOPQRSTUVWXYZ
Maps to: DEFGHIJKLMNOPQRSTUVWXYZABC

Example 1

Use the Caesar cipher with shift of 3 to encrypt the message: "We ride at noon"

We use the mapping above to replace each letter. W gets replaced with Z, and so forth, giving the encrypted message: ZH ULGH DW QRRQ.

Notice that the length of the words could give an important clue to the cipher shift used. If we saw a single letter in the encrypted message, we would assume it must be an encrypted A or I, since those are the only single letters than form valid English words.

To obscure the message, the letters are often rearranged into equal sized blocks. The message ZH ULGH DW QRRQ could be written in blocks of three characters as ZHU LGH DWQ RRQ.

---

[1] http://en.wikipedia.org/w/index.php?title=File:Caesar3.svg&page=1. PD

Example 2

Decrypt the message GZD KNK YDX MFW JXA if it was encrypted using a shift cipher with shift of 5.

We start by writing out the character mapping by shifting the alphabet, with A mapping to F, five characters later in the alphabet.

Original: ABCDEFGHIJKLMNOPQRSTUVWXYZ

Maps to: FGHIJKLMNOPQRSTUVWXYZABCDE

We now work backwards to decrypt the message. The first letter G is mapped to by B, so B is the first character of the original message. Continuing, our decrypted message is BUY FIF TYS HAR ESA.

Removing spaces we get BUYFIFTYSHARESA. In this case, it appears an extra character was added to the end to make the groups of three come out even, and that the original message was "Buy fifty shares."

Try it Now 1

Decrypt the message BNW MVX WNH if it was encrypted using a shift cipher with shift 9 (mapping A to J).

Notice that in both the ciphers above, the extra part of the alphabet wraps around to the beginning. Because of this, a handy version of the shift cipher is a cipher disc, such as the Alberti cipher disk shown here[2] from the 1400s. In a cipher disc, the inner wheel could be turned to change the cipher shift. This same approach is used for "secret decoder rings."

The security of a cryptographic method is very important to the person relying on their message being kept secret. The security depends on two factors:
1. The security of the method being used
2. The security of the encryption key used

In the case of a shift cipher, the method is "a shift cipher is used." The encryption key is the specific amount of shift used.

Suppose an army is using a shift cipher to send their messages, and one of their officers is captured by their enemy. It is likely the method and encryption key could become compromised. It is relatively hard to change encryption methods, but relatively easy to change encryption keys.

---

[2] http://en.wikipedia.org/wiki/File:Alberti_cipher_disk.JPG

During World War II, the Germans' Enigma encryption machines were captured, but having details on the encryption method only slightly helped the Allies, since the encryption keys were still unknown and hard to discover. Ultimately, the security of a message cannot rely on the method being kept secret; it needs to rely on the key being kept secret.

Encryption Security
The security of any encryption method should depend only on the encryption key being difficult to discover. It is not safe to rely on the encryption method (algorithm) being kept secret.

With that in mind, let's analyze the security of the Caesar cipher.

Example 3.

Suppose you intercept a message, and you know the sender is using a Caesar cipher, but do not know the shift being used. The message begins EQZP. How hard would it be to decrypt this message?

Since there are only 25 possible shifts, we would only have to try 25 different possibilities to see which one produces results that make sense. While that would be tedious, one person could easily do this by hand in a few minutes. A modern computer could try all possibilities in under a second.

Shift	Message	Shift	Message	Shift	Message	Shift	Message
1	DPYO	7	XJSI	13	RDMC	19	LXGW
2	COXN	8	WIRH	14	QCLB	20	KWFV
3	BNWM	9	VHQG	15	PBKA	21	JVEU
4	AMVL	10	UGPF	16	OAJZ	22	IUDT
5	ZLUK	11	TFOE	17	NZIY	23	HTCS
6	YKTJ	**12**	**SEND**	18	MYHX	24	GSBR
						25	FRAQ

In this case, a shift of 12 (A mapping to M) decrypts EQZP to SEND. Because of this ease of trying all possible encryption keys, the Caesar cipher is not a very secure encryption method.

Brute Force Attack
A brute force attack is a method for breaking encryption by trying all possible encryption keys.

To make a brute force attack harder, we could make a more complex substitution cipher by using something other than a shift of the alphabet. By choosing a random mapping, we could get a more secure cipher, with the tradeoff that the encryption key is harder to describe; the key would now be the entire mapping, rather than just the shift amount.

Example 4

Use the substitution mapping below to encrypt the message "March 12 0300"

Original: ABCDEFGHIJKLMNOPQRSTUVWXYZ0123456789
Maps to: 2BQF5WRTD8IJ6HLCOSUVK3A0X9YZN1G4ME7P

Using the mapping, the message would encrypt to 62SQT ZN Y1YY

Try it Now 2

Use the substitution mapping from Example 4 to decrypt the message C2SVX2VP

While there were only 25 possible shift cipher keys (35 if we had included numbers), there are about 10^{40} possible substitution ciphers[3]. That's much more than a trillion trillions. It would be essentially impossible, even with supercomputers, to try every possible combination. Having a huge number of possible encryption keys is one important part of key security.

Unfortunately, this cipher is still not secure, because of a technique called frequency analysis, discovered by Arab mathematician Al-Kindi in the 9th century. English and other languages have certain letters than show up more often in writing than others.[4] For example, the letter E shows up the most frequently in English. The chart to the right shows the typical distribution of characters.

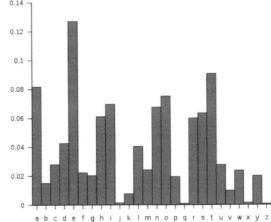

Example 5

The chart to the right shows the frequency of different characters in some encrypted text. What can you deduce about the mapping?

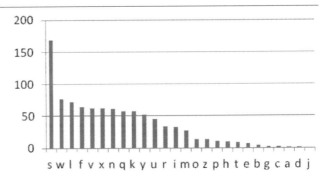

Because of the high frequency of the letter S in the encrypted text, it is very likely that the substitution maps E to S. Since W is the second most frequent character, it likely that T or A maps to W. Because C, A, D, and J show up rarely in the encrypted text, it is likely they are mapped to from J, Q, X, and Z.

---

[3] There are 35 choices for what *A* maps to, then 34 choices for what *B* maps to, and so on, so the total number of possibilities is 35*34*33*...*2*1 = 35! = about 10^{40}

[4] http://en.wikipedia.org/w/index.php?title=File:English_letter_frequency_(alphabetic).svg&page=1 PD

In addition to looking at individual letters, certain pairs of letters show up more frequently, such as the pair "th." By analyzing how often different letters and letter pairs show up an encrypted message, the substitution mapping used can be deduced[5].

Transposition Ciphers

Another approach to cryptography is **transposition cipher.**

> **Transposition Ciphers**
> A transposition cipher is one in which the order of characters is changed to obscure the message.

An early version of a transposition cipher was a Scytale[6], in which paper was wrapped around a stick and the message was written. Once unwrapped, the message would be unreadable until the message was wrapped around a same-sized stick again.

One modern transposition cipher is done by writing the message in rows, then forming the encrypted message from the text in the columns.

Example 6

Encrypt the message "Meet at First and Pine at midnight" using rows 8 characters long.

We write the message in rows of 8 characters each. Nonsense characters are added to the end to complete the last row.

```
MEETATFI
RSTANDPI
NEATMIDN
IGHTPXNR
```

We could then encode the message by recording down the columns. The first column, reading down, would be MRNI. All together, the encoded message would be MRNI ESEG ETAH TATT ANMP TDIX FPDN IINR. The spaces would be removed or repositioned to hide the size of table used, since that is the encryption key in this message.

Example 7

Decrypt the message CEE IAI MNL NOG LTR VMH NW using the method above with a table with rows of 5 characters.

Since there are total of 20 characters and each row should have 5 characters, then there will be 20/5 = 4 rows.

---

[5] For an example of how this is done, see http://en.wikipedia.org/wiki/Frequency_analysis
[6] http://en.wikipedia.org/wiki/File:Skytala%26EmptyStrip-Shaded.png

We start writing, putting the first 4 letters, CEEI, down the first column.

```
CALLM
EINTH
EMORN
INGVW
```

We can now read the message: CALL ME IN THE MORNING VW. The VW is likely nonsense characters used to fill out the message.

More complex versions of this rows-and-column based transposition cipher can be created by specifying an order in which the columns should be recorded. For example, the method could specify that after writing the message out in rows that you should record the third column, then the fourth, then the first, then the fifth, then the second. This adds additional complexity that would make it harder to make a brute-force attack.

To make the encryption key easier to remember, a word could be used. For example, if the key word was "MONEY", it would specify that rows should have 5 characters each. The order of the letters in the alphabet would dictate which order to read the columns in. Since E, the 4th letter in the word, is the earliest letter in the alphabet from the word MONEY, the 4th column would be used first, followed by the 1st column (M), the 3rd column (N), the 2nd column (O), and the 5th column (Y).

Example 8

Encrypt the message BUY SOME MILK AND EGGS using a transposition cipher with key word MONEY.

Writing out the message in rows of 5 characters:
```
BUYSO
MEMIL
KANDE
GGSPK
```

We now record the columns in order 4 1 3 2 5:
SIDP BMKG YMNS UEAG OLEK
As before, we'd then remove or reposition the spaces to conceal evidence of the encryption key.

Try it Now 3
Encrypt the message "Fortify the embassy" using a transposition cipher with key word HELP

To decrypt a keyword-based transposition cipher, we'd reverse the process. In the example above, the keyword MONEY tells us to begin with the 4th column, so we'd start by writing SIDP down the 4th column, then continue to the 1st column, 3rd column, etc.

Example 9

Decrypt the message RHA VTN USR EDE AIE RIK ATS OQR using a row-and-column transposition cipher with keyword PRIZED.

The keyword PRIZED tells us to use rows with 6 characters. Since D comes first in the alphabet, we start with 6th column. Since E is next in the alphabet, we'd follow with the 5th column. Continuing, the word PRIZED tells us the message was recorded with the columns in order 4 5 3 6 2 1.

For the decryption, we set up a table with 6 characters in each row. Since the beginning of the encrypted message came from the last column, we start writing the encrypted message down the last column.

					R
					H
					A
					V

The 5th column was the second one the encrypted message was read from, so is the next one we write to.

				T	R
				N	H
				U	A
				S	V

Continuing, we can fill out the rest of the message.

A	I	R	S	T	R
I	K	E	O	N	H
E	A	D	Q	U	A
R	T	E	R	S	V

Reading across the rows gives our decrypted message: AIRSTRIKEONHEADQUARTERSV

Unfortunately, since the transposition cipher does not change the frequency of individual letters, it is still susceptible to frequency analysis, though the transposition does eliminate information from letter pairs.

Advanced shared symmetric-key methods

Both the substitution and transposition methods discussed so far are shared **symmetric-key** methods, meaning that both sender and receiver would have to have agreed upon the same secret encryption key before any methods could be sent.

All of the methods so far have been susceptible to frequency analysis since each letter is always mapped to the same encrypted character. More advanced methods get around this weakness. For example, the Enigma machines used in World War II had wheels that rotated. Each wheel was a substitution cipher, but the rotation would cause the substitution used to shift after each character.

For a simplified example, in the initial setup, the wheel might provide the mapping

```
Original:    ABCDEFGHIJKLMNOPQRSTUVWXYZ0123456789
Maps to:     2BQF5WRTD8IJ6HLCOSUVK3A0X9YZN1G4ME7P
```

After the first character is encrypted, the wheel rotates, shifting the mapping one space, resulting in a new shifted mapping:

Original: `ABCDEFGHIJKLMNOPQRSTUVWXYZ0123456789`

Maps to: `P2BQF5WRTD8IJ6HLCOSUVK3A0X9YZN1G4ME7`

Using this approach, no letter gets encrypted as the same character over and over.

Example 10

Encrypt the message "See me". Use a basic Caesar cipher with shift 3 as the initial substitution, but shift the substitution one place after each character.

The initial mapping is

Original: `ABCDEFGHIJKLMNOPQRSTUVWXYZ`

Maps to: `DEFGHIJKLMNOPQRSTUVWXYZABC`

This would map the first letter, S to V. We would then shift the mapping by one.

Original: `ABCDEFGHIJKLMNOPQRSTUVWXYZ`

Now maps to: `EFGHIJKLMNOPQRSTUVWXYZABCD`

Now the next letter, E, will map to I. Again we shift the cipher

Original: `ABCDEFGHIJKLMNOPQRSTUVWXYZ`

Now maps to: `FGHIJKLMNOPQRSTUVWXYZABCDE`

The next letter, E, now maps to J. Continuing this process, the final message would be VIJSL.

Notice that frequency analysis is much less useful now, since the character E has been mapped to three different characters due to the shifting of the substitution mapping.

Try it Now 4

Decrypt the message KIQRV if it was encrypted using a basic Caesar cipher with shift 3 as the initial substitution, but shifting the substitution one place after each character.

The actual Engima machines used in WWII were more complex. Each wheel consisted of a complex substitution cipher, and multiple wheels were used in a chain[7]. The specific wheels used, order of the wheels, and starting position of the wheels formed the encryption key. While captured Engima devices provided the Allied forces details on the encryption method, the keys still had to be broken to decrypt messages.

---

[7] http://en.wikipedia.org/wiki/File:Enigma_rotors_with_alphabet_rings.jpg

These code breaking efforts led to the development of some of the first electronic computers by Alan Turing at Bletchley Park in the United Kingdom. This is generally considered the beginnings of modern computing[8].

In the 1970s, the U.S. government had a competition and ultimately approved an algorithm deemed DES (Data Encryption Standard) to be used for encrypting government data. It became the standard encryption algorithm used. This method used a combination of multiple substitution and transposition steps, along with other steps in which the encryption key is mixed with the message. This method uses an encryption key with length 56 bits, meaning there are 2^{56} possible keys.

This number of keys make a brute force attack extremely difficult and costly, but not impossible. In 1998, a team was able to find the decryption key for a message in 2 days, using about $250,000 worth of hardware. However, the price and time will go down as computer power increases.

From 1997 to 2001 the government held another competition, ultimately adopting a new method, deemed AES (Advanced Encryption Standard). This method uses encryption keys with 128, 192, or 256 bits, providing up to 2^{256} possible keys, making brute force attacks essentially impossible.

Public Key Cryptography

Suppose that you are connecting to your bank's website. It is possible that someone could intercept any communication between you and your bank, so you'll want to encrypt the communication. The problem is that all the encryption methods we've discussed require than both parties have already agreed on a shared secret encryption key. How can you and your bank agree on a key if you haven't already?

This becomes the goal of public key cryptography – to provide a way for two parties to agree on a key without a snooping third party being able to determine the key. The method relies on a one-way function; something that is easy to do one way, but hard to reverse. We will explore the Diffie-Hellman-Merkle key exchange method.

As an example, let's consider mixing paint. It's easy to mix paint to make a new color, but much harder to separate a mixed paint into the two original colors used.[9][10]

---

[8] For a good overview, see http://www.youtube.com/watch?v=5nK_ft0Lf1s
[9] http://en.wikipedia.org/w/index.php?title=File:Diffie-Hellman_Key_Exchange.svg&page=1
[10] For a video overview of this process, see http://www.youtube.com/watch?v=YEBfamv-_do

Using this analogy, Alice and Bob publically agree on a common starter color. Each then mixes in some of their own secret color. They then exchange their mixed colors.

Since separating colors is hard, even if a snooper were to obtain these mixed colors, it would be hard to obtain the original secret colors.

Once they have exchanged their mixed colors, Alice and Bob both add their secret color to the mix they obtained from the other person. In doing so, both Alice and Bob now have the same common secret color, since it contains a mix of the original common color, Alice's secret color, and Bob's secret color.

They now have a common secret color they can use as their encryption key, even though neither Alice nor Bob knows the other's secret color.

Likewise, there is no way for a snooper to obtain the common secret color without separating one of the mixed colors.

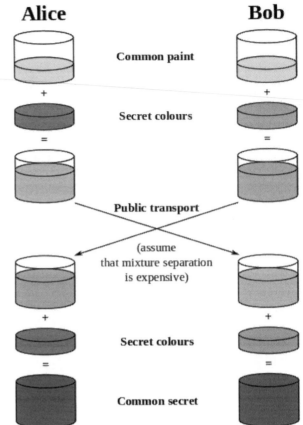

To get this process to work for computer communication, we need to have the process result in a share common number to act as the common secret encryption key. For this, we need a numerical one-way function.

Modular arithmetic

If you think back to doing division with whole numbers, you may remember finding the whole number result and the remainder after division.

> **Modulus**[11]
> The **modulus** is another name for the remainder after division.
> For example, 17 mod 5 = 2, since if we divide 17 by 5, we get 3 with remainder 2.

Modular arithmetic is sometimes called clock arithmetic, since analog clocks wrap around times past 12, meaning they work on a modulus of 12. If the hour hand of a clock currently points to 8, then in 5 hours it will point to 1. While 8+5 = 13, the clock wraps around after 12, so all times can be thought of as modulus 12. Mathematically, 13 mod 12 = 1.

---

[11] Sometime, instead of seeing 17 mod 5 = 2, you'll see 17 ≡ 2 (mod 5). The ≡ symbol means "congruent to" and means that 17 and 2 are equivalent, after you consider the modulus 5.

Example 11

Compute: a) 10 mod 3   b) 15 mod 5   c) 2^7 mod 5

a) Since 10 divided by 3 is 3 with remainder 1, 10 mod 3 = 1
b) Since 15 divided by 5 is 3 with no remainder, 15 mod 5 = 0
c) 2^7 = 128. 128 divide by 5 is 25 with remainder 3, so 2^7 mod 5 = 3

Try it Now 5

Compute: a) 23 mod 7   b) 15 mod 7   c) 2034 mod 7

Recall that when we divide 17 by 5, we could represent the result as 3 remainder 2, as the mixed number $3\frac{2}{5}$, or as the decimal 3.4. Notice that the modulus, 2, is the same as the numerator of the fractional part of the mixed number, and that the decimal part 0.4 is equivalent to the fraction $\frac{2}{5}$. We can use these conversions to calculate the modulus of not-too-huge numbers on a standard calculator.

> **Modulus on a Standard Calculator**
> To calculate a mod n on a standard calculator
> 1) Divide a by n
> 2) Subtract the whole part of the resulting quantity
> 3) Multiply by n to obtain the modulus

Example 12

Calculate 31345 mod 419

31345 / 419 = 74.8090692   Now subtract 74 to get just the decimal remainder
74.8090692 − 74 = 0.8090692   Multiply this by 419 to get the modulus

0.8090692 * 419 = 339   This tells us 0.8090692 was equivalent to $\frac{339}{419}$

In the text above, only a portion of the decimal value was written down. In practice, you should try to avoid writing down the intermediary steps, and instead allow your calculator to retain as many decimal values as it can.

The one-way function

When you use a prime number p as a modulus, you can find a special number called a generator, g, so that g^n mod p will result in all the values from 1 to $p - 1$.

In the table to the right, notice that when we give values of n from 1 to 6, we get out all values from 1 to 6. This means 3 is a generator when 7 is the modulus.

n	3^n	3^n mod 7
1	3	3
2	9	2
3	27	6
4	81	4
5	243	5
6	729	1

This gives us our one-way function. While it is easy to compute the value of g^n mod p when we know n, it is difficult to find the exponent n to obtain a specific value.

For example, suppose we use $p = 23$ and $g = 5$. If I pick n to be 6, I can fairly easily calculate 5^6 mod 23 = 15625 mod 23 = 8.

If someone else were to tell you 5^n mod 23 = 7, it is much harder to find n. In this particular case, we'd have to try 22 different values for n until we found one that worked – there is no known easier way to find n other than brute-force guessing.

While trying 22 values would not take too long, when used in practice much larger values for p are used, typically with well over 500 digits. Trying all possibilities would be essentially impossible.

The key exchange

Before we can begin the key exchange process, we need a couple more important facts about modular arithmetic.

Modular Exponentiation Rule
$(a^b$ mod $n) = (a$ mod $n)^b$ mod n

Example 13

Compute 12^5 mod 7 using the exponentiation rule.

Evaluated directly: $12^5 = 248,832$, so 12^5 mod 7 = 248,823 mod 7 = 3.

Using the rule above, 12^5 mod 7 = (12 mod 7)^5 mod 7 = 5^5 mod 7 = 3125 mod 7 = 3.

You may remember a basic exponent rule from algebra: $(a^b)^c = a^{bc} = a^{cb} = (a^c)^b$
For example: $64^2 = (4^3)^2 = 4^6 = (4^2)^3 = 16^3$

We can combine the modular exponentiation rule with the algebra exponent rule to define the modular exponent power rule.

Modular Exponent Power Rule
$(a^b$ mod $n)^c$ mod $n = (a^{bc}$ mod $n) = (a^c$ mod $n)^b$ mod n

---

Example 14

Verify the rule above if $a = 3$, $b = 4$, $c = 5$, and $n = 7$

$(3^4 \bmod 7)^5 \bmod 7 = (81 \bmod 7)^5 \bmod 7 = 4^5 \bmod 7 = 1024 \bmod 7 = 2$

$(3^5 \bmod 7)^4 \bmod 7 = (243 \bmod 7)^4 \bmod 7 = 5^4 \bmod 7 = 625 \bmod 7 = 2$, the same result.

Try it Now 6

Use the modular exponent rule to calculate 10000 mod 7, by noting $10000 = 10^4$.

This provides us the basis for our key exchange.   While it will be easier to understand in the following example, here's the process:

1. Alice and Bob agree publically on values a prime p and generator g.

2. Alice picks some secret number a, while Bob picks some secret number b.

3. Alice computes $A = g^a \bmod p$ and sends it to Bob.

4. Bob computes $B = g^b \bmod p$ and sends it to Alice.

5. Alice computes $B^a \bmod p$, which is $(g^b \bmod p)^a \bmod p$

6. Bob computes $A^b \bmod p$, which is $(g^a \bmod p)^b \bmod p$

The modular exponent power rule tells us $(g^a \bmod p)^b \bmod p = (g^b \bmod p)^a \bmod p$, so Alice and Bob will arrive at the same shared value to use as a key, even though neither knows the other's secret number, and no eavesdropper can determine this value knowing only g, p, A, and B.

Example 15

Alice and Bob publically share a generator and prime modulus. In this case, we'll use 3 as the generator and 17 as the prime.

(This example is continued on the next page)

	Alice		**Bob**
Alice and Bob publically share a generator and prime modulus.	$g = 3$, $p = 17$	Common info	$g = 3$, $p = 17$
Each then secretly picks a number n of their own.	$n = 8$	secret number	$n = 6$
Each calculates $g^n \bmod p$	$3^8 \bmod 17 = 16$		$3^6 \bmod 17 = 15$
They then exchange these resulting values.	$A = 16$ $B = 15$		$B = 15$ $A = 16$
Each then raises the value they received to the power of their secret $n \bmod p$.	$B^n \bmod p =$ $15^8 \bmod 17 = 1$	mix in secret number	$A^n \bmod p =$ $16^6 \bmod 17 = 1$
The result is the shared secret key.	1	shared secret key	1

The shared secrets come out the same because of the modular exponent power rule $(a^b \bmod n)^c \bmod n = (a^c \bmod n)^b \bmod n$. Alice computed $(3^6 \bmod 17)^8 \bmod 17$ while Bob computed $(3^8 \bmod 17)^6 \bmod 17$, which the rule says will give the same results.

Notice that even if a snooper were to obtain both exchanged values $A = 16$ and $B = 15$, there is no way they could obtain the shared secret key from these without having at least one of Alice or Bob's secret numbers. There is no easy way to obtain the secret numbers from the shared values, since the function was a one-way function.

Using this approach, Alice and Bob can now use the shared secret key obtained as the key for a standard encryption algorithm like DES or AES.

Try it Now 7
Suppose you are doing a key exchange with Kylie using generator 5 and prime 23. Your secret number is 2. What number do you send to Kylie? If Kylie sends you the value 8, determine the shared secret key.

RSA

There are several other public-key methods used, including RSA, which is very commonly used. RSA involves distributing a public encryption key, which anyone can use to encrypt messages to you, but which can only be decrypted using a separate private key. You can think of this as sending an open padlock to someone – they can lock up information, but no one can unlock it without the key you kept secret.

RSA's security relies on the difficulty of factoring large numbers. For example, it's easy to calculate that 53 times 59 is 3127, but given the number 12,317 that is a product of two primes, it's much harder to find the numbers that multiply to give that number. It's exponentially harder when the primes each have 100 or more digits. Suppose we find two primes p and q and multiply them to get $n = pq$. This number will be very hard to factor. If we also know p and q, there are shortcuts to find two numbers e and d so that $m^{ed} \bmod n = m \bmod n$ for all numbers m. Without knowing the factorization of n, finding these values is very hard.

To use RSA, we generate two primes p and q and multiply them to get $n = pq$. Since we know the factorization, we can easily find e and d so $m^{ed} = m \bmod n$. Now, we lock away p, q, and d. We then send the values e and n out publically. To encrypt a message m, the sender computes $S = m^e \bmod n$. As we saw earlier, the modulus is a one-way function which makes the original message very hard to recover from S. However, we have our private key d we can use to decrypt the message. When we receive the secret message S, we compute $S^d \bmod n = \left(m^e\right)^d \bmod n = m^{ed} \bmod n = m \bmod n$, recovering the original message[12].

Example 16

Suppose that Alice has computed $n = 3127$, $e = 3$, and $d = 2011$. Show how Bob would encrypt the message 50 and how Alice would then decrypt it.

Bob would only know his message, $m = 50$ and Alice's public key: $n = 3127$ and $e = 3$. He would encrypt the message by computing $m^e \bmod n$: $50^3 \bmod 3127 = 3047$.

Alice can then decrypt this message using her private key d by computing $S^d \bmod n$: $3047^{2011} \bmod 3127 = 50$.

This method differs from Diffie-Hellman-Merkle because no exchange process is needed; Bob could send Alice an encrypted message using Bob's public key without having to communicate with Alice beforehand to determine a shared secret key. This is especially handy for applications like encrypting email, where both parties might not be online at the same time to perform a Diffie-Hellman-Merkle style key exchange.

---

[12] Many details have been left out, including how e and d are determined, and why this all works. For a bit more detail, see http://www.youtube.com/watch?v=wXB-V_Keiu8, or http://doctrina.org/How-RSA-Works-With-Examples.html

Other secret keeping methods

While this chapter has focused on cryptography methods for keeping a message secure, there is another area of secret keeping called **steganography** which focuses on hiding the existence of a message altogether. Historically, steganography included techniques like invisible ink, watermarks, or embedding a secret message inside a longer document that appeared unimportant. The digital age has provided many new ways to hide messages in digital photos or in the background noise of a music file.

Resources

There are several calculation tools available based on the material in this chapter available at: http://www.opentextbookstore.com/mathinsociety/apps/

Try it Now Answers

1. SEND MONEY 2. PARTY AT 9

3. HELP gives column order 2 1 3 4.
   FORT
   IFYT
   HEEM
   BASS
   YPLR
   Encrypted text: OFE APF IHB YRY ESL TTM SR

4. The initial mapping was:
   Original: ABCDEFGHIJKLMNOPQRSTUVWXYZ
   Now maps to: DEFGHIJKLMNOPQRSTUVWXYZABC

   Using this, we can see the first character of the encrypted message, K, can be decrypted to the letter H. We now shift the mapping by one character.
   Original: ABCDEFGHIJKLMNOPQRSTUVWXYZ
   Now maps to: EFGHIJKLMNOPQRSTUVWXYZABCD

   The second character in the message, I, can be decrypted to the letter E. Continuing this process of shifting and decrypting, KIQRV decrypts to HELLO.

5. a) 2 b) 1 c) 4

6. $10000 \bmod 7 = 10^4 \bmod 7 = (10 \bmod 7)^4 \bmod 7 = 3^4 \bmod 7 = 81 \bmod 7 = 4$

7. To compute the number we'd send to Kylie, we raise the generator to the power of our secret number modulus the prime: $5^2 \bmod 23 = 25 \bmod 23 = 2$.

   If Kylie sends us the value 8, we determine the shared secret by raising her number to the power of our secret number modulus the prime: $8^2 \bmod 23 = 64 \bmod 23 = 18$. 18 would be the shared secret.

Exercises

Substitution ciphers

In the questions below, if it specifies an alphabetic cipher, then the original map used letters only: ABCDEFGHIJKLMNOPQRSTUVXYZ. If it specifies an alphanumeric cipher, then the original map used letters and numbers:
ABCDEFGHIJKLMNOPQRSTUVXYZ0123456789

1. Encrypt the message "SEND SUPPLIES" using an alphabetic Caesar cipher with shift 7 (mapping A to H).

2. Encrypt the message "CANCEL CONTRACT" using an alphanumeric Caesar cipher with shift 16 (mapping A to Q).

3. Decrypt the message "2R1 ONO 5SN OXM O" if it was encrypted using an alphanumeric Ceasar cipher with shift 10 (mapping A to K).

4. Decrypt the message "RJJY NSAJ SNHJ" if it was encrypted using an alphabetic Ceasar cipher with shift 5 (mapping A to F).

For questions 5-8 use this substitution mapping:
Original: ABCDEFGHIJKLMNOPQRSTUVWXYZ0123456789
Maps to: HLCO2BQF5WRTZN1G4D8IJ6SUVK3A0X9YME7P

5. Use the substitution mapping to encrypt the message "DEAR DIARY"

6. Use the substitution mapping to encrypt the message "ATTACK AT SUNRISE"

7. Use the substitution mapping to decrypt the message "Z2DQ 2D1N"

8. Use the substitution mapping to decrypt the message "Z22 IHI3 YX3"

Transposition ciphers

9. Encrypt the message "Meet in the library at ten" using a tabular transposition cipher with rows of length 5 characters.

10. Encrypt the message "Fly surveillance over the northern county" using a tabular transposition cipher with rows of length 8 characters.

11. Decrypt the message "THE VHI NIE SAN SHT STI MQA DAN SDR S" if it was encrypted using a tabular transposition cipher with rows of length 7 characters.

12. Decrypt the message "DOLR UTIR INON KVEY AZ" if it was encrypted using a tabular transposition cipher with rows of length 6 characters.

13. Encrypt the message "Buy twenty million" using a tabular transposition cipher with the encryption keyword "RENT".

14. Encrypt the message "Attack from the northeast" using a tabular transposition cipher with the encryption keyword "POWER".

15. Decrypt the message "RYL OEN ONI TPM IEE YTE YDH WEA HRM S" if it was encrypted using a tabular transposition cipher with the encryption keyword "READING".

16. Decrypt the message "UYH SRT ABV HLN SEE L" if it was encrypted using a tabular transposition cipher with the encryption keyword "MAIL".

Shifting substitution ciphers

17. Encrypt the message "SEND SUPPLIES" using an alphabetic Caesar cipher that starts with shift 7 (mapping A to H), and shifts one additional space after each character is encoded.

18. Encrypt the message "CANCEL CONTRACT" using an alphabetic Caesar cipher that starts with shift 5 (mapping A to F), and shifts one additional space after each character is encoded.

Modular arithmetic

19. Compute
    a. 15 mod 4
    b. 10 mod 5
    c. 257 mod 11

20. Compute
    a. 20 mod 4
    b. 14 mod 3
    c. 86 mod 13

21. Determine if 4 is a generator modulus 11

22. Determine if 2 is a generator modulus 13

23. Use the modular exponent rule to calculate 157^{10} mod 5

24. Use the modular exponent rule to calculate 133^8 mod 6

Diffie-Hellman-Merkle key exchange

25. Suppose you are doing a key exchange with Marc using generator 5 and prime 23. Your secret number is 7. Marc sends you the value 3. Determine the shared secret key.

26. Suppose you are doing a key exchange with Jen using generator 5 and prime 23. Your secret number is 4. Jen sends you the value 8. Determine the shared secret key.

RSA

27. Suppose that Alice has computed $n = 33$, $e = 7$, and $d = 3$. Show how Bob would encrypt the message 5 and how Alice would then decrypt it.

28. Suppose that Alice has computed $n = 55$, $e = 7$, and $d = 13$. Show how Bob would encrypt the message 8 and how Alice would then decrypt it.

Extensions

29. To further obscure a message, sometimes the usual alphabet characters are replaced with other symbols. Design a new set of symbols, and use it to encode a message. Exchange with a friend and see if they can decode your message.

30. To make an encryption harder to break, sometimes multiple substitution and transposition ciphers are used in sequence. For example, a method might specify that the first letter of the encryption keyword be used to determine the initial shift for a Caesar cipher (perhaps with a rotating cipher), and also be used for a transposition cipher. Design your own sequence of encryption steps and encrypt a message. Exchange with a friend and see if they can follow your process to decrypt the message.

31. When using large primes, computing values like 67^{24} mod 83 can be difficult on a calculator without using additional tricks, since 67^{24} is a huge number. We will explore an approach used.
    a. Notice that 67^2 mod 83 is fairly easy to calculate: 67^2 mod 83 = 4489 mod 83 = 7.

       Since 67^4 mod 83 = $(67^2)^2$ mod 83 can be rewritten using the modular exponent rule as $(67^2 \text{ mod } 83)^2$ mod 83, this is also easy to evaluate: 67^4 mod 83 = $(67^2 \text{ mod } 83)^2$ mod 83 = 7^2 mod 83 = 49.

       This process can be continued to find 67^8 mod 83 as $(6^4)^2$ mod 83. Find this value, then find 67^{16} mod 83 and 67^{32} mod 83.

    b. There is a rule that (ab) mod $n = (a \text{ mod } n)(b \text{ mod } n)$ mod n
       Noting that $17000 = 170*100$, calculate 17000 mod 83 using the rule above.

    c. Note that $67^5 = 67^4 67$. Use this, along with the rule from above and the results from part a to compute 67^5 mod 83.

    d. Note that $67^7 = 67^{4+2+1} = 67^4 \, 67^2 \, 67^1$. Compute 67^7 mod 83.

    e. Write 67^{24} as a product of powers of 67, and use this to compute 67^{24} mod 83

32. Use the process from the previous question to evaluate 23^{34} mod 37

33. To encrypt text messages with RSA, the words are first converted into a string of numbers, and then encrypted. Several characters are usually combined together to produce a message number smaller than the modulus, but approximately the same size. Look up an ASCII table to convert the message "SCALE THE WALLS" to numbers, then encrypt it using the RSA public key $n = 10823$, $e = 5$. Since ASCII characters are two digits, pair up characters to form four-digit numbers before encoding. For example A is 65 and B is 66, so the character pair AB could be treated as the number 6566 and encrypted as 10148

34. Explore approaches to steganography that don't require specialized software. Attempt to hide a message using one of these techniques, and see if a fellow student can detect the message.

35. When you visit a secure website, your web browser will report that the site's identity has been verified by a third party, called a certificate authority. This is meant to assure you that you are visiting the actual company's website. Research how these certificates work.

Solutions to Selected Exercises

Problem Solving

1. $18/230 = 0.07826 =$ about 7.8%

3. €250(0.23) =€ 57.50 in VAT

5. $15000(5.57) = \$83,550$

7. absolute increase: 1050. Relative: $1050/3250 = 0.323 = 32.3\%$ increase

9. a. $2200 - 2200(0.15) = 2200(0.85) = \$1870$
   b. Yes, their goal was to decrease by at least 15%. They exceeded their goal.

11. Dropping by 6% is the same as keeping 94%. $a(0.94) = 300$. $a = 319.15$. Attendance was about 319 before the drop.

13. a) Kaplan's enrollment was 64.3% larger than Walden's. 30510
    b) Walden's enrollment was 39.1% smaller than Kaplan's.
    c) Walden's enrollment was 60.9% of Kaplan's.

15. If the original price was $100, the basic clearance price would be $100 - \$100(0.60) =$ $40. The additional markdown would bring it to $40 - \$40(0.30) = \$28$. This is 28% of the original price.

17. These are not comparable; "a" is using a base of all Americans and is talking about health insurance from any source, while "b" is using a base of adults and is talking specifically about health insurance provided by employers.

21. These statements are equivalent, if we assume the claim in "a" is a percentage point increase, not a relative change. Certainly these messages are phrased to convey different opinions of the levy. We are told the new rate will be $9.33 per $1000, which is 0.933% tax rate. If the original rate was 0.833% (0.1 percentage point lower), then this would indeed be a 12% relative increase.

23. 20% of 30% is $30\%(0.20) = 6\%$, a 6 percentage point decrease.

25. Probably not, unless the final is worth 50% of the overall class grade. If the final was worth 25% of the overall grade, then a 100% would only raise her average to 77.5%

27. $4/10 pounds = \$0.40$ per pound   (or 10 pounds/$4 = 2.5 pounds per dollar)

29. $x = 15$ 31. 2.5 cups   33. 74 turbines

35. 96 inches   37. $6000   39. 55.6 meters

43. The population density of the US is 84 people per square mile. The density of India is about 933 people per square mile. The density of India is about 11 times greater than that of the U.S.

49. The oil in the spill could produce 93.1 million gallons of gasoline. Each car uses about 600 gallons a year. That would fuel 155,167 cars for a year.

53. An answer around 100-300 gallons would be reasonable

57. 156 million miles

59. The time it takes the light to reach you is so tiny for any reasonable distance that we can safely ignore it. 750 miles/hr is about 0.21 miles/sec. If the sound takes 4 seconds to reach you, the lightning is about 0.84 miles away. In general, the lightning will be $0.21n$ miles away, which is often approximated by dividing the number of seconds by 5.

61. About 8.2 minutes

63. Four cubic yards (or 3.7 if they sell partial cubic yards)

Voting Theory

1.

Number of voters	3	3	1	3	2
1st choice	A	A	B	B	C
2nd choice	B	C	A	C	A
3rd choice	C	B	C	A	B

3. a. 9+19+11+8 = 47
   b. 24 for majority; 16 for plurality (though a choice would need a minimum of 17 votes to actually win under the Plurality method)
   c. Atlanta, with 19 first-choice votes
   d. Atlanta 94, Buffalo 111, Chicago 77. Winner: Buffalo
   e. Chicago eliminated, 11 votes go to Buffalo. Winner: Buffalo
   f. A vs B: B. A vs C: A. B vs C: B. B gets 2 pts, A 1 pt. Buffalo wins.

5. a. 120+50+40+90+60+100 = 460
   b. 231 for majority; 116 for plurality
   c. A with 150 first choice votes
   d. A 1140, B 1060, C 1160, D 1240. Winner: D
   e. B eliminated, votes to C. D eliminated, votes to A. Winner: A
   f. A vs B: B. A vs C: A. A vs D: D. B vs C: C. B vs D: D. C vs D: C
      A 1pt, B 1pt, C 2pt, D 2pt. Tie between C and D.
      Winner would probably be C since C was preferred over D

7. a. 33
   b. 17

9. Yes, B

11. B, with 17 approvals

13. Independence of Irrelevant Alternatives Criterion

15. Condorcet Criterion

Weighted Voting

1. a. 9 players
   b. 10+9+9+5+4+4+3+2+2 = 48
   c. 47

3. a. 9, a majority of votes
   b. 17, the total number of votes
   c. 12, which is 2/3 of 17, rounded up

5. a. P1 is a dictator (can reach quota by themselves)
   b. P1, since dictators also have veto power
   c. P2, P3, P4

7. a. none
   b. P1
   c. none

9. a. 11+7+2 = 20
   b. P1 and P2 are critical

11. Winning coalitions, with critical players underlined:
    {$\underline{P1},\underline{P2}$} {$\underline{P1},P2,P3$} {$\underline{P1},P2,P4$} {$\underline{P1},P2,P3,P4$} {$\underline{P1},\underline{P3}$} {$\underline{P1},P3,P4$}
    P1: 6 times, P2: 2 times, P3: 2 times, P4: 0 times. Total: 10 times
    Power: P1: 6/10 = 60%, P2: 2/10 = 20%, P3: 2/10 = 20%, P4: 0/10 = 0%

13. a. {$\underline{P1}$} {$\underline{P1},P2$} {$\underline{P1},P3$} {$\underline{P1},P4$} {$\underline{P1},P2,P3$} {$\underline{P1},P2,P4$} {$\underline{P1},P3,P4$} {$\underline{P1},P2,P3,P4$}
       P1: 100%, P2: 0%, P3: 0%, P4: 0%
    b. {$\underline{P1},\underline{P2}$} {$\underline{P1},\underline{P3}$} {$\underline{P1},\underline{P4}$} {$\underline{P1},P2,P3$} {$\underline{P1},P2,P4$} {$\underline{P1},P3,P4$} {$\underline{P1},P2,P3,P4$}
       P1: 7/10 = 70%, P2: 1/10 = 10%, P3: 1/10 = 10%, P4: 1/10 = 10%
    c. {$\underline{P1},\underline{P2}$} {$\underline{P1},\underline{P3}$} {$\underline{P1},P2,P3$} {$\underline{P1},P2,P4$} {$\underline{P1},P3,P4$} {$\underline{P1},P2,P3,P4$}
       P1: 6/10 = 60%, P2: 2/10 = 20%, P3: 2/10 = 20%, P4: 0/10 = 0%

15. P3 = 5. P3+P2 = 14. P3+P2+P1 = 27, reaching quota. P1 is critical.

17. Sequential coalitions with pivotal player underlined
    <P1,$\underline{P2}$,P3> <P1,$\underline{P3}$,P2> <P2,$\underline{P1}$,P3> <P2,P3,$\underline{P1}$> <P3,$\underline{P1}$,P2> <P3,$\underline{P2}$,P1>
    P1: 2/6 = 33.3%, P2: 2/6 = 33.3%, P3: 2/6 = 33.3%

410

19. a. 6, 7
    b. 8, given P1 veto power
    c. 9, given P1 and P2 veto power

21. If adding a player to a coalition could cause it to reach quota, that player would also be critical in that coalition, which means they are not a dummy. So a dummy cannot be pivotal.

23. We know P2+P3 can't reach quota, or else P1 wouldn't have veto power.
    P1 can't reach quota alone.
    P1+P2 and P1+P3 must reach quota or else P2/P3 would be dummy.
    a. {P1,P2} {P1,P3} {P1,P2,P3}. P1: 3/5, P2: 1/5, P3: 1/5
    b. <P1,P2,P3> <P1,P3,P2> <P2,P1,P3> <P2,P3,P1> <P3,P1,P2> <P3,P2,P1>
       P1: 4/6, P2: 1/6, P3: 1/6

25. [4: 2, 1, 1, 1] is one of many possibilities

27. [56: 30, 30, 20, 20, 10]

29. [54: 10, 10, 10, 10, 10, 1, 1, 1, 1, 1, 1, 1, 1, 1, 1] is one of many possibilities

Fair Division
1. Chance values the veggie half at $7.50 and pepperoni half at $2.50.
    A full pepperoni slice is ¼ of the pepperoni half. Value $2.50/4 = $0.625
    A full veggie slice is ¼ of the veggie half. Value $7.50/4 = $1.875
    A slice that is ½ pepperoni ½ veggie is value $0.3125+$0.9375 = $1.25

3. Erin: Bowl 1, Catherine: Bowl 2, Shannon: Bowl 3

5. a. 25 Snickers @ $0.01 each, 20 Milky Ways @ $0.05 each, 60 Reese's @ $0.02 each
      Value: $0.25 + $1.00 + $1.20 = $2.45
   b. No. Dustin values the whole bag at $8, so a fair share would be $4.
   c. Lots of possibilities. Here's a couple:
      80 Milky Ways, 0 Snickers, 0 Reese's
      50 Snickers, 50 Milky Ways, 50 Reese's

7. a. Zoe
   b. Maggie: s2, s3. Meredith: s1, s2. Holly: s3
   c. Maggie: s2, Meredith: s1, Holly: s3, Zoe: s4

9. a. P5
   b. $6.50 (doesn't need to trim it much since they're last)
   c. P4 would receive it, with value $6.00 (since P4 would trim it)

11. a. (320+220)/4 = $135

   b. Desk and Vanity both go to A. A pays $320 + $220 - $135 = $405 to estate
      B gets $95, C gets $125, D gets $110.

   c. Surplus of $405 - $95 - $125 - $110 = $75 gets split, $18.75 each.
      A gets desk and vanity, pays $386.25 to estate
      B gets $113.75, C gets $143.75, D gets $128.75

13. Fair shares: Abby: 10.333, Ben: 9, Carla: 7.667
    Motorcycle to Abby, Car to Ben, Tractor to Abby, Boat to Abby
    Initial: Abby pays $10.667, Ben pays $2, Carla gets $7.667
    Surplus: $5; $1.667 each
    Final: Abby gets Motorcycle, Tractor and Boat, pays $9
           Ben gets Car, pays $0.333
           Carla gets $9.334

15. Fair shares: Sasha: $135, Megan: $140
    Sasha gets: Couch, detail cleaning. Value $80
    Megan gets: TV, Stereo, carpets. Value: $260
    Initial: Sasha gets $55, Megan pays $120.
    Surplus: $65; $32.50 each
    Final: Sasha gets Couch and does detail cleaning, gets $87.50
            Megan gets TV and stereo, and cleans carpets, pays $87.50

17. a. s3, worth $270

    b. s1 and s4 have combined value $440 for Greedy, so piece would be worth $220

Apportionment

1. a. Math: 6, English: 5, Chemistry: 3, Biology: 1
   b. Math: 7, English: 5, Chemistry: 2, Biology: 1
   c. Math: 6, English: 5, Chemistry: 3, Biology: 1
   d. Math: 6, English: 5, Chemistry: 3, Biology: 1
   e. Math: 6, English: 5, Chemistry: 2, Biology: 2

3. a. Morning: 1, Midday: 5, Afternoon: 6, Evening: 8
   b. Morning: 1, Midday: 4, Afternoon: 7, Evening: 8
   c. Morning: 1, Midday: 5, Afternoon: 6, Evening: 8
   d. Morning: 1, Midday: 5, Afternoon: 6, Evening: 8
   e. Morning: 2, Midday: 5, Afternoon: 6, Evening: 7

5. a. Alice: 18, Ben: 14, Carlos: 4
   b. Alice: 19, Ben: 14, Carlos: 3
   c. Alice: 19, Ben: 14, Carlos: 3
   d. Alice: 19, Ben: 14, Carlos: 3
   e. Alice: 18, Ben: 14, Carlos: 4

7. a. A: 40, B: 24, C: 15, D: 30, E: 10
b. A: 41, B: 24, C: 14, D: 30, E: 10
c. A: 40, B: 24, C: 15, D: 30, E: 10
d. A: 40, B: 24, C: 15, D: 30, E: 10
e. A: 40, B: 24, C: 15, D: 29, E: 11

Graph Theory

1.

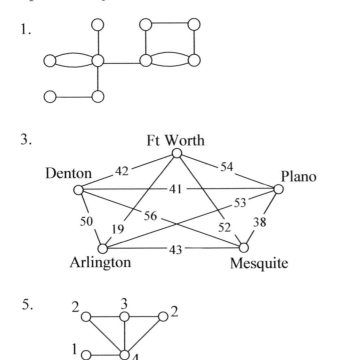

3.

5.

7. The first and the third graphs are connected

9. Bern to Frankfurt to Munchen to Berlin: 12hrs 50 min. (Though trip through Lyon, Paris and Amsterdam only adds 30 minutes)

11. The first graph has an Euler circuit. The last two graphs each have two vertices with odd degree.

13. One of several possible eulerizations requiring 5 duplications:

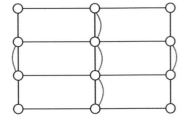

17. Only the middle graph has a Hamiltonian circuit.

19. a. Ft Worth, Arlington, Mesquite, Plano, Denton, Ft Worth: 183 miles
    b. Same as part a
    c. Same as part a

21. a. ABDCEA
    b. ACEBDA
    c. ADBCEA

23.

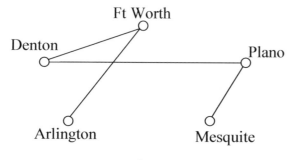

25.

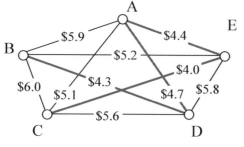

Scheduling

1.

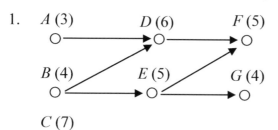

3.

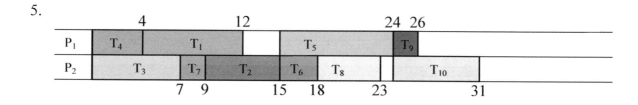

5.

7.

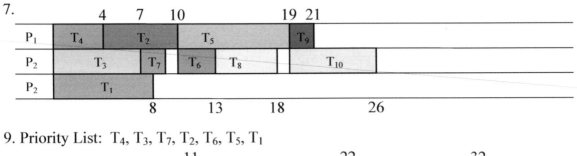

9. Priority List: T_4, T_3, T_7, T_2, T_6, T_5, T_1

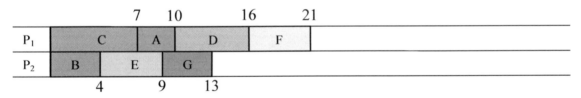

11. Priority List: T_5, T_1, T_3, T_{10}, T_2, T_8, T_4, T_6, T_7, T_9

13. Priority List: C, D, E, F, B, G, A

15. a.

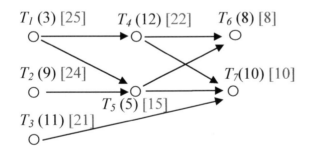

b. Critical path: T_1, T_4, T_7. Minimum completion time: 25
c. Critical path priority list: T_1, T_2, T_4, T_3, T_5, T_7, T_6

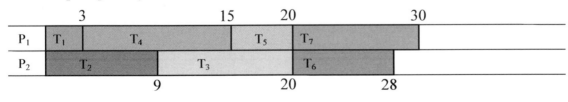

17. a.

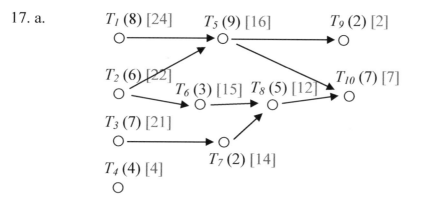

T_1 (8) [24] T_5 (9) [16] T_9 (2) [2]

T_2 (6) [22] T_6 (3) [15] T_8 (5) [12] T_{10} (7) [7]

T_3 (7) [21]

T_4 (4) [4] T_7 (2) [14]

b. Critical path: T_1, T_5, T_{10}. Minimum completion time: 24
c. Critical path priority list: T_1, T_2, T_3, T_5, T_6, T_7, T_8, T_{10}, T_4, T_9

19. Critical path priority list: B, A, D, E, C, F, G

	4		15	20
P_1	B	D	E	F
P_2	A	C		G
	3		10	19

Growth Models

1. a. $P_0 = 20$. $P_n = P_{n-1} + 5$
   b. $P_n = 20 + 5n$

3. a. $P_1 = P_0 + 15 = 40+15 = 55$. $P_2 = 55+15 = 70$
   b. $P_n = 40 + 15n$
   c. $P_{10} = 40 + 15(10) = 190$ thousand dollars
   d. $40 + 15n = 100$ when $n = 4$ years.

5. Grew 64 in 8 weeks: 8 per week
   a. $P_n = 3 + 8n$
   b. $187 = 3 + 8n$. $n = 23$ weeks

7. a. $P_0 = 200$ (thousand), $P_n = (1+.09) P_{n-1}$ where n is years after 2000
   b. $P_n = 200(1.09)^n$
   c. $P_{16} = 200(1.09)^{16} = 794.061$ (thousand) $= 794,061$
   d. $200(1.09)^n = 400$. $n = \log(2)/\log(1.09) = 8.043$. In 2008.

9. Let $n=0$ be 1983. $P_n = 1700(2.9)^n$. 2005 is $n=22$. $P_{22} = 1700(2.9)^{22} = 25,304,914,552,324$ people. Clearly not realistic, but mathematically accurate.

11. If n is in hours, better to start with the explicit form. $P_0 = 300$. $P_4 = 500 = 300(1+r)^4$
$500/300 = (1+r)^4$. $1+r = 1.136$. $r = 0.136$
  a. $P_0 = 300$. $P_n = (1.136)P_{n-1}$
  b. $P_n = 300(1.136)^n$
  c. $P_{24} = 300(1.136)^{24} = 6400$ bacteria
  d. $300(1.136)^n = 900$. $n = \log(3)/\log(1.136) =$ about 8.62 hours

13. a. $P_0 = 100$ $P_n = P_{n-1} + 0.70 (1 - P_{n-1} / 2000) P_{n-1}$
  b. $P_1 = 100 + 0.70(1 - 100/2000)(100) = 166.5$
  c. $P_2 = 166.5 + 0.70(1 - 166.5/2000)(166.5) = 273.3$

15. To find the growth rate, suppose n=0 was 1968. Then P_0 would be 1.60 and $P_8 = 2.30 = 1.60(1+r)^8$, $r = 0.0464$. Since we want n=0 to correspond to 1960, then we don't know P_0, but P_8 would $1.60 = P_0(1.0464)^8$. $P_0 = 1.113$.
  a. $P_n = 1.113(1.0464)^n$
  b. $P_0 = \$1.113$, or about $1.11
  c. 1996 would be n=36. $P_{36} = 1.113(1.0464)^{36} = \$5.697$. Actual is slightly lower.

17. The population in the town was 4000 in 2005, and is growing by 4% per year.

Finance
  1. $A = 200 + .05(200) = \$210$

  3. I=200. $t = 13/52$ (13 weeks out of 52 in a year). $P_0 = 9800$
$200 = 9800(r)(13/52)$ $r = 0.0816 = 8.16\%$ annual rate

  5. $P_{10} = 300(1 + .05/1)^{10(1)} = \$488.67$

  7. a. $P_{20} = 2000(1 + .03/12)^{20(12)} = \$3641.51$ in 20 years
  b. $3641.51 - 2000 = \$1641.51$ in interest

  9. $P_8 = P_0(1 + .06/12)^{8(12)} = 6000$. $P_0 = \$3717.14$ would be needed

  11. a. $P_{30} = \dfrac{200\left((1 + 0.03/12)^{30(12)} - 1\right)}{0.03/12} = \$116,547.38$
  b. $200(12)(30) = \$72,000$
  c. $\$116,547.40 - \$72,000 = \$44,547.38$ of interest

  13. a. $P_{30} = 800,000 = \dfrac{d\left((1 + 0.06/12)^{30(12)} - 1\right)}{0.06/12}$ $d = \$796.40$ each month
  b. $\$796.40(12)(30) = \$286,704$
  c. $\$800,000 - \$286,704 = \$513,296$ in interest

15. a. $P_0 = \dfrac{30000\left(1 - (1+0.08/1)^{-25(1)}\right)}{0.08/1} = \$320,253.29$

    b. $30000(25) = \$750,000$

    c. $\$750,000 - \$320,253.29 = \$429,756.71$

17. $P_0 = 500,000 = \dfrac{d\left(1 - (1+0.06/12)^{-20(12)}\right)}{0.06/12}$ $d = \$3582.16$ each month

19. a. $P_0 = \dfrac{700\left(1 - (1+0.05/12)^{-30(12)}\right)}{0.05/12} = $ a $\$130,397.13$ loan

    b. $700(12)(30) = \$252,000$

    c. $\$252,200 - \$130,397.13 = \$121,602.87$ in interest

21. $P_0 = 25,000 = \dfrac{d\left(1 - (1+0.02/12)^{-48}\right)}{0.02/12} = \$542.38$ a month

23. a. Down payment of 10% is $\$20,000$, leaving $\$180,000$ as the loan amount

    b. $P_0 = 180,000 = \dfrac{d\left(1 - (1+0.05/12)^{-30(12)}\right)}{0.05/12}$ $d = \$966.28$ a month

    c. $P_0 = 180,000 = \dfrac{d\left(1 - (1+0.06/12)^{-30(12)}\right)}{0.06/12}$ $d = \$1079.19$ a month

25. First we find the monthly payments:

$P_0 = 24,000 = \dfrac{d\left(1 - (1+0.03/12)^{-5(12)}\right)}{0.03/12}$. $d = \$431.25$

Remaining balance: $P_0 = \dfrac{431.25\left(1 - (1+0.03/12)^{-2(12)}\right)}{0.03/12} = \$10,033.45$

27. $6000(1+0.04/12)^{12N} = 10000$

$(1.00333)^{12N} = 1.667$

$\log\left((1.00333)^{12N}\right) = \log(1.667)$

$12N\log(1.00333) = \log(1.667)$

$N = \dfrac{\log(1.667)}{12\log(1.00333)} = $ about 12.8 years

29. $3000 = \dfrac{60\left(1 - (1 + 0.14/12)^{-12N}\right)}{0.14/12}$

$3000(0.14/12) = 60\left(1 - (1.0117)^{-12N}\right)$

$\dfrac{3000(0.14/12)}{60} = 0.5833 = 1 - (1.0117)^{-12N}$

$0.5833 - 1 = -(1.0117)^{-12N}$

$-(0.5833 - 1) = (1.0117)^{-12N}$

$\log(0.4167) = \log\left((1.0117)^{-12N}\right)$

$\log(0.4167) = -12N \log(1.0117)$

$N = \dfrac{\log(0.4167)}{-12\log(1.0117)} = $ about 6.3 years

31. First 5 years: $P_5 = \dfrac{50\left((1 + 0.08/12)^{5(12)} - 1\right)}{0.08/12} = \$3673.84$

Next 25 years: $3673.84(1 + .08/12)^{25(12)} = \$26,966.65$

33. Working backwards, $P_0 = \dfrac{10000\left(1 - (1 + 0.08/4)^{-10(4)}\right)}{0.08/4} = \$273,554.79$ needed at retirement. To end up with that amount of money, $273,554.70 = \dfrac{d\left((1 + 0.08/4)^{15(4)} - 1\right)}{0.08/4}$.

He'll need to contribute $d = \$2398.52$ a quarter.

Statistics

1. a. Population is the current representatives in the state's congress
   b. 106
   c. the 28 representatives surveyed
   d. 14 out of 28 = ½ = 50%
   e. We might expect 50% of the 106 representatives = 53 representatives

3. This suffers from leading question bias

5. This question would likely suffer from a perceived lack of anonymity

7. This suffers from leading question bias

9. Quantitative

11. Observational study

13. Stratified sample

15. a. Group 1, receiving the vaccine
   b. Group 2 is acting as a control group. They are not receiving the treatment (new vaccine).
   c. The study is at least blind. We are not provided enough information to determine if it is double-blind.
   d. This is a controlled experiment

17. a. Census
   b. Observational study

Describing Data

1. a. Different tables are possible

Score	Frequency
30	1
40	0
50	4
60	3
70	6
80	5
90	2
100	3

b. This is technically a bar graph, not a histogram:

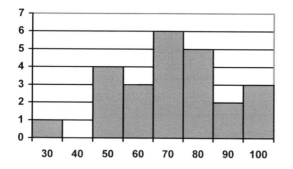

c.

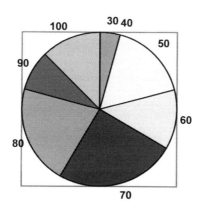

420

3. a. 5+3+4+2+1 = 15
   b. 5/15 = 0.3333 = 33.33%

5. Bar is at 25%. 25% of 20 = 5 students earned an A

7. a. (7.25+8.25+9.00+8.00+7.25+7.50+8.00+7.00)/8 = $7.781
   b. In order, 7.50 and 8.00 are in middle positions. Median = $7.75
   c. 0.25*8 = 2. Q1 is average of 2nd and 3rd data values: $7.375
      0.75*8 = 6. Q3 is average of 6th and 7th data values: $8.125
      5-number summary: $7.00, $7.375, $7.75, $8.125, $9.00

9. a. (5*0 + 3*1 + 4*2 + 2*3 + 1*5)/15 = 1.4667
   b. Median is 8th data value: 1 child
   c. 0.25*15 = 3.75. Q1 is 4th data value: 0 children
      0.75*15 = 11.25. Q3 is 12th data value: 2 children
      5-number summary: 0, 0, 1, 2, 5

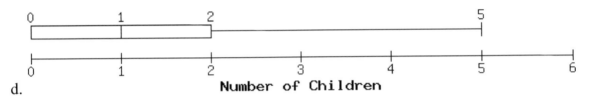

   d.

11. Kendra makes $90,000. Kelsey makes $40,000. Kendra makes $50,000 more.

Probability

1. a. $\dfrac{6}{13}$ b,. $\dfrac{2}{13}$

3. $\dfrac{150}{335} = 44.8\%$

5. $\dfrac{1}{6}$

7. $\dfrac{26}{65}$

9. $\dfrac{3}{6} = \dfrac{1}{2}$

11. $\dfrac{4}{52} = \dfrac{1}{13}$

13. $1 - \dfrac{1}{12} = \dfrac{11}{12}$

15. $1 - \dfrac{25}{65} = \dfrac{40}{65}$

17. $\dfrac{1}{6} \cdot \dfrac{1}{6} = \dfrac{1}{36}$

19. $\dfrac{1}{6} \cdot \dfrac{3}{6} = \dfrac{3}{36} = \dfrac{1}{12}$

21. $\dfrac{17}{49} \cdot \dfrac{16}{48} = \dfrac{17}{49} \cdot \dfrac{1}{3} = \dfrac{17}{147}$

23. a. $\dfrac{4}{52} \cdot \dfrac{4}{52} = \dfrac{16}{2704} = \dfrac{1}{169}$

   b. $\dfrac{4}{52} \cdot \dfrac{48}{52} = \dfrac{192}{2704} = \dfrac{12}{169}$

   c. $\dfrac{48}{52} \cdot \dfrac{48}{52} = \dfrac{2304}{2704} = \dfrac{144}{169}$

   d. $\dfrac{13}{52} \cdot \dfrac{13}{52} = \dfrac{169}{2704} = \dfrac{1}{16}$

   e. $\dfrac{48}{52} \cdot \dfrac{39}{52} = \dfrac{1872}{2704} = \dfrac{117}{169}$

25. $\dfrac{4}{52} \cdot \dfrac{4}{51} = \dfrac{16}{2652}$

27. a. $\dfrac{11}{25} \cdot \dfrac{14}{24} = \dfrac{154}{600}$

   b. $\dfrac{14}{25} \cdot \dfrac{11}{24} = \dfrac{154}{600}$

   c. $\dfrac{11}{25} \cdot \dfrac{10}{24} = \dfrac{110}{600}$

   d. $\dfrac{14}{25} \cdot \dfrac{13}{24} = \dfrac{182}{600}$

   e. no males = two females. Same as part d.

29. P(F and A) = $\dfrac{10}{65}$

31. P(red or odd) = $\dfrac{6}{14} + \dfrac{7}{14} - \dfrac{3}{14} = \dfrac{10}{14}$. Or 6 red and 4 odd-numbered blue marbles is 10 out of 14.

33. P(F or B) = $\dfrac{26}{65} + \dfrac{22}{65} - \dfrac{4}{65} = \dfrac{44}{65}$. Or P(F or B) = $\dfrac{18+4+10+12}{65} = \dfrac{44}{65}$

35. P(King of Hearts or Queen) = $\dfrac{1}{52} + \dfrac{4}{52} = \dfrac{5}{52}$

37. a. P(even | red) = $\dfrac{2}{5}$ b. P(even | red) = $\dfrac{2}{6}$

39. P(Heads on second | Tails on first) = $\dfrac{1}{2}$. They are independent events.

41. P(speak French | female) $= \dfrac{3}{14}$

43. Out of 4,000 people, 10 would have the disease. Out of those 10, 9 would test positive, while 1 would falsely test negative. Out of the 3990 uninfected people, 399 would falsely test positive, while 3591 would test negative.

    a. P(virus | positive) $= \dfrac{9}{9+399} = \dfrac{9}{408} = 2.2\%$

    b. P(no virus | negative) $= \dfrac{3591}{3591+1} = \dfrac{3591}{3592} = 99.97\%$

45. Out of 100,000 people, 300 would have the disease. Of those, 18 would falsely test negative, while 282 would test positive. Of the 99,700 without the disease, 3,988 would falsely test positive and the other 95,712 would test negative.

    P(disease | positive) $= \dfrac{282}{282+3988} = \dfrac{282}{4270} = 6.6\%$

47. Out of 100,000 women, 800 would have breast cancer. Out of those, 80 would falsely test negative, while 720 would test positive. Of the 99,200 without cancer, 6,944 would falsely test positive.

    P(cancer | positive) $= \dfrac{720}{720+6944} = \dfrac{720}{7664} = 9.4\%$

49. $2 \cdot 3 \cdot 8 \cdot 2 = 96$ outfits

51. a. $4 \cdot 4 \cdot 4 = 64$ b. $4 \cdot 3 \cdot 2 = 24$

53. $26 \cdot 26 \cdot 26 \cdot 10 \cdot 10 \cdot 10 = 17{,}576{,}000$

55. $_4P_4$ or $4 \cdot 3 \cdot 2 \cdot 1 = 24$ possible orders

57. Order matters. $_7P_4 = 840$ possible teams

59. Order matters. $_{12}P_5 = 95{,}040$ possible themes

61. Order does not matter. $_{12}C_4 = 495$

63. $_{50}C_6 = 15{,}890{,}700$

65. $_{27}C_{11} \cdot 16 = 208{,}606{,}320$

67. There is only 1 way to arrange 5 CD's in alphabetical order. The probability that the CD's are in alphabetical order is one divided by the total number of ways to arrange 5 CD's. Since alphabetical order is only one of all the possible orderings you can either use permutations, or simply use 5!. P(alphabetical) $= 1/5! = 1/(5 \text{ P } 5) = \dfrac{1}{120}$.

69. There are $_{48}C_6$ total tickets. To match 5 of the 6, a player would need to choose 5 of those 6, $_6C_5$, and one of the 42 non-winning numbers, $_{42}C_1$. $\dfrac{6 \cdot 42}{12271512} = \dfrac{252}{12271512}$

71. All possible hands is $_{52}C_5$. Hands will all hearts is $_{13}C_5$. $\dfrac{1287}{2598960}$.

73. $\$3\left(\dfrac{3}{37}\right) + \$2\left(\dfrac{6}{37}\right) + (-\$1)\left(\dfrac{28}{37}\right) = -\$\dfrac{7}{37} = -\$0.19$

75. There are $_{23}C_6 = 100{,}947$ possible tickets.

Expected value $= \$29{,}999\left(\dfrac{1}{100947}\right) + (-\$1)\left(\dfrac{100946}{100947}\right) = -\$0.70$

77. $\$48(0.993) + (-\$302)(0.007) = \$45.55$

Sets

1. {m, i, s, p}

3. One possibility is: Multiples of 3 between 1 and 10

5. Yes

7. True

9. True

11. False

13. $A \cup B = \{1, 2, 3, 4, 5\}$

15. $A \cap C = \{4\}$

17. $A^c = \{6, 7, 8, 9, 10\}$

19. $D^c \cap E = \{t, s\}$

21. $(D \cap E) \cup F = \{k, b, a, t, h\}$

23. $(F \cap E)^c \cap D = \{b, c, k\}$

25.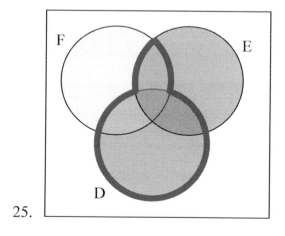

27.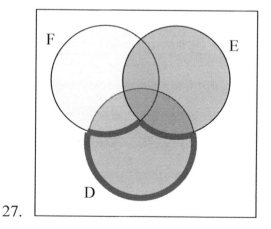

29. One possible answer: $(A \cap B) \cup (B \cap C)$

31. $(A \cap B^c) \cup C$

33. 5

35. 6

37. $n(A \cap C) = 5$

39. $n(A \cap B \cap C^c) = 3$

41. $n(G \cup H) = 45$

43. 136 use Redbox

45. 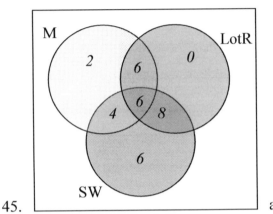 a) 8 had seen exactly one b) 6 had only seen SW

Historical Counting

1. Partial answer: Jars: 3 singles, 3 @ x2, 2 @ x6, 1 @ x12. 3+6+12+12 = 33

3. 113

5. 3022

7. 53

9. 1100100

11. 332

13. 111100010

15. 7,1,10 base 12 = 1030 base 10

17. 6,4,2 base 12 = 914 base 10

19. 175 base 10 = 1,2,7 base 12 = ♉♊♏

21. 10000 base 10 = 5,9,5,4 base 12 = ♍♑♍♌

23. 135 = 6,15 base 20 =

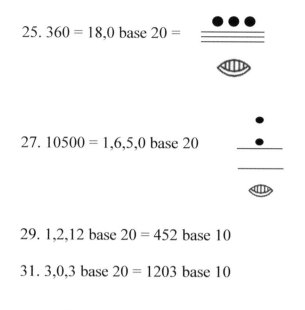

25. 360 = 18,0 base 20 =

27. 10500 = 1,6,5,0 base 20

29. 1,2,12 base 20 = 452 base 10

31. 3,0,3 base 20 = 1203 base 10

33.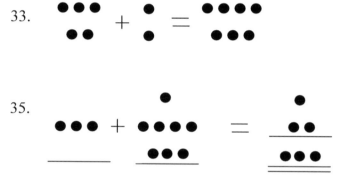

35.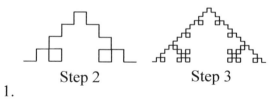

Fractals

Step 2   Step 3

1.

Step 2   Step 3

3.

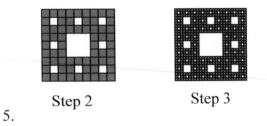

Step 2   Step 3

5.

9. Four copies of the Koch curve are needed to create a curve scaled by 3.

$$D = \frac{\log(4)}{\log(3)} \approx 1.262$$

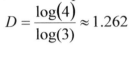

1

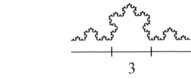

3

11. Eight copies of the shape are needed to make a copy scaled by 3. $D = \dfrac{\log(8)}{\log(3)} \approx 1.893$

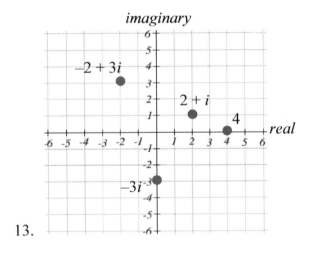

13.

15. a) $5 - i$ b) $5 - 4i$

17. a) $6 + 12i$ b) $10 - 2i$ c) $14 + 2i$

19. $(2 + 3i)(1 - i) = 5 + i$. It appears that multiplying by $1 - i$ both scaled the number away from the origin, and rotated it clockwise about $45°$.

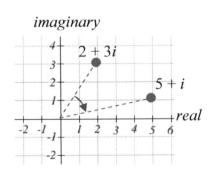

21.
$$z_1 = iz_0 + 1 = i(2) + 1 = 1 + 2i$$
$$z_2 = iz_1 + 1 = i(1 + 2i) + 1 = i - 2 + 1 = -1 + i$$
$$z_3 = iz_2 + 1 = i(-1 + i) + 1 = -i - 1 + 1 = -i$$

$$z_0 = 0$$

$$z_1 = z_0{}^2 - 0.25 = 0 - 0.25 = -0.25$$

23. $z_2 = z_1{}^2 - 0.25 = (-0.25)^2 - 0.25 = -0.1875$

$$z_3 = z_2{}^2 - 0.25 = (-0.1875)^2 - 0.25 = -0.21484$$

$$z_4 = z_3{}^2 - 0.25 = (-0.21484)^2 - 0.25 = -0.20384$$

25. attracted, to approximately $-0.37766 + 0.14242i$

27. periodic 2-cycle   29. Escaping   31. periodic 3-cycle

33. a) Yes, periodic 3-cycle b) Yes, periodic 3-cycle c) No

Cryptography

1. ZLU KZB WWS PLZ   3. SHRED EVIDENCE

5. O2H DO5 HDV   7. MERGER ON

9. MNB AET RTE HAT TLR EII YN

11. THE STASH IS HIDDEN AT MARVINS QNS

13. UEM IYN IOB WYL TTL N

15. HIRE THIRTY NEW EMPLOYEES MONDAY

17. ZMW NDG CDA YVK

19. a) 3   b) 0   c) 4

21. We test out all n from 1 to 10

n	4^n	4^n mod 11
1	4	4
2	16	5
3	64	9
4	256	3
5	1024	1
6	4096	4
7	16384	5
8	65536	9
9	262144	3
10	1048576	1

Since we have repeats, and not all values from 1 to 10 are produced (for example, there is no n is 4^n mod 11 = 7), 4 is *not* a generator mod 11.

23. $157^{10} \bmod 5 = (157 \bmod 5)^{10} \bmod 5 = 2^{10} \bmod 5 = 1024 \bmod 5 = 4$

25. $3^7 \bmod 23 = 2$

27. Bob would send $5^7 \bmod 33 = 14$. Alice would decrypt it as $14^3 \bmod 33 = 5$

31. a. $67^8 \bmod 83 = (67^4 \bmod 23)^2 \bmod 83 = 49^2 \bmod 83 = 2401 \bmod 83 = 77$
    $67^{16} \bmod 83 = (67^8 \bmod 23)^2 \bmod 83 = 77^2 \bmod 83 = 5929 \bmod 83 = 36$

   b. $17000 \bmod 83 = (100 \bmod 83)*(170 \bmod 83) \bmod 83 = (17)(4) \bmod 83 = 68$

   c. $67^5 \bmod 83 = (67^4 \bmod 83)(67 \bmod 83) \bmod 83 = (49)(67) \bmod 83 = 3283 \bmod 83 = 46$

   d. $67^7 \bmod 83 = (67^4 \bmod 83)\ (67^2 \bmod 83)(67 \bmod 83) \bmod 83 = (49)(7)(67) \bmod 83 = 22981 \bmod 83 = 73$.

   e. $67^{24} = 67^{16}67^8$ so
   $67^{24} \bmod 83 = (67^{16} \bmod 83)(67^8 \bmod 83) \bmod 83 = (77)(36) \bmod 83 = 2272 \bmod 83 = 33$

66472108R00239